FOOD PROCESSING AND PRESERVATION TECHNOLOGY

Advances, Methods, and Applications

FOOD PROCESSING AND PRESERVATION TECHNOLOGY

Advances, Methods, and Applications

Edited by
Megh R. Goyal, PhD, P.E.
Santosh K. Mishra, PhD
Preeti Birwal, PhD

AAP APPLE
ACADEMIC
PRESS

First edition published 2022

Apple Academic Press Inc.
1265 Goldenrod Circle, NE,
Palm Bay, FL 32905 USA

4164 Lakeshore Road, Burlington,
ON, L7L 1A4 Canada

CRC Press
6000 Broken Sound Parkway NW,
Suite 300, Boca Raton, FL 33487-2742 USA

2 Park Square, Milton Park,
Abingdon, Oxon, OX14 4RN UK

Library and Archives Canada Cataloguing in Publication

Title: Food processing and preservation technology : advances, methods, and applications / edited by Megh R. Goyal, PhD, P.E., Santosh K. Mishra, PhD, Preeti Birwal, PhD.

Names: Goyal, Megh R., editor. | Mishra, Santosh K., editor. | Birwal, Preeti, editor.

Series: Innovations in agricultural and biological engineering.

Description: First edition. | Series statement: Innovations in agricultural and biological engineering | Includes bibliographical references and index.

Identifiers: Canadiana (print) 20210336137 | Canadiana (ebook) 2021033617X | ISBN 9781771889957 (hardcover) | ISBN 9781774639467 (softcover) | ISBN 9781003153184 (ebook)

Subjects: LCSH: Food—Preservation—Technological innovations. | LCSH: Food industry and trade—Technological innovations. | LCSH: Food industry and trade—Safety measures. | LCSH: Food—Quality.

Classification: LCC TP371.2 .F66 2022 | DDC 664/.028—dc23

Library of Congress Cataloging-in-Publication Data

Names: Goyal, Megh R., editor. | Mishra, Santosh K., editor. | Birwal, Preeti, editor.

Title: Food processing and preservation technology : advances, methods, and applications / Megh R. Goyal, Santosh K. Mishra, Preeti Birwal.

Other titles: Innovations in agricultural and biological engineering.

Description: First edition. | Palm Bay, FL, USA : Apple Academic Press, [2022] | Series: Innovations in agricultural and biological engineering | Includes bibliographical references and index. | Summary: "Food Processing and Preservation Technology: Advances, Methods, and Applications confronts the challenges of food preservation by providing new research and information on the use of novel processing and preservation technologies during production, processing, and transportation in the food industry for the improvement of shelf life and the safety of foods. The book is organized in two main parts. The first section focuses on novel and nonthermal processing of food and food products. It looks at dielectric heating and ohmic heating as well as three-dimensional printing of foods and ozonization of food products. Part two delves into process interventions for food processing and preservations, discussing the applications of diverse novel food processing. The authors discuss drying technologies, advances in food fermentation technologies, mechanization of traditional indigenous products for preservation of food and safety, and different properties and concepts of bakery products. Key features: Examines different properties and attributes of some bakery foods, etc. Elucidates on novel nonthermal processing techniques and their mechanisms of actions for minimal loss of food nutrients and for food safety Discusses a variety of modern technologies that aim to reduce the spoilage of food products This volume presents valuable research on food processing, quality control, and safety measures for food products by means of novel processing and preservation technologies during production, processing, and transportation in the food industry"-- Provided by publisher.

Identifiers: LCCN 2021049572 (print) | LCCN 2021049573 (ebook) | ISBN 9781771889957 (hardcover) | ISBN 9781774639467 (paperback) | ISBN 9781003153184 (ebook)

Subjects: LCSH: Food--Preservation--Technological innovations. | Food industry and trade--Technological innovations. | Food industry and trade--Safety measures.

Classification: LCC TP371.2 .F68 2022 (print) | LCC TP371.2 (ebook) | DDC 664/.028--dc23/eng/20211109

LC record available at https://lccn.loc.gov/2021049572

LC ebook record available at https://lccn.loc.gov/2021049573

ISBN: 978-1-77188-995-7 (hbk)
ISBN: 978-1-77463-946-7 (pbk)
ISBN: 978-1-00315-318-4 (ebk)

ABOUT THE BOOK SERIES: INNOVATIONS IN AGRICULTURAL AND BIOLOGICAL ENGINEERING

Under this book series, Apple Academic Press, Inc. is publishing book volumes over a span of 8–10 years in the specialty areas defined by the American Society of Agricultural and Biological Engineers (<asabe.org>). Apple Academic Press, Inc. aims to be a principal source of books in agricultural and biological engineering (ABE). We welcome book proposals from readers in areas of their expertise.

The mission of this series is to provide knowledge and techniques for agricultural and biological engineers (ABEs). The book series offers high-quality reference and academic content on ABE that is accessible to academicians, researchers, scientists, university faculty and university-level students, and professionals around the world.

Agricultural and biological engineers ensure that the world has the necessities of life, including safe and plentiful food, clean air and water, renewable fuel and energy, safe working conditions, and a healthy environment by employing knowledge and expertise of the sciences, both pure and applied, and engineering principles. Biological engineering applies engineering practices to problems and opportunities presented by living things and the natural environment in agriculture.

ABE embraces a variety of the following specialty areas (<asabe.org>): aquaculture engineering, biological engineering, energy, farm machinery and power engineering, food and process engineering, forest engineering, information and electrical technologies, soil and water conservation engineering, natural resources engineering, nursery and greenhouse engineering, safety and health, and structures and environment.

For this book series, we welcome chapters or book proposals on the following specialty areas (but not limited to):

1. Academia to industry to end-user loop in agricultural and biological engineering
2. Agricultural mechanization
3. Aquaculture engineering
4. Biological engineering in agriculture

5. Biotechnology applications in agricultural & biological engineering
6. Energy source engineering
7. Food and process engineering
8. Forest engineering
9. Hill land agriculture
10. Human factors in engineering
11. Information and electrical technologies
12. Irrigation and drainage engineering
13. Nanotechnology applications in agricultural & biological engineering
14. Natural resources engineering
15. Nursery and greenhouse engineering
16. Potential of phytochemicals from agricultural and wild plants for human health
17. Power systems and machinery design
18. GPS and remote sensing potential in agricultural and biological engineering
19. Robot engineering and drones in agriculture
20. Simulation and computer modeling
21. Smart engineering applications in agriculture
22. Soil and water engineering
23. Micro-irrigation engineering
24. Structures and environment engineering
25. Waste management and recycling
26. Rural electrification.
27. Sanitary engineering
28. Farm to fork technologies in agriculture
29. Impact of global warming and climatic change on agriculture economy
30. Any other focus areas

For more information on this series, readers may contact:

Megh R. Goyal, PhD, PE
Book Series Senior Editor-in-Chief:
Innovations in Agricultural and Biological Engineering
E-mail: goyalmegh@gmail.com

OTHER BOOKS ON AGRICULTURAL & BIOLOGICAL ENGINEERING BY APPLE ACADEMIC PRESS, INC.

Management of Drip/Trickle or Micro Irrigation
Megh R. Goyal, PhD, PE, Senior Editor-in-Chief

Evapotranspiration: Principles and Applications for Water Management
Megh R. Goyal, PhD, PE and Eric W. Harmsen, Editors

Book Series: RESEARCH ADVANCES IN SUSTAINABLE MICRO IRRIGATION
Senior Editor-in-Chief: Megh R. Goyal, PhD, PE
Volume 1: Sustainable Micro Irrigation: Principles and Practices
Volume 2: Sustainable Practices in Surface and Subsurface Micro Irrigation
Volume 3: Sustainable Micro Irrigation Management for Trees and Vines
Volume 4: Management, Performance, and Applications of Micro Irrigation Systems
Volume 5: Applications of Furrow and Micro Irrigation in Arid and Semi-Arid Regions
Volume 6: Best Management Practices for Drip Irrigated Crops
Volume 7: Closed Circuit Micro Irrigation Design: Theory and Applications
Volume 8: Wastewater Management for Irrigation: Principles and Practices
Volume 9: Water and Fertigation Management in Micro Irrigation
Volume 10: Innovation in Micro Irrigation Technology

Book Series: INNOVATIONS AND CHALLENGES IN MICRO IRRIGATION
Senior Editor-in-Chief: Megh R. Goyal, PhD, PE
- Engineering Interventions in Sustainable Trickle Irrigation: Water Requirements, Uniformity, Fertigation, and Crop Performance
- Fertigation Technologies for Micro Irrigated Crops: Performance, Requirements, and Efficiency
- Management of Drip/Trickle or Micro Irrigation
- Management Strategies for Water Use Efficiency and Micro Irrigated Crops: Principles, Practices, and Performance

- Micro Irrigation Engineering for Horticultural Crops: Policy Options, Scheduling and Design
- Micro Irrigation Management: Technological Advances and Their Applications
- Micro Irrigation Scheduling and Practices
- Performance Evaluation of Micro Irrigation Management: Principles and Practices
- Potential of Solar Energy and Emerging Technologies in Sustainable Micro Irrigation
- Principles and Management of Clogging in Micro Irrigation
- Sustainable Micro Irrigation Design Systems for Agricultural Crops: Methods and Practices

Book Series: INNOVATIONS IN AGRICULTURAL & BIOLOGICAL ENGINEERING
Senior Editor-in-Chief: Megh R. Goyal, PhD, PE
- Advances in Food Process Engineering: Novel Processing, Preservation and Decontamination of Foods
- Biological and Chemical Hazards in Food and Food Products: Prevention, Practices, and Management
- Bioremediation and Phytoremediation Technologies in Sustainable Soil Management, Volume 2: Microbial Approaches and Recent Trends
- Bioremediation and Phytoremediation Technologies in Sustainable Soil Management, Volume 3: Inventive Techniques, Research Methods, and Case Studies
- Bioremediation and Phytoremediation Technologies in Sustainable Soil Management, Volume 4: Degradation of Pesticides and Polychlorinated Biphenyls
- Dairy Engineering: Advanced Technologies and Their Applications
- Developing Technologies in Food Science: Status, Applications, and Challenges
- Emerging Technologies in Agricultural Engineering
- Engineering Interventions in Agricultural Processing
- Engineering Interventions in Foods and Plants
- Engineering Practices for Agricultural Production and Water Conservation: An Interdisciplinary Approach
- Engineering Practices for Management of Soil Salinity: Agricultural, Physiological, and Adaptive Approaches

- Engineering Practices for Milk Products: Dairyceuticals, Novel Technologies, and Quality
- Field Practices for Wastewater Use in Agriculture: Future Trends and Use of Biological Systems
- Flood Assessment: Modeling and Parameterization
- Food Engineering: Emerging Issues, Modeling, and Applications
- Food Process Engineering: Emerging Trends in Research and Their Applications
- Food Processing and Preservation Technology: Advances, Methods, and Applications
- Food Technology: Applied Research and Production Techniques
- Functional Dairy Ingredients and Nutraceuticals: Physicochemical, Technological, and Therapeutic Aspects
- Handbook of Research on Food Processing and Preservation Technologies, Volume 2: Nonthermal Food Preservation and Novel Processing Strategies
- Handbook of Research on Food Processing and Preservation Technologies, Volume 3: Computer-Aided Food Processing and Quality Evaluation Techniques
- Handbook of Research on Food Processing and Preservation Technologies, Volume 4: Design and Development of Specific Foods, Packaging Systems, and Food Safety
- Handbook of Research on Food Processing and Preservation Technologies, Volume 5: Emerging Techniques for Food Processing, Quality, and Safety Assurance
- Handbook of Research on Food Processing and Preservation Technologies: Volume 1: Nonthermal and Innovative Food Processing Methods
- Modeling Methods and Practices in Soil and Water Engineering
- Nanotechnology and Nanomaterial Applications in Food, Health, and Biomedical Sciences
- Nanotechnology Applications in Agricultural and Bioprocess Engineering: Farm to Table
- Nanotechnology Applications in Dairy Science: Packaging, Processing, and Preservation
- Nanotechnology Horizons in Food Process Engineering, Volume 1: Food Preservation, Food Packaging, and Sustainable Agriculture
- Nanotechnology Horizons in Food Process Engineering, Volume 2: Scope, Biomaterials, and Human Health

- Nanotechnology Horizons in Food Process Engineering, Volume 3: Trends, Nanomaterials, and Food Delivery
- Novel Dairy Processing Technologies: Techniques, Management, and Energy Conservation
- Novel Processing Methods for Plant-Based Health Foods: Extraction, Encapsulation and Health Benefits of Bioactive Compounds
- Novel Strategies to Improve Shelf-Life and Quality of Foods: Quality, Safety, and Health Aspects
- Processing of Fruits and Vegetables: From Farm to Fork
- Processing Technologies for Milk and Milk Products: Methods, Applications, and Energy Usage
- Quality Control in Fruit and Vegetable Processing: Methods and Strategies
- Scientific and Technical Terms in Bioengineering and Biological Engineering
- Soil and Water Engineering: Principles and Applications of Modeling
- Soil Salinity Management in Agriculture: Technological Advances and Applications
- State-of-the-Art Technologies in Food Science: Human Health, Emerging Issues and Specialty Topics
- Sustainable Biological Systems for Agriculture: Emerging Issues in Nanotechnology, Biofertilizers, Wastewater, and Farm Machines
- Sustainable Nanomaterials for Biosystem Engineering: Impacts, Challenges, and Future Prospects
- Technological Interventions in Dairy Science: Innovative Approaches in Processing, Preservation, and Analysis of Milk Products
- Technological Interventions in Management of Irrigated Agriculture
- Technological Interventions in the Processing of Fruits and Vegetables
- Technological Processes for Marine Foods, From Water to Fork: Bioactive Compounds, Industrial Applications, and Genomics

ABOUT SENIOR-EDITOR-IN-CHIEF

Megh R. Goyal, PhD, PE, is, currently a retired professor of agricultural and biomedical engineering from the General Engineering Department at the College of Engineering at the University of Puerto Rico–Mayaguez Campus (UPRM); and Senior Acquisitions Editor and Senior Technical Editor-in-Chief for Agricultural and Biomedical Engineering for Apple Academic Press Inc.

During his long career, he has worked as a Soil Conservation Inspector; Research Assistant at Haryana Agricultural University and Ohio State University; Research Agricultural Engineer/Professor at the Department of Agricultural Engineering of UPRM; and Professor of Agricultural and Biomedical Engineering in the General Engineering Department of UPRM. He spent a one-year sabbatical leave in 2002–2003 at the Biomedical Engineering Department of Florida International University, Miami, USA.

Dr. Goyal was the first agricultural engineer to receive the professional license in agricultural engineering from the College of Engineers and Surveyors of Puerto Rico. In 2005, he was proclaimed the "Father of Irrigation Engineering in Puerto Rico for the Twentieth Century" by the American Society of Agricultural and Biological Engineers, Puerto Rico Section, for his pioneering work on micro irrigation, evapotranspiration, agroclimatology, and soil and water engineering.

During his professional career of 51 years, he has received many awards, including Scientist of the Year, Membership Grand Prize for the American Society of Agricultural Engineers Campaign, Felix Castro Rodriguez Academic Excellence Award, Man of Drip Irrigation by the Mayor of Municipalities of Mayaguez/Caguas/Ponce and Senate/Secretary of Agriculture of ELA, Puerto Rico, and many others. He has been recognized as one of the experts "who rendered meritorious service for the development of [the] irrigation sector in India" by the Water Technology Centre of Tamil Nadu Agricultural University in Coimbatore, India, and ASABE who bestowed on him the 2108 Netafim Microirrigation Award.

Dr. Goyal has authored more than 200 journal articles and edited more than 95 books.

Dr. Goyal received his BSc degree in Engineering from Punjab Agricultural University, Ludhiana, India, and his MSc and PhD degrees from the Ohio State University, Columbus, Ohio, USA. He also earned a Master of Divinity degree from the Puerto Rico Evangelical Seminary, Hato Rey, Puerto Rico, USA.

Readers may contact him at goyalmegh@gmail.com.

ABOUT THE EDITORS

Santosh Kumar Mishra, PhD

Santosh Kumar Mishra, PhD, an Assistant Professor in the Department of Dairy Microbiology, College of Dairy Science and Technology, Guru Angad Dev Veterinary and Animal Sciences University, Ludhiana, Punjab, India. He is working presently on areas of functional foods and dairy products incorporating live probiotics and technology of functional lactic cultures for fermented and nonfermented dairy products. He served the dairy industry as Quality Assurance Executive at Mother Dairy, New Delh, India. He is also handling externally funded projects by DST, MoFPI, and UGC as PI or Co-PI. He has received several awards for best papers and posters/presentations. He is the recipient of junior and senior research fellowship during his master and doctoral programs at the National Dairy Research Institute, Karnal, Haryana, India. Recently, he received an award of honor at an international conference sponsored by Partap College of Education, Ludhiana, in association with the International Professionals Development Association, UK. He is the member of various scientific societies: life member of SASNET-Fermented Foods, Anand, and member of the Indian Dairy Associations, New Delhi. He has published several research, review, and popular articles in national and international journals as well several book chapters and teaching reviews in various training programs. He has recently completed the a young scientist project by DST, SEED Department, Government of India, New Delhi, on isolation and characterization of novel oxalate degrading lactic acid bacteria for potential probiotic management of kidney stone.

Dr Mishra received his BTech degree in Dairy Technology from Maharashtra Animal and Fisheries Sciences University, Nagpur, India, and his MSc and PhD degrees from the National Dairy Research Institute, Karnal, Haryana, India.

Readers may contact him at: skmishra84@gmail.com.

Preeti Birwal, PhD

Preeti Birwal, PhD, is a Scientist (Processing and Food Engineering) in the Department of Processing and Food Engineering, College of Agricultural Engineering and Technology, Punjab Agricultural Univer-sity, Ludhiana, Punjab, India.

She is currently working in the area of nonthermal food preservation, fermented beverages, food pack-aging, and technology of millet-based beer. She has served Jain Deemed to be University, Bangalore, as an assistant, where she has served as a member of the Board of Examiners and Placements. She has participated in many national and international conferences and seminars and won prizes for her oral and poster presentations. She has delivered lectures as a resource person on doubling farmers' income through dairy technology in training sponsored by the Directorate of Extension, Ministry of Agriculture and Farmers Welfare, Government of India. She is also serving as editor and reviewer of several journals and has been named "outstanding reviewer of the month" by the journal Current Research in Nutrition and Food Science. Recently she has organized a national conference. She has 18 research papers, an edited book, several book chapters, over 28 popular articles, several conference papers and abstracts, and several editorial opinions to her credit. She has successfully guided five postgraduate students for their dissertation work. She is also serving as external examiner for various Indian state agricultural universities.

Dr. Birwal earned her PhD (Dairy Engineering) on nonthermal pres-ervation of milk from ICAR-NDRI, Bangalore, and has received a merit certificate. She graduated with a degree in Dairy Technology from ICAR-National Dairy Research Institute (NDRI), Karnal, and master's degree in Food Process Engineering and Management from NIFTEM, Haryana, India. She is recipient of a university bronze medal under undergraduate program. She is the recipient of MHRD (2008), Nestle India (2009), GATE (2012–2014), UGC-RGN fellowships (2014–2018). She has successfully completed AUTOCAD 2D & 3D certification.

Readers may contact her at: preetibirwal@gmail.com

CONTENTS

Contributors ....*xvii*

Abbreviations ....*xxi*

Symbols ....*xxv*

Preface ....*xxvii*

PART I: NOVEL AND NONTHERMAL PROCESSING OF FOOD AND FOOD PRODUCTS 1

1. **Dielectric Heating: Recent Trends and Application in Food Processing** 3

   Mohan Naik, D. Lavanya, Suka Thangaraju, Nikitha Modupalli, and Venkatachalapathy Natarajan

2. **Ohmic Heating in Food Processing: A Futuristic Technology** 37

   Swastika Das, Robina Rai, and Shyam K. Singh

3. **Power Ultrasound: A Green Technology for Processing of Foods** 69

   Yashini Muthukrishnan, Chikkaballur K. Sunial, and Ashish Rawson

4. **Three-Dimensional (3D) Printing of Foods** 101

   Aswin S. Warrier

5. **Ozonization of Food Products** 113

   Subhashini Sundaramoorthy, Niveadhitha Sundramoorthy, Rakesh Ramalingam, and Madhumitha Maran

PART II: PROCESS INTERVENTIONS FOR FOOD PROCESSING AND PRESERVATION 139

6. **Pretreatments and Drying of Food Products** 141

   Ritesh B. Watharkar and Sadhana Sharma

7. **Principles of Foaming and Foam Mat Drying Technology: Fruits and Vegetables** 163

   Ritesh B. Watharkar

8. **Mechanization of Manufacturing Processes for Traditional Indian Dairy Products** 181

   Gajanan P. Deshmukh, Rekha R. Menon, and Naveen Jose

9. **Advances in Food Fermentation**.. 203

Barinderjeet S. Toor, Ankita Kataria, Amarjeet Kaur, and Savita Sharma

10. **Physicochemical and Thermal Properties of Bakery Products** 245

Sudharshan R. Ravula, Divyasree Arepally, A. K. Datta, and T. K. Goswami

Index... *275*

CONTRIBUTORS

Divyasree Arepally
PhD Research Scholar, Department of Agricultural and Food Engineering,
Indian Institute of Technology, Kharagpur 721302, West Bengal, India; Mobile: +91-9966995212;
E-mail: divyasreearepally@gmail.com

Preeti Birwal
Scientist, Department of Processing and Food Engineering, College of Agricultural Engineering and
Technology, Punjab Agricultural University, Ludhiana 141004, Punjab, India;
Mobile: +91-9896649633; E-mail: preetibirwal@gmail.com

Swastika Das
Guest Faculty, Assam University Silchar, Dorgakona, Silchar 788003, Assam, India;
Mobile: +91-9735836444; E-mail: swastika.5396@gmail.com

Gajanan P. Deshmukh
PhD Research Scholar, ICAR- National Dairy Research Institute, SRS, Bangalore 560030, India;
Mobile: 91-8147208662; E-mail: gajanannnn@gmail.com

A.K. Datta
Professor, Department of Agricultural and Food Engineering, Indian Institute of Technology,
Kharagpur 721302, West Bengal, India; Mobile: +91-9051309591; E-mail: akd@agfe.iitkgp.ac.in

T.K. Goswami
Professor, Department of Agricultural and Food Engineering, Indian Institute of Technology,
Kharagpur 721302, West Bengal, India; Mobile: +91-9647485515; E-mail: tkg@agfe.iitkgp.ac.in

Megh R. Goyal
Retired Faculty in Agricultural and Biomedical Engineering from College of Engineering at University
of Puerto Rico—Mayaguez Campus; and Senior Technical Editor-in-Chief in Agricultural and
Biomedical Engineering for Apple Academic Press, Inc.; PO Box 86, Rincon, PR 006770086, USA;
E-mail: goyalmegh@gmail.com

Naveen Jose
PhD Research Scholar, ICAR- National Dairy Research Institute, SRS, Bangalore 560030, India;
Mobile: 91-9400569023; E-mail: naveenjose50@gmail.com

Ankita Kataria
PhD Research Scholar, Department of Food Science and Technology, Punjab Agricultural University,
Ludhiana 141004, India; Mobile: +91-8197247436; E-mail: ankitakataria92@gmail.com

Amarjeet Kaur
Senior Milling Technologist, Department of Food Science and Technology, Punjab Agricultural
University, Ludhiana 141004, India; Mobile: +91-9888466677; E-mail: foodtechak@gmail.com

D. Lavanya
PhD Research Scholar, Department of Food Engineering, Indian Institute of Food Processing
Technology, Pudukkottai Road, Thanjavur 613005, Tamil Nadu, India; Mobile: +91-9686771314;
E-mail: lavanya.devraj91@gmail.com

Madhumitha Maran
MTech. Research Scholar, Department of Food Technology, FET, Jain (Deemed-to-be-University), Jain Global Campus, Jakkasandra Post, Kanakapura Taluk, Ramanagara District, Karnataka 562 112, India; Mobile: +91-8056374858; E-mail: madhumithaaquarius@gmail.com

Rekha R. Menon
Principal Scientist, ICAR- National Dairy Research Institute, SRS, Bangalore 560030, India; Mobile: +91-9916703069; E-mail: rekhmn@gmail.com

Santosh K. Mishra
Assistant Professor, Department of Dairy Microbiology, College of Dairy Science and Technology, Guru Angad Dev Veterinary and Animal Sciences University (GADVASU), Ludhiana 141004, Punjab, India; Mobile: +91-9464995049; E-mail: skmishra84@gmail.com

Nikitha Modupalli
PhD Research Scholar, Department of Food Engineering, Indian Institute of Food Processing Technology, Pudukkottai Road, Thanjavur 613005, Tamil Nadu; India; Mobile: +91-8531081060; E-mail: nikitha.modupalli93@gmail.com

Yashini Muthukrishnan
MTech. Research Scholar, Indian Institute of Food Processing Technology, Thanjavur 613005, Tamil Nadu; Mobile: +91-8012571112; E-mail: yashinim97@gmail.com

Mohan Naik
PhD Research Scholar, Department of Food Engineering, Indian Institute of Food Processing Technology, Pudukkottai Road, Thanjavur 613005, Tamil Nadu, India; Mobile: +91-7353615709; E-mail: mohannaik023@gmail.com

Venkatachalapathy Natarajan
Professor & Head, Department of Food Engineering, Indian Institute of Food Processing Technology, Pudukkottai Road, Thanjavur 613005, Tamil Nadu, India; Mobile: +91-9750968403; E-mail: venkat@iifpt.edu.in

Robina Rai
Guest Faculty, Assam University Silchar, Dorgakona, Silchar: 788003, Assam, India; Mobile: +91-7086125023; E-mail: robinarai95@gmail.com

Rakesh Ramalingam
PhD Research Scholar, Department of Management Studies, Dr. M.G.R Educational and Research Institute (Deemed to be University), Periyar E.V.R. High Road, Maduravoyal, Chennai 600095, India; Mobile: +91-9600711588; E-mail: rakesh10mba@gmail.com

Sudharshan R. Ravula
PhD Research Scholar, Department of Agricultural and Food Engineering, Indian Institute of Technology, Kharagpur 721302, West Bengal, India; Mobile: +91-9492883137; E-mail: r.sudharshanreddy@gmail.com

Ashish Rawson
Associate Professor, Indian Institute of Food Processing Technology, Thanjavur 613005, Tamil Nadu; Mobile: +91-7373068426; E-mail: ashish.rawson@iifpt.edu.in

Sadhana Sharma
PhD Research Scholar, Department of Food Engineering & Technology, Tezpur University, Sonitpur, Napaam 784028, Assam, India; Mobile: +91-7503685332; E-mail: sadhana.foodtech@gmail.com

Savita Sharma
Senior Dough Rheologist, Department of Food Science and Technology, Punjab Agricultural University, Ludhiana 141004, India; Mobile: +91-9814769992; E-mail: savitasharmans@yahoo.co.in

Shyam K. Singh
Guest Faculty, Assam University Silchar, Dorgakona, Silchar: 788003, Assam, India;
Mobile: +91-8876877706; E-mail: shyamsingh.iitkgp@gmail.com

Niveadhitha Sundramoorthy
Assistant Professor, Department of Food Technology, Sri Shakthi Institute of Engineering and
Technology, Sri Sakthi Nagar, L & T Bypass, Coimbatore 641 062, India; Mobile: +91-9962913510;
E-mail: s.nivi333@gmail.com

Subhashini Sundaramoorthy
Assistant Professor, Department of Food Technology, FET, Jain (Deemed-to-be-University),
Jain Global Campus, Jakkasandra Post, Kanakapura Taluk 562112, Ramanagara District, Karnataka,
India; Mobile: +91-9962610287; E-mail: s.subhashini10@gmail.com

Chikkaballpur K. Sunial
Assistant Professor, Indian Institute of Food Processing Technology, Thanjavur 613005, Tamil Nadu;
Mobile: +91-9750968423; E-mail: sunil.ck@iifpt.edu.in

Suka Thangaraju
PhD Research Scholar, Department of Food Engineering, Indian Institute of Food Processing
Technology, Pudukkottai Road, Thanjavur 613005, Tamil Nadu; India;
Mobile: +91-9488111443; E-mail: sukathangaraj@gmail.com

Barinderjeet S. Toor
PhD Research Scholar, Department of Food Science and Technology, Punjab Agricultural University,
Ludhiana 141004, India; Mobile: +91-8054829495, E-mail: barinderjeet-fst@pau.edu

Aswin S. Warrier
Assistant Professor (Dairy Engineering), KVASU Dairy Plant, Kerala Veterinary and Animal Sciences
University, Mannuthy Campus, Thrissur, Kerala 680651, India. Mobile: +91-7559067959.
E-mail: aswinswarrier@kvasu.ac.in

Ritesh B. Watharkar
Assistant Professor, Department of Food Processing and Technology, Karunya Institute of Science and
Technology, Coimbatore 641003, Tamil Nadu, India; Mobile: +91- 9370614686;
E-mail: watharkarritesh2019@gmail.com

ABBREVIATIONS

$W_{c+s+sample}$	Weight of (container + seeds + sample)
W_{c+s}	Weight of (container + seeds)
ΔT	Temperature increase
2D	Two-Dimensional
3D	Three Dimensional
3DP	Three-Dimensional Printing
4D	Four-Dimensional
x	Distance below a Surface (m)
A	Area of the product
AACC	American Association of Cereal Chemists
AFD	Atmospheric freeze-drying
AM	Additive Manufacturing
ANN	Artificial Neural Network
ASTM	American Society of Testing Materials
AYBWEP	African yam bean water-extractable proteins
BM	Buffalo milk
BOD	Biological oxygen demand
BP	Banana Peel
BWOSS	Batch wash ozone sanitation system
Ca	Calcium
CAD	Computer Aided Design
CCP	Critical control point
CCT	Capped Column Test
C_{fc}	Crust concentration
cfu/g	colony-forming units per gram
CIE	Commission Internationale de I'Eclairage model
CO	carbon monoxide
COD	chemical oxygen demand
DDVP	Dichlorvos
DES	Deep eutectic solvent
DIC	Detente instantanee controlee
DIY	Do It Yourself
DNA	Deoxyribonucleic Acid
DON	Deoxynivalenol

DS	Dry solids
DSC	Differential Scanning Calorimeter
D_T	Decimal Reduction Time (S)
EHD	Electro hydrodynamic drying
EIS	Electrical Impedance Spectroscopy
EMT	Effective Medium Theory
FAO	Food and Agriculture Organization
FCC	Federal Communications and Commissions
FD	Foam density
FDA	Food and Drug Administration
FDM	Fused Deposition Modeling
FE	Foam expansion
Fe	Iron
FFDM	Fat free dry matter
FMD	Foam mat drying
FT-NIR	Fourier transform near infrared
GFA	Guar foaming albumin
GRAS	Generally recognized as safe
HACCP	Hazard analysis critical control point
H_f	Heat capacity of calorimeter
HPH	High pressure homogenization
HPP	High pressure processing
HSD	Horizontal spray drying
IBS	Irritable bowel syndrome
ICMSF	International Commission on Microbiological Specifications for foods
IoT	Internet of Things
ISSHE	Inclined scraped surface heat exchanger
K	Potassium
K	Thermal conductivity
k_{eff}	Effective thermal conductivity
k_i	Conductivity of the ith phase
kJ/mol	Kilo Joule per mole
L	Length of the product
LAB	Lactic acid bacteria
M	Molecular weight
MC	Moisture content
MDSC	Modulated Differential Scanning Calorimeter
MFMD	Microwave assisted FMD

Mg	Magnesium
mg/L	Milligram/liter
Mn	Manganese
MNV	Murine norovirus
MRE	Magnitude of Relative Error
MS	Mild steel
MT	Million tons
MTS	Mano thermo-sonication
MW	Microwave
MW	Microwave
Na	Sodium
NaOH	Sodium hydroxide
NASA	National Aeronautics and Space Administration
NDDB	National Dairy Development Board
NDRI	National Dairy Research Institute
NF	Nanofiltration
O_2	Oxygen
O_3	Ozone
OH	Ohmic Heating
OTA	Ochratoxin A
P	Phosphorous
PBs	Plateau border
PEF	Pulsed Electric Field
PEF	Pulsed electric field
PGA	Propylene glycol alginate
POD	Peroxidase
PP	Polyethylene pouches
ppm	Parts per million
PPO	Polyphenol oxidase
PPO	Polyphenol oxidase
PUFA	Polyunsaturated fatty acid
PUP	Poultice Up Process
PV	Peroxide value
RCOG	Radiochemical Ozone Generation
r_d	Distance from the probe heater
RF	Radio Frequency
RH	Relative humidity
RO	reverse osmosis
ROS	Reactive oxygen species

RW	Refractance Window
SL	Stereolithography
SLS	Selective Laser Sintering
SPI	Soya protein isolate
SPU	Stephan processing unit
SSC	Soluble solids content
SSHE	Scrapped surface heat exchanger
STL	Stereolithography (file format)
t_1, t_2	Initial and final time
T_{cw}	Temperatures of the cold water
T_d	Temperatures of the dough
T_{eq}	Temperatures of the equilibrium of mixture
TFC	Total fungal count
TFSSHE	Thin film scrapped surface heat exchanger
T_{hw}	Temperatures of the hot water
TIDPs	Traditional Indian dairy products
TSS	Total soluble solids
UAC	Ultrasound assisted cutting
UAE	Ultrasound assisted extraction
USFDA	United States Food and Drug Administration
UV	Ultraviolet
VFD	Variable frequency drive
V_s	Volume of sample
WHO	World Health Organization
WPM	Water displacement method
W_s	Weight of sample
WWF	Whole wheat flour
X	Sample thickness
Zn	Zinc
μL/L	Microliter/liter
μmol/mol	Micromole/mole

SYMBOLS

ρ_c	Density of carbohydrates
ρ_f	Density of fat
ρ_p	Density of protein
ρ_w	Density of water
ρ_0	Initial density of dough
C_p	Specific Heat (J/kg°C)
C_p	Specific heat of dough
C_w	Specific heat of water
ε	Porosity
ε_i	Volume fraction of the ith phase
q	Heat
Q	Heat flux
R	Universal gas constant
T	Temperature (°C)
t	Duration of a Single Pulse (s)
v	Wave Frequency (Hz)
α	Thermal diffusivity
Λ	Wavelength (nm)
ρ	Density
ρ_b	Apparent density

PREFACE

Tropical climate average ambient temperature throughout the year is very high in most parts of the world, especially in developing nations, which is very favorable for the growth of spoilage microorganisms. However, due to acute power shortages, it is difficult to maintain the cold chain right from food production at crucial farms to food processing plants and finally to distribution outlets of foods. Therefore, food preservation prior to distribution and sale is a major problem in the tropical climates of most of the developing nations. In order to assure the consumer that the product is safe for human consumption, due importance is given to the quality and safety part of the production, processing, and distribution.

The primary aim of this book is intended to be used as a textbook for those students who are taking a college- or university-level food processing, safety or quality assurance course for the first time. The objective was to compile information that dairy and food science students are expected to be familiar with as part of their college or university program before they seek career positions in the food industry. This book will be further useful to food industry quality practitioners or employees who need to become familiar with updated information pertaining to their routine work.

The book will encourage the preservation of traditional knowledge of nonthermal processing and preservation techniques along with their advanced counterparts. Moreover, these novel processing techniques draw attention of researchers/policymakers because of their demonstrated beneficial effects by enhancing the shelf life of food along with maintenance of proper quality control and food safety.

By searching the literature, one can find volumes of books and specialized publications on nonthermal preservation techniques, but only few of them have targeted quality, safety and novel preservation methods of different foods using these novel techniques. Unfortunately, most of these publications have dealt with theoretical aspects of these strategies and technologies with little emphasis on real application in consumer and food products.

The book contains two main parts: Part I: Novel and Nonthermal Processing of Food and Food Products: Here we discuss the application of novel technologies for the improvement of food shelf-life and safety concepts are elucidated. Part II: Process Interventions for Food Processing and Preservations: The application of novel food processing technologies like different types of drying technologies, advances in food fermentation technologies, mechanization of traditional indigenous products for preservation of food and safety and different properties and concept of bakery products are elucidated.

We introduce this book volume under book series *Innovations in Agricultural & Biological Engineering*. This book volume is a treasure house of information and an excellent reference material for researchers, scientists, students, growers, traders, processors, industries, and others for quality control and safety of food products during production, processing, and transportation in any food industry and boost their confidence in the area of safety and quality aspects of food products.

This book has surpassed our vision and expectations due to the contributions by all cooperating authors to this book volume who have been most valuable in this compilation. Their names are mentioned in each chapter and in the list of contributors. We are grateful to all of them for their expertise, commitment, and dedication.

We would like to thank editorial staff at Apple Academic Press, Inc. for their valuable help and advice throughout this project.

We request the reader to offer your constructive suggestions that may help to improve the next edition.

Also, we would like to thank our families, who have taught us the importance of working hard, having clear goals, and standing for what we believe is right. It is a lesson that guides us in everything we do. Last but not the least we wish to thank our spouses, for their understanding and patience throughout this project.

—Editors

PART I
Novel and Nonthermal Processing of Food and Food Products

CHAPTER 1

DIELECTRIC HEATING: RECENT TRENDS AND APPLICATIONS IN FOOD PROCESSING

MOHAN NAIK, D. LAVANYA, SUKA THANGARAJU, NIKITHA MODUPALLI, and VENKATACHALAPATHY NATARAJAN*

ABSTRACT

In food processing industries, dielectric heating has been widely using for drying, sterilization, pasteurization, cooking, blanching, baking, cooling, thawing, insect and microbial control, and assisted technology for the extraction of valuable compounds from cell-matrix. Recent advances in dielectric heating include microwave freeze drying, microwave vacuum drying, microwave hot air assisted drying, and microwave-radio frequency plasma processing has emerged as novel processing techniques. Compared to conventional heating methods, dielectric heating has the capacity to inactivate harmful microorganisms and could destroy the active enzymes. It moderately affects the texture, color, and other bioactive components in foods. Numerous efforts have been made toward the nonuniformity of heating. Recent trends include the application of mathematical modeling, plasma processing, freeze-drying, and other drying techniques coupled with dielectric heating. This chapter summarizes recent trends and developments in the application of dielectric heating in food processing.

1.1 INTRODUCTION

It is surprising to know that dielectric heating has been in use for quite a long time. It may appear to many engineers that these are new forms of heating

*Corresponding author. E-mail: venkat@iifpt.edu.in

but the first application of dielectric heating has started during the World War II. Dielectric heating is termed the "workhorse" heating method which has a wide range of applications in wood, paper, plastic, textile industries. The market has pressure to produce healthy and safe food products that are preservative-free, chemical-free, tastier, and cheaper. Dielectric heating may play an important role in terms of environmental impact perspective, energy-efficient, uniform and proper heating of food products, etc.

Dielectric heating can be widely employed in food processing, which includes drying and dehydration, sterilization, pasteurization, baking, blanching, microbial, and pest control. When an alternating electric field or electromagnetic radiation heats the dielectric material, the process is termed as dielectric heating. When an insulating material is exposed to electromagnetic radiation, the atoms get stressed and because of the interatomic friction, heat is produced. This heating process is termed as dielectric heating.

Dielectric heating can be applied to all electromagnetic radiation frequencies which range from 300 kHz to 300 GHz. The dielectric heating frequencies can be broadly classified into radio frequency (RF) (300 kHz–300 MHz) and microwave (MW) (300 MHz–300 GHz), which can be further classified and it is given in Table 1.1. The important characteristics which differentiate RF and MW are that in RF, an electric field is built-up within the electrodes of the apparatus and in MW, a wave is propagated and reflected. RF performs adeptly with a huge quantity of product with good ionic conductivity. Whereas MW goes well with a small quantity of product with dipolar nature [14, 91, 95].

TABLE 1.1   Dielectric Heating Frequency Ranges

Frequency Range	Frequency Type
300–3000 kHz	Middle frequency
3–30 MHz	High frequency
30–300 MHz	Very high frequency
300–3000 MHz	Ultrahigh frequency
3–30 GHz	Super high frequency
30–300 GHz	Extremely high frequency

Before selecting the frequency range for dielectric heating, the following four points are to be considered: (1) dielectric properties of the material which varies according to the frequency along with the temperature and moisture content [30]; (2) quantity of material to be processed [48]; (3) the moisture level of the materials to be processed; (4) conditions required for the material to be processed (e.g., pressure, temperature, etc.) [10]. Though there is a wide range of frequencies that comes under dielectric heating, the practical application of dielectric frequency includes 13.56, 27.12, 40.68, 896, 915, and 2450 MHz. In the recent years, it is seen that there has been a rise of interest in the application of dielectric heating for industrial purposes. This may be due to the energy crisis in the world and the wide acceptance of MW oven. The purpose of this chapter is to give an idea about dielectric heating, the principle involved in it, and utilization of dielectric heating in the food manufacturing sector.

1.2   PRINCIPLES OF OPERATION

In a dielectric heating system, an alternating magnetic field is introduced in between the electrodes. An important factor to remember in terms of dielectric heating is that they are not forms of heat but they produce different combinations of energy, which are manifested as the interaction of heat with materials. There are two important phenomena which are responsible for dielectric heating: ionic conduction and dipole rotation.

1.2.1   IONIC CONDUCTION

Food always has millions of ions in it like sodium, hydronium, chloride, hydroxyl ions, etc. Since ions are charged molecules, they are accelerated by the electric field strength. These cause ions to make movement in the opposite direction opposite of their polarity. And thus, they get hit with other nonpolar molecules and liberates kinetic-energy that makes them accelerate and hit other water molecules in the billiard ball model. When the polarity changes, the ions are accelerated in the opposite direction. This happens a million times per second where a large number of collisions and energy transfer happens. Hence, here happens the energy transfer in two steps: (1) electrical energy is converted into induced order kinetic energy, (2) this kinetic energy is further altered into disordered kinetic energy, which is also termed as heat. This type of heating is neither dependent on temperature nor frequency. The power developed per unit volume (P_v) by ionic conduction is given by the following formula:

$$P_v = E^2 q n \mu \tag{1.1}$$

where q is the amount of electric charge on each of the ions, n is ion density, and μ is the level of mobility of the ions.

1.2.2 DIPOLAR ROTATION

Water is one best example of dipolar molecules, which means the molecules having an asymmetric charge center. Due to the stress created by the electric field, the other molecules may also become dipolar in nature which is also known as induced dipoles. These dipoles are affected by the rapidly changing polarity of the electric field intensity. Though the molecules are randomly aligned, the electric field intensity applied to the molecules tries to pull them into a proper arrangement. But when the field relaxes to zero, the dipoles get back to their random alignment. This accumulation and deterioration of the electric field happen a million times per second. It results in energy transfer from electric energy to potential energy and then to kinetic energy or thermal energy in the material. Dipolar rotation is dependent on temperature and molecular size for accumulation and deterioration of frequency known as relaxation frequency [50]. The power developed per unit volume (P_v) by dipolar rotation is given by the following formula:

$$P_v = kE^2 f\varepsilon' \tan \delta \qquad (1.2)$$

or

$$P_v = kE^2 f\varepsilon'' \qquad (1.3)$$

where k is the constant which is dependent on units of measurement; E is the electric field strength; f is frequency; ε' is the relative dielectric constant or relative permittivity; $\tan \delta$ is dissipation factor; and ε'' is the loss factor.

The relative dielectric constant gives details about the degree to which an electric field may accumulate within a material. The loss tangent measures how much of the electric field will be converted to heat. From the above equation, it is revealed that E and f are functions of the equipment, and ε', ε'', and $\tan \delta$ are functions related to material that is being heated. Another significant point is that to maintain the particular power level P_v, and the frequency f is altered to increase the electric field strength E.

1.3 FACTORS AFFECTING INTERACTION OF ELECTROMAGNETIC WAVES WITH MATERIALS

Any material can be categorized into four different groups according to the way they interact with electromagnetic radiations. They are conductors, insulators, dielectrics, and magnetic compounds. Conductors are materials with free electrons present in them that reflect the electromagnetic radiations just like a mirror reflecting light. These materials can be used as applicators or waveguide. Insulators are nonconductive materials like air, ceramics, glass, etc. They are used for containing materials that are to be heated by electromagnetic radiations.

These materials are also termed "nonlossy dielectrics." Dielectrics are those materials with properties between conductors and insulators that come under the term "lossy dielectrics" which can absorb electromagnetic radiation and convert it into heat. Any material which contains moisture is termed as dielectrics, for example, food, wood, oils, etc. Magnetic compounds are materials that react with magnetic components of electromagnetic radiation and will heat the material. They are used as a smothering or protecting device to prevent leakage of electromagnetic radiations.

As mentioned earlier, the properties that decide whether the material can be effectively heated by dielectrics are the dielectric properties. They are relative dielectric constant ε', dissipation factor, tan δ, and the loss factor, ε''. The complex dielectric constant may be given as

$$\varepsilon = \varepsilon' - j\varepsilon'' \qquad (1.4)$$

where, $j = \sqrt{-1}$ which means there is a 90° phase between the real and imaginary complex dielectric constants. And dissipation factor can be expressed as

$$\tan \delta = \frac{\varepsilon''}{\varepsilon'} \qquad (1.5)$$

These factors are affected by several parameters.

1.3.1 MOISTURE CONTENT

Water is an important dielectric component that absorbs electromagnetic radiation. The quantity of free moisture present in food greatly affects the dielectric constant. The dielectric constant of the water at room temperature is approximately around 78. Therefore, with a large quantity of water present in the material has a higher dielectric constant. Few thumb rules that can be applied:

1. higher will be the dielectric constant, with higher moisture content,
2. with increase in moisture content, dielectric loss also increases and level values in the range of 20%–30% and decrease at still higher water content, and
3. the dielectric constant of the mixture of components stands in the range of the individual components.

The components of the food are dielectrically inert ($\varepsilon' < 3$, $\varepsilon'' < 0.1$) in nature when it is compared with ionic fluids or water. When the moisture content of food is very low, the remaining water molecules are intact and remain uninfluenced by the electric field and the low specific heat capacity turns into the essential element to heat the product. For high-carbohydrate foods like alcoholic beverages, bakery products, the dissolved sugars, and alcohol are important interactive materials [85].

1.3.2 TEMPERATURE

The temperature complexity of the dielectric property increase or decrease depending on the composition of any food product. Generally, when the material is below its freezing point, it shows reduced dielectric constant and dielectric loss. MW heating mostly depends on the amount of moisture present in it. While the RF heating depends on food compositions which influence the heating distribution. Nevertheless, with adequate heating duration, the heat transfers and thermal conductivity lowered the gradient in temperature throughout the product [30].

1.3.3 DENSITY

The dielectric constant value of air is 1, which makes the material transparent to electromagnetic radiations; therefore, the incorporation of air reduces the dielectric constant. Hence, lower density food products get heated faster at a given power level than the food with higher density and similar composition. Porosity is probably having a considerable effect on the dielectric constant, but it has less effect on the dielectric loss factor. In spite of low dielectric loss and less absorption, the food with low specific heat capacity is capable of being heated through electromagnetic waves because of the less requirement of energy required per unit weight to increase the temperature [33].

1.3.4 FREQUENCY

Dielectric properties of the material are dependent on the frequency of the electromagnetic radiation applied to the product. The frequency of electromagnetic radiation for any unit operation is crucial to choose since the dielectric performance and heating capacity of food differ according to temperature and frequency that are considerably affected by moisture and salt concentrations. The ionic loss for any food product is huge and dipole losses are reduced at the frequency 915 MHz than at 2450 MHz. For the industrial, medical and scientific field 5800 MHz is the most commonly used frequency which has higher dipole losses and negligible ionic losses [75, 93].

1.3.5 THERMAL CONDUCTIVITY

Thermal conductivity does not play a major role in the dielectric heating of the product compared to conventional heating. This is because of the speed with the former heat, thus decreasing the time in which thermal conductivity can be effective. There are cases in which thermal conductivity has an important function in dielectric heat. For example, when the penetration depth of electromagnetic radiation is small, then the thermal conductivity may be dependent on the heat transfer to the interior of the product. Thermal conductivity also affects the homogeneity of the heating process. If the product size is large when compared to the wavelength, then the superficial way of heat is preferred, whilst for product dimensions similar to wavelength, the temperature could be greater in the center. Thinner parts of the product might be overheated in comparison to the thicker portion. This effect can be checked by lowering the input power and prolonging the duration of heat [92].

1.3.6 PENETRATION DEPTH

Penetration depth is not the property of the material, but it rather results in various properties of the material. It is important that the electromagnetic radiation should penetrate as deeply as possible during bulk heating of the product. If it does not happen then the heating is limited to the external surface of the material. The parameters affecting the penetration depth of the product are wavelength, dielectric loss factor, and dielectric constant which is according to the following formula:

$$D = \frac{\lambda_0 \sqrt{2}}{2\pi} \left[\varepsilon' \sqrt{1 + (\varepsilon''/\varepsilon')^2} - 1 \right]^{-1/2} \tag{1.6}$$

If the value of ε'' is low, then the Eq. (1.6) can be written as

$$D = \frac{\lambda_0 \sqrt{\varepsilon'}}{2\pi\varepsilon''} \tag{1.7}$$

From the above equations, it is clear that materials with high dielectric constant and loss factor will have smaller depths of penetration [59, 75].

1.3.7 PERMITTIVITY

Permittivity is the evaluation of material capacity to store electric energy. The permittivity of any material is given by the following formula:

$$\frac{D}{E} = \varepsilon_{abs} = \varepsilon_0 \varepsilon = \varepsilon_0 \left(\varepsilon' - j\varepsilon'' \right) \tag{1.8}$$

where ε_0 is the absolute permittivity of vacuum ($\varepsilon_0 = 8.855 \times 10^{-12}$ F/m).

Lumped circuit method is used to establish the intricate permittivity over the range of frequency from 0 to 200 MHz. At a frequency range above 10 MHz, a capacitance bridge is used to measure the capacitance and dissipation factor. Resonant circuits are used for frequencies ranges from 10^4 to 10^8 Hz with fixed inductors and variable capacitors with resonance. The resonance method can also be used to measure the capacitance of the material [75].

1.4   APPLICATIONS IN FOOD PROCESSING

Dielectric heating has a wide range of applications in the area of food processing industries as it is safe in maintaining the quality of food products through swift and evenly heating. By varying the frequency range of electric current, dielectric heating has been designed in various applications like mainly in the area of crop production, to kill some pests in food, for uniform and quick heating of food materials. The following section of this chapter details the various food applications of dielectric heating.

1.4.1   DRYING AND DEHYDRATION

Drying and dehydration is a thermophysical and physicochemical unit operation. Drying is the process of removal of moisture to a predetermined period of time, whereas dehydration is the process of removal of moisture to a bone-dry condition with pores on the surface that can be rehydrated by the addition of moisture. Drying is an energy-intensive method as it uses the latent heat of vaporization to remove the excess moisture present in the food products by decreasing the water activity and extending the shelf life. The major challenges that the food processing industries facing in the field of conventional dryings like hot air are longer drying periods with prolonged exposure to elevated thermal energy leads to the deprivation of overall food quality (color, texture, and nutritional composition) [54].

Apart from other conventional heating methods where the transfer of energy takes place from a hotter to the cooler medium of material is modified through conduction, convection, and radiation. In the case of dielectric heating due to dissipation of electromagnetic energy directly into the food material leads to molecular interaction with the electromagnetic field generates the thermal energy required for the evaporation of moisture inside the dried material to attain uniform heating throughout the product [88].

Dielectric heating comprises both RF and MW heating which utilizes electromagnetic spectrum as an energy source. Figure 1.1 proposes the allotment of MW and RF in the electromagnetic spectrum [74]. In RF, the applied frequencies were in the range of 10–300 MHz with a frequency range of 300–30,000 MHz in the case of MW heating. As per the US Federal Communications and Commissions, the recommended frequency range for RF heating of 13.56 MHz ± 6.68 kHz, 27.12 MHz ± 160.00 kHz, 40.68 MHz ± 20.00 kHz, and MW processing of 915 ± 25 MHz, 2450 ± 50 MHz, 5800 ± 75 MHz, and 24,125 ± 125 MHz for industrial purpose.

In 1950, the usage of MW energy for the generation of thermal energy in agricultural commodities came into existence. The interest in the field of MW drying technique was broadened in the early 1960s for industrial and other scientific applications [45]. Due to the crucial economic and technical obstacles, MW devices have been approved for the drying of agro-food and other biological articles [60].

Application of RF drying in the field of food processing started in the year 1940 to cook processed meat, to dehydrate bakery products and vegetables [40] followed by the thawing of frozen products in 1960 [34]. RF processed bottled fruit juices such as orange, peach, and quince showed better retention of organoleptic properties with a reduced load of bacteria compared to other conventional thermal processing techniques [16]. In the 1980s, RF heating was done to investigate the postbaking properties of crackers and cookies [5]. Thus, dielectric heating through RF and MW helps not only to shorten the drying time but also improves the end quality of the dried products. Studies conducted by few researchers in dielectric drying of food materials using both RF and MW heating are presented in Table 1.2.

In conjunction with dielectric heating to make the MW and RF drying more effective, certain agriculture products need to be pretreated, for example, immersion of seedless grapefruits in the hot alkaline solution of ethyl oleate not only imparts better quality but also reduces the drying time both in convection and MW drying.

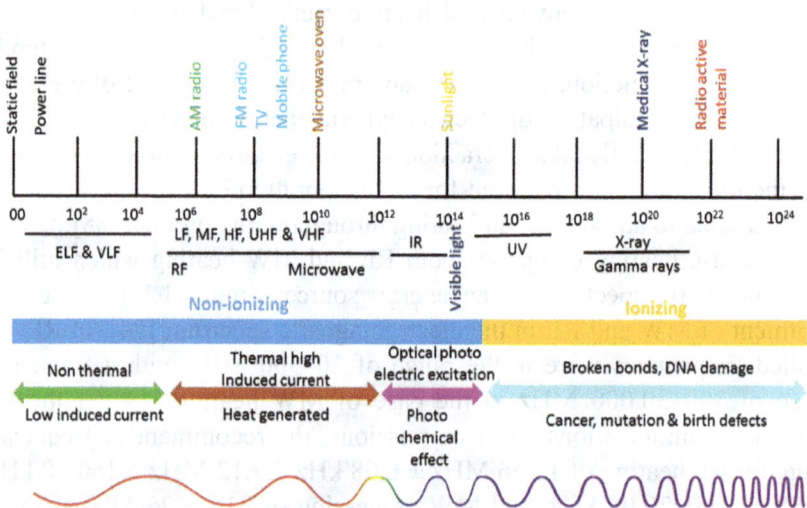

FIGURE 1.1   Allotment of MW and RF frequencies in the electromagnetic spectrum.

TABLE 1.2 Recent Applications of Dielectric Heating in Drying of Food Materials

Mode of Treatment	Product Category	Frequency Levels and Exposure Interval	Improvements After Treatment	References
RF cooking	Meat and meat products (ground, communited, and noncommunited)	27.12 MHz at 72 °C	• Reduction in cooking time up to 1/25th of conventional cooking time • Faster heating rates at 10–20 °C • Less loss of juiciness • Retains color and water holding capacity • The decrease in hardness with an increases springiness in case of RF cooked communited meat • Suitable for communited and ground meat products • Reduces *E. coli* and enhances product stability	[43]
RF-assisted pasteurization	Vacuum-packed ham slices	600 W at 27.12 MHz at 75–85 °C for 5 min	• Improves the storage stability of packed hams • Decreases bacterial load, reduces moisture losses • Helps the overall acceptability and sensory characteristics of ham slices	[58]
MW-assisted drying	Apples	915 MHz or 2.45 GHz	• Removal of moisture from 87%–4% • Decreases dielectric constant and loss factor	[21]

TABLE 1.2 *(Continued)*

Mode of Treatment	Product Category	Frequency Levels and Exposure Interval	Improvements After Treatment	References
MW-assisted drying	Thompson seedless grapes	2450 MHz	• Removes the excess moisture • Reduces the microbial load • Even drying • Better retention of color and texture compared to convection drying	[90]
MW-assisted convective air drying	Carrot cubes	600 W at 45–60 °C at 2–4 MW power levels	• Considerable reduction in drying time up to 25%–90% • Better retention of color at a lower power level compared to the higher power level	[70]
MW drying	Corn	• 0.25, 0.50, and 0.75 W/g • Inlet air velocity: 0.2, 0.5, and 0.8 m/s • Inlet air temperature: 30, 35, and 40 °C	• Reduced drying rate at higher MW power level, irrespective of air inlet temperature • Can be effectively utilized in the drying of seed, food and animal feed • >92% germination can be attained at MW power level of 0.25 W/g.	[82]
Combined of infrared and microwave drying	Banana and Kiwi fruit	Infrared power: 600 W, microwave power: 320 or 420 W	• Reduction in drying time (98%) • Shorter drying time of 0.75–2.25 min in MW heating compared to hot air drying which requires 200 min to reduce the final moisture	[63]

TABLE 1.2 (Continued)

Mode of Treatment	Product Category	Frequency Levels and Exposure Interval	Improvements After Treatment	References
Combined infrared and microwave drying	Raspberry	IRP (675 W)+MWP (600 W), vacuum pressure 85 kPa	• Time-saving process up to (55.56%) • Greater rehydration capacity (25.63%) • 2.5 times enhancement in product crispiness • Better retention of anthocyanins	[83]
Combined infrared and microwave drying	Green pepper	IRP (1500 W)+MWP (100 W), air temperature 65 °C	• Minimal energy intake • 57%–78% deprivation in drying time • Decreases water activity (20%–30%) • Higher rehydration efficiencies	[55]
Combined infrared and microwave drying	Eggplants	IRP (1500 W)+MWP (630 W),	• Effective diffusivity • Decreased drying time from 22% to 12% • 54% reduction in shrinkage • Higher rehydration ratios up to 37%	[7]

1.4.2   OTHER MW DRIED PRODUCTS

In food processing, MW heating has shown a wide range of application due to its adaptability. Pasta is one of the most preferred products in the western countries, which is prepared by using coarsely ground flour called semolina (durum wheat) by hydrating, mixing, kneading, and extruding through an extruder using a suitable die. The last and foremost step in pasta making is drying, to remove the excess moisture present on the surface. Drying moisture from pasta is a slow dehydration process that consumes excess time through hot air drying and gives rise to inferior product quality. This issue can be effectively solved through MW drying where the moisture is removed from the interior to the surface by electromagnetic waves. Drying of pasta can be made still more effective when MW drying is combined with hot air drying [28]. Another outcome of MW drying is the of MW energy along with fluidized-bed drying in the drying of macaroni beads at household level resulted in a 50% reduction of fluidized bed drying time [28, 31]. The efficiency of air drying through convectional and MW-assisted hot air rotary driers was compared and evaluated on the drying time and final product quality of penne short cut pasta. The study revealed that the assistance of MW energy along with any other driers helps in the rapid cut down of drying time without affecting the final product quality [4]. Hence, MW drying can be effectively utilized in the drying of agricultural commodities, fruits–vegetables, meat and meat products, and baked products to remove the excess moisture present on the outer surface.

MW-assisted dielectric heating has its own advantages and limitations. Dielectric drying has shown some limitations in the case of drying fruits and vegetables because of the high set-up, initial investment made this technology complicated compared to other conventional drying techniques.

1.4.3   BAKING

In general, baking is the terminology used to describe a cooking process that utilizes dry heat. This type of cooking is mainly carried out in an oven that uses MW energy and the products obtained after baking are categorized as "baked goods or products." This includes bread, cookies, biscuits, buns pizzas, etc. Dielectric heating is generally matched-up with MW and RF which generally utilizes the electromagnetic spectrum in generating the heat inside the products. It is also called a volumetric heat transfer technique which poses certain advantages in food processing sectors over conventional heating. Dielectric technology has been widely utilized in the drying of textile packages, in drying of biscuits, and other products after baking to remove the surface moisture, to hinder the activity of harmful microorganisms, and to enhance the storage life of the product [36]. As per [41], RF heating and dielectric heating also called as postbaking that is considered as one of the oldest technology which has been known for more than 40 years in the baking industries. There are various unique advantages of MW heating in baking industries as less labor requirement, consumes less processing time and space. The other applications of MW heating include proofing, baking, and pasteurization of baked goods [89].

There is a great demand for advanced technology which consumes less time along with nutrition retention in certain food products due to an 85% change in eating patterns among the growing population. The unique efficiency of MW heating has been not only being confined to the drying and dehydration but also adopted in various unit operations like thawing and tempering of frozen foods products including meat, dairy products, fish, sausage, hams, and bacon, drying of pasta and vegetable, etc. [12].

In conventional baking, the outer surface of the dough is heated by a combination of heat transfer, that is, conduction, convection, and radiation, which causes structural changes followed by gelatinization of starch, denaturation of protein, volume expansion, CO_2 gas liberation from the leavening agents, evaporation of moisture, the formation of crust, and nonenzymatic browning. In MW baking at ambient heating conditions, the heat is generated by the interaction between water molecules coupled with other dissolved solutes and ions to causes structural changes within the product. The mode of heat transfer in the case of industrial ovens is by radiation, conduction, and convection which depends mainly on the design of the oven, configuration, and operation. Studies conducted by several researchers in the field of heat transfer in the baking of biscuits, bread, and cake confirmed that baking was usually done by the utilization of electrical energy, forced convection, and gas-fired ovens [98]. An about 50%–80% utilizes radiation mode of heat transfer compared to baking through conventional ovens. In industrial baking ovens, control of heat through radiation is much difficult than in conventional heating [51].

Most of the researchers found out the undesirable changes in the texture of bread baked in nonconventional ovens due to altered patterns in heat and mass transfer with a shorter baking time linked to MW radiation. Due to uneven heat distribution in the crust part in MW-baked breads resulted in the formation of crustless products with a coarser and less firm texture. The other deformities caused due to MW baking include inadequate gelatinization of starch, changes in gluten structure due to MW-induced heating caused by the rapid steam, and gas generation might be the earliest causes for deprived quality losses in bread. Though surface browning in the products after baking was very negligible in the case of MW-baked goods than conventionally baked ones but MW baking yields superior and shelf-stable products. Most of the studies confirmed that bread-keeping quality can be improved through air impingement convention ovens. The main difference between a convention and MW heating is in the case of MW baking heat is transferred from the walls of the oven to the surface of the product through radiation, whereas in convention baking, heat is transferred from hot air flowing inside the oven to the product. For instance, biscuits baked in a band oven comprises all three modes of heat transfer namely conduction, convection, and radiation. Out of the total heat transfer in the case of band types of ovens from bands to the biscuits 43% of heat transfer was occupied by radiation followed by 37% by convection and conduction shares 20%. Likewise, 70% of the bread baked in an electric oven can be done through radiative heat transfer [87].

The process of baking involves a composite chemical reaction in the removal of moisture which imparts the desirable texture and structure to the final product. In all the baked products, moisture content judges the final quality of the product. For example, in bread, loss of moisture is considered as negligible as it helps in the formation of the loaf structure and the crust after baking. In the case of biscuits, loss of moisture is important in transferring the crispiness character to the biscuits after baking. In baking, when RF heating is assisted with conventional heating leads to the change in a wide array of process parameters like in simple drying. Some of the quality changes happened in baked goods can be measured through physical observations by judging the taste, texture of the crumb, marketability, etc.

The main objective for the introduction of RF along with other conventional drying systems in baking bread causes moisture to redistribute which leads to the formation of open structures in bread dough is considered an undesirable strategy. The major parameter to be considered during the baking of bread is the moisture distribution which is markedly nonuniform between the two zones, that is, the outside crust and the soft spongy part inside the bread. When the baking process is combined with the RF, it leads to the rapid rise in the temperature up to 100 °C, which causes the product to pasteurize followed by the chemical reactions in the outer crust to cause surface heat transfer. Incorporation of dielectric heating along with MW heating in baking noted a considerable reduction in baking time with a rapid temperature rise causes the product pasteurization. The major drawback with MW heating is the use of containers which inevitably changes the color of the product making it undesirable in the market. Works done by EA technology in the field of dielectric heating of bread through MW and RF heating have been come up with a cheap compatible container for better heat transfer to the bread loaf surface to obtain reliable crust structure and to retain the color of the product after baking. There are various other substantial benefits of RF and MW technology in the field of bread baking, the major among them are reduction of baking time and surface pasteurization without giving up the desired product quality.

Work done by previous researchers found that due to rapid rise in temperature at 90 °C leads to an increase in -amylase in European bread and other work done by FMBRA in the area of MW heating of bread observed that due to rapid and uniform heating of bread loaf at 90 °C ceases the enzyme activity which imparts an undesirable flavor and texture. The other advantage of RF heating over bread baking is it reduces the proving time and quality of the yeast followed by the rapid expansion of the dough caused by the water vapor production due to the carbon dioxide released by the yeast within the closed cell of the dough.

The major attribute to be taken into the consideration in case of biscuits after baking is crispiness, texture, and color. The combination of RF and MW-assisted dielectric heating of biscuits has proven the better color, texture without giving up the organoleptic properties. RF baking is similar to that of MW baking where the food products are exposed to a high voltage signal of RF through a set of a parallel electrode where the food is placed in between the electrodes. When the current is passed in between the electrodes, the food particles get heated up as a result of the rotation of polar molecules. The baking rates can be increased from 30% to 50% by installing RF units along with the conventional ovens. Through MW baking in combination with other conventional baking, techniques help not only in maintaining the better quality of the final products but also helps in the effective reduction in the microbial and enzymatic activities by uniformly removing the excess free moisture available for the microbial multiplication, reduces product discoloration and damage due to thermal processing. Certain studies conducted by researchers in the field of baking through the combination of MW and RF heating along with dielectric heating have concluded that the combination of the technologies is cost-effective, reduces more power consumption, and preserves the final product after baking like fluffiness in cakes, crispiness in crackers and biscuits, etc.

1.4.3.1   UTILIZATION OF MW-ASSISTED SYSTEMS IN BAKING OF VARIOUS FOOD PRODUCTS

MW treatment has been recognized in various food processing industries over other conventional heating due to its lower processing time, better product quality, and comfortability in producing various dried products. The drying quality in MW drying can be improved in combination with other conventional drying techniques for better processing conditions. The application of MW heating in the pasteurization of packed bread was done [11] at 14–17 MHz and can be further proceeded up to 2450 MHz. Apart from bread baking, MW energy especially MW proofing has been successfully utilized as a supplementary heating source in the production of yeast-raised doughnuts. The major advantage is its cut-off the production cost and time of proofing from approximately 35 to 4 min [77]. Moreover, MW reduces the frying time and improves the quality of food products [76]. Commercially, the adoption of MW heating is used in the drying of pasta along with hot air and also in the formulation of food products like pancakes [79].

When certain cereal-based products like bread, biscuits, and cakes were subjected to the baking process various physicochemical and sensory changes would occur. These cooking-induced changes taking place in baked goods are highly important for the product acceptability and digestibility of the consumers. A combination of dielectric heating in MW baking brings certain changes in baked goods like caramelization or nonenzymatic browning due to the complex reaction between the reducing sugars and amino acids followed by lipid peroxidation. The mixture of a chemical reaction inside the product is mainly influenced by various factors such as heat or temperature, pH, moisture content followed by the presence or absence of metallic cations, and the structure of the sugar. Table 1.3 depicts the MW baking of cookies, biscuits, and pound cakes.

Apart from baking, MW energy can be effectively utilized in reducing the proofing time of doughnut dough to 4 min that was comparatively consuming a proofing time of 40–60 min in case of conventional drying. The baking efficiency can be successfully achieved in deep fat fried of cake doughnuts through MW baking. Deep fat frying helps in attaining the uniform brown color to the product, whereas MW baking removes the excess moisture to maintain the final organoleptic properties of the cake-doughnut. MW heating can also be successfully introduced at the final stages after baking of biscuits, crackers, fried potatoes, chips, etc., to maintain the final product structure [2, 9, 78].

TABLE 1.3 Application of Microwave in Baking Process

Products	MW Baking (Exposure Temperature and Time)	Quality Changes	Ref.
Cookies	617.3 and 745.5 W after conventional heating at 240 °C for 4 min	• Reduction in total moisture and moisture gradient • Reduced cracks • Slight darker color • Increased volume expansion in control samples	[9]
Biscuits	MW heating at 700 W for 30 s immediately after subjecting to convection and reel ovens	• 5% reduction in checking compared to conventional ovens (61%) • Less vulnerable to checking upon exposure to high ambient humidity	[2]
Pound cakes	MW baking: 240 W for 5 min Conventional baking using swing oven: 180 °C for 40 min	• Structure of MW baked and swing oven-baked pound cakes revealed similar structure • Microwave baking is found to be suitable in baking of pound cake dough consisting of high fat, sugar, and moisture	[78]
Legume cake	Combination of infrared baking (1500 W) and MW power (700 W) for 4 min	• Softer and voluminous cake texture • Larger pores	[61]
Gluten-free bread	Infrared power (150 W, 20%) + microwave power (20 min, 30%) using 3 halogen lamps	• Lowers the hardness level • Higher volume with desirable color	[62]
Soy cake	Combination of Infrared power 1500 W + MWP (705 W for 9.5 min)	• Reduction in baking time up to 1.37 m³/kg • Specific volume increases • Reduction in hardness value up to 0.1646 N	[1]

1.4.4 PASTEURIZATION AND STERILIZATION

Conventional methods of heating for achieving pasteurization in solids and liquid foods require a large amount of heat generation, which needs to be effectively transferred onto the foods by conduction, convection, and radiation with as minimal losses as possible. This method is expensive, time-consuming, and often results in nonuniform heating [99]. Dielectric heating offers an efficient, rapid, and economical solution to this problem, and allows uniform heating throughout the medium. Also, the application of conventional pasteurization method can be hardly applicable to low-moisture foods due to the lower heat transfer rate of solids [47].

Several previous studies indicated that the dielectric heating mechanism has largely been used for low moisture foods effectively. Li et al. [44] developed a dielectric pasteurization technique for almonds using RF. They have effectively pasteurized and dried the presoaked and tempered almonds using RF treatment with a 10.5 cm electrode gap for 1.5 min, after which drying for 8.5 min at an electrode gap on 8.5 min, without adversely affecting the quality parameters and moisture content. Geveke et al. [24] described a rapid reduction in pasteurization time of shell eggs by 60% in RF treatment for 23.5 min, in comparison with 60 min of hot water immersion treatment. They concluded that the maximum temperatures of albumen and yolk were 50 and 61 °C, respectively, thus not affecting the foaming quality of yolk, along with achieving a 6.5 log reduction of *E. coli*, in response to 6.6 log reduction using hot water.

An investigation conducted by Dev et al. [19] reported the dielectric properties of different components of egg and relevance of MWs in pasteurization of eggs. Another study on tomato elaborated on the effect of dielectric properties (dielectric constant and loss factor) at different MW power range from 915 to 2450 MHz, which can be effectively applied to develop MW pasteurization and sterilization system for tomato [66]. Peng et al. [67] also demonstrated the applicability of 915 MHz single-mode MW-assisted pasteurization and sterilization technique for prepacked carrots in the brine solution, by using Gellan gel systems as a model. Two processes, namely F90 °C = 3 min and F90 °C = 10 min, were successfully established, which reduced the total time for processing by half and two-thirds, respectively.

Ozturk et al. [64] determined that RF heating of dried vegetable powders like broccoli powder, ranging from 0.56 to 2.12 °C/s were directly correlated to moisture content and dielectric losses, and can be used as an efficient method to sterilize dried vegetable powders without potentially deteriorating its quality. A study on pasteurization/sterilization of fermented red chili paste using ohmic heating demonstrated 99.7% effective deactivation in *Bacillus* strains, as compared with 81.9% inactivation by conduction heating [13]. These studies excellently demonstrate the applicability of dielectric heating techniques in both solid and liquid foods, without compromising of final product quality and safety.

1.4.5 EXTRACTION

In recent years, extraction of essential oils, phytochemicals, organic and aromatic compounds from plant-based foods has been profoundly important in many agriculture-allied industries. Traditional techniques like steam distillation, hydro-distillation, or Soxhlet extraction techniques consume a large amount of time and energy and require higher quantities of suitable solvents, thus creating an environmental concern [17]. These conventional methods also require heating of the components which can cause a loss in either quality or quantity of the extracted compounds [27]. One more disadvantage in conventional extraction methods is that elevation of temperature and presence of oxygen during extraction can trigger chemical modifications, compromising the purity of the extract. Thus, dielectric, more precisely MW-assisted extraction techniques form an economical, and highly effective solution using minimal exposure times to give better quality yield [84].

The added advantage for MW-assisted processes is that they can be used as a treatment on its own or can be used to assist any extraction process to improve quality and yield. Périno-Issartier et al. [68] described that the MW-assisted hydrodiffusion technique extracted a maximum yield of 5.4% of higher quality lavandin essential oils from its flowers at a minimal exposure time of 15 min at 1.3 kWh power consumption, against 120 min and 8.06 kWh power for steam distillation method. MW-assisted negative pressure cavitation system has been employed to increase the output of bioactive compound chimaphilin, 20-O-aglloyl, and hyperin from pyrola, showed an overall reduction in extraction time by 40% [96]. MW-assisted extraction process for pumpkin oil from seeds resulted in a better quality of oil with a higher concentration of polyunsaturated omega-6 fatty acid and antioxidant capacity than soxhlet extracted oil [35]. Studies by Cravotto et al. [15] described elevated oil yield using MW-assisted and ultrasound and MW-assisted extraction techniques from plant foods like seaweed, in comparison with conventional extraction techniques. One of the studies stated that MW exposure of pepper for 2 min resulted in the enhanced recovery of piperine up to 95%, contradictory to 20% recovery in unexposed pepper with even vigorous agitation of 1 h [73].

Pettinato et al. [69] evaluated the use of MW as a preheating medium for extraction of solids from spent coffee grounds and elaborated that the highest yield was obtained when heated using 408 K power level for 10 min and extracted for 120 min. This study emphasized the importance of heating time in MW-assisted extraction processes, as it regulated the time required for extraction and efficiency of extraction. Another study on the extraction of chitosan from *Rhizopus oryzae* NRRL 1526 strain biomass, set MW assistance parameters for 300 W and 22 min as optimum extraction conditions [80]. It was also observed that MW-assisted extraction increased the yield by almost 2-folds compared to conventional systems (13.43% vs 6.67% (w/w) of biomass taken). MW aided extraction techniques to offer an improved alternative to the conventional extraction techniques, as it is economical, power-saving, lesser extraction time, and better yield. It also can extract better quality extracts with better properties and without oxidation of the compounds.

1.4.6   MICROBIAL AND INSECT CONTROL

Microbial inactivation by volumetric heating techniques like MW and RF is primarily due to internal heat generation. RF and MW processes generate electromagnetic energy generates internal heat in the cells. This causes denaturation of proteins, inactivation of enzymes, and changes in nucleic acids composition. These changes cause the deformation of genetic material like DNA, by splitting them into single strands by breaking the hydrogen bonds between strands. This directly leads to inactivation and apoptosis of microorganisms [18].

Studies on nonthermal RF technique for inactivation of microbial inactivation showed a 1.3 ± 0.2 log and 4.8 ± 0.2 log reductions of *E. coli* at 50 °C and 60 °C, respectively, much greater than thermal inactivation [25]. A similar study on apple cider treated with RF at 50.5 ± 1 °C with 0.5 kV/cm field strength for 50 min exhibited a 4.7 log reduction in *Listeria innocua* [26]. Another advantage of volumetric heating techniques is that they can be effectively used for prepacked foods for microbial inactivation without adversely affecting the package or contents. A study on cooked, prepacked and RF treated ham showed that at the treatment of 85 °C, the ham quality was maintained intact for 28 days of shelf storage under vacuum and refrigerated conditions [57]. RF heating ranging from 70 to 90 °C for 5 min on disinfection of fish meal resulted in >5 log reduction of *Salmonella spp.* and *E. coli* efficiently, stating the total elimination of bacterial pathogens [42]. Kim et al. [39] concluded that the utilization of RF for treating black and red pepper samples achieved 5 log reduction of *S. typhi* and *E. coli* effectively, without altering the quality attributes. Liu et al. [46] successfully established that *E. faecium* can be used as a surrogate for *Salmonella* in wheat flour to simulate the process design required to achieve a 5 log reduction of food-borne pathogens in low-moisture foods. Similarly, by use of dielectric barrier discharge, treatment has shown a significant reduction in psychrotrophic and lactic acid bacteria along with *Pseudomonas* counts effectively in mackerel fish [3].

Low-temperature MW treatment of hazelnuts contaminated with *Aspergillus parasiticus* has shown that for 120 s treatment at 2.450 GHz power level has achieved a log reduction of 2.5 CFU/mL. An increase in exposure time has caused the development of undesirable burning signs on the surface of hazelnuts [8]. Another study on microbial inactivation on oak barrels used for wine fermentation has revealed that short MW exposure for 3 min at 3000 W resulted in complete disinfection of the barrels, as it is important to decontaminate them to avoid the growth of spoilage yeast species like *Brettanomyces bruxellenxis* [29].

Insect infestation is another problem in the storage and safety of several food components, especially grains and spices. The usage of dielectric heating for disinfestation applies the principle of selective heating as the dielectric properties of insects and pests differ from those of grains and other food components. This causes the building up of excess heating in the insect or pest leading to death, though the product will be at moderate temperature itself [49]. One of the key advantages of the use of dielectric heating as a disinfestation technique is that it is environment-friendly, green technique, and does not leave any pesticide or chemical traces. Power level-time and temperature-time relationships can be used to establish an effective protocol to achieve 100% mortality using RF and MW techniques, without adversely affecting the quality characteristics of host food materials. Previous studies on MW-assisted disinfestation for 100 W for the 30 s have successfully achieved 100% mortality in rice weevil in raw rice grain [32].

For few products, to achieve a 100% mortality rate without excessive quality degradation, prescribed holding time after exposure time is required. In case of disinfestation of milled rice, 100% insect mortality was achieved after exposure to RF treatment at 11 cm electrode gap and 50 °C hot air surface heating, followed by a treatment time of 5 min at 50 °C [100]. A study on almonds treated with RF at 6kW power till 63 °C and treating for two minutes in hot air, followed by forced air cooling on a single layer for shelled and 5-cm layer for in-shell product, resulted in 100% insect mortality without no significant effect on the kernel moisture content and overall quality [22].

Previous studies on complete disinfestation of red flour beetle in rape-seeds by adopting RF heating up to 60 °C without holding time, showing that thermal death kinetics follow the first-order reaction with 100 kJ/mol activation energy [97]. Similarly, MW exposure for 3 min at 50 °C, followed by a cooling period of 3.5 min effectively eliminated 100% of *Sitotroga cerealella* in larval stages in white corn grain [23]. MW treatment of 2450 MHz for 2 min has reportedly inhibited *Botrytis cinerea* and *Penicillium expansum,* two major pathogens in peaches [38]. The study has also deduced that in fruits inoculated with the pathogens, the rate of occurrence of infected wounds reduced significantly after both storage and shelf life. Purohit et al. [72] investigated the effect of MW heating on disinfestation of *Callosobruchus maculatus* in mung bean and stated that 100% mortality was achieved for all life stages at the power level of 400 W for 28 s of exposure, with a surface temperature of 68.1 °C. Usage of RF and MW heating treatments ensured centralized heating in insects alone, but not in food components due to differences in dielectric properties. This concept has effectively been used in several food components to achieve complete or partial disinfestation or disinfection.

1.5   RECENT TRENDS AND FUTURE ADVANCES IN DIELECTRIC HEATING OF FOODS

From the past two decades, various developments and advances in dielectric heating (MW and RF) was noticed. Nowadays, dielectric heating has demonstrated practical application in cooking, disinfection, drying, baking, roasting, sterilization, pasteurization, and thawing. Still, nonuniformity and overheating are the most important challenge with dielectric heating. Many methods have been suggested to combat these limitations by including rotation, mixing, sample moving, forced air, vacuum, and modifying the composition of the surrounding media. However, research still needed to improve the quality of dielectric treated foods to maintain uniformity. In the case of

MW heating, rotation or mode of stirrer is not sufficient to achieve uniform heating. Phase control heating, variable frequency ovens, and combining ovens with other heating methods are being studied.

Adoption of phase control (constructive interface) technique to allow more control of heating. This can be attained by varying the relative phase of two MW signals to control the MW field and the consequent temperature distribution. Thus, under these conditions more uniform heating is possible. Apart from this, the use of a variable frequency oven in order to achieve uniform heating is possible. This can be used to control the heating pattern in a spherical product whose geometry dominates the heating pattern when heated in a fixed frequency applicator [86]. MW heating has combined with vacuum and with turnable to achieve uniform heating with similar characteristics with those produced from the freeze-drying process [52]. On the other hand, homogeneous heating pattern is achieved by coupling MW with a fluidized bed dryer. Thus, enhanced drying rate and samples were dried in lower time can be achieved [6].

Although MW combined with hot air or infrared or heat pump heating to reduce the quality problems in foodstuffs, such as uneven crust color formation, loss of firmness, excessive moisture migration, and nonuniform heating. Recent developments showed that MW freeze-drying or MW vacuum-freeze drying is a prominent emerging dehydration technique, most commonly applied particularly fruits and vegetables, soups, and seafood. MW freeze dryers offer numerous advantages over conventional freeze-drying like shorter drying time, energy saving, flexibility, and improved quality. However, currently, its application is limited to handle small scale due to complexity in designing equipment for large volume [20]. Using an MW-assisted pulse-spouted-vacuum dryer is possible to produce high-quality dried food products. Enhanced drying rate, retention in color, highest rehydration capacity, and the acceptable sensory appeal was observed in the case of dried snack [53].

RF heating offers better heating uniformity over MW heating due to greater penetration depth. RF heating system has proven numerous real-time applications in the case of food processing, such as drying, disinfection, extraction, sterilization, thawing, cooking, and roasting. However, nonuniform heating and runaway heating are major challenges behind RF heating. To combat these problems, numerous efforts have been made with the use of other drying methods. A recent study reported that hot air-RF drying shown an advantage over hot air drying in terms of quality (aroma, color) of dried products, moreover, took shorter duration of time and lesser power [81]. Zhou et al. [102]

assessed the quality of RF with vacuum and osmatic dehydrated Kiwifruit. Less deterioration in quality and physicochemical properties were noticed. Similarly, RF-vacuum-hot-air dying shown rapid drying rate and uniform drying moisture distribution within the sample. Thus, improved quality attributes like color, rehydration capacity, and shrinkage ratio were noticed.

Recent development and application have shown the use of carbon dioxide argon gas RF cold plasma in modifying characteristics of starches (waxy rice, maize, potato). Treated samples showed increased enthalpy of gelatinization, resistant starch content, and decrease in starch crystallinity. Hence, carbon dioxide argon gas RF cold plasma could serve as an effective tool in modifying the properties of starches [56]. Microbial inactivation is one of the thrust areas of food safety and preservation. MW and RF-powered cold plasma possess antimicrobial potential against a broad range of microorganisms and heat-resistant spores. Due to nonthermal effect, it ensures the safety and microbiological quality of foodstuff [71]. Application of computer simulation modeling for validation, optimization, and evaluation of RF drying/heating performance is one of the novel applications to predict the moisture–temperature distribution, heating rate, and uniformity in foods [65]. However, an in-depth study is essential to apply the computational modeling and simulation to predict the drying characteristics of foodstuff.

1.6 SUMMARY

Extension of shelf life, deliver safe, and high quality food and food ingredients are a key challenge for food processors. For processing and preservation, most commonly thermal energy is preferred to inactivate enzymes, micro-organisms, and insects. Most of the traditional thermal methods use high-temperature steam, water, or air as the source of heat. Moreover, longer treatment duration degrades the quality of foods. Hence, there is a trend in the exploitation of interest in search of alternative advanced heating technologies for food processing application. Dielectric heating is preferred over traditional methods of heating because of uniformity in heating, shorter duration, depth of penetration, and high quality of the product. Dielectric heating collectively includes MW and RF heating. The practical application of dielectric heating demonstrates a wide range of applications in food processing like drying, sterilization, pasteurization, thawing, roasting, extraction, and freezing. Nonuniform heating and runaway heating are the major challenges with dielectric heating. To combat the problems, dielectric heating coupled with vacuum, pressure, infrared, or rotation and combined with other methods is the feasible solution. The application of computational modeling helps to predict the moisture temperature distribution, uniformity, and heating rate. However, further research and development are essential to develop equipment for industrial applications.

KEYWORDS

- **dielectric heating**
- **drying**
- **extraction**
- **food quality**
- **microwave heating**
- **nonuniformity**
- **pasteurization**
- **radio frequency heating**
- **sterilization**

REFERENCES

1. Ackiyan, O. Optimization of Formulation of Soy-Cakes Baked in Infrared-Microwave Combination Oven by Response Surface Methodology. Journal of Food Science and Technology, 2015, 52 (5), 2910–2917.
2. Ahmad, S. S.; Morgan, M. T.; Okos, M. R. Effects of Microwave on the Drying, Checking and Mechanical Strength of Baked Biscuits. *Journal of Food Engineering*, **2001**, *50* (2), 63–75.
3. Albertos, I.; Martín-Diana, A. B.; Cullen, P. J.; Tiwari, B. K.; Ojha, S. K.; Bourke, P.; Álvarez, C.; Rico, D. Effects of Dielectric Barrier Discharge (DBD) Generated Plasma on Microbial Reduction and Quality Parameters of Fresh Mackerel (*Scomber Scombrus*) Fillets. *Innovative Food Science and Emerging Technologies*, **2017**, *44*, 117–122.
4. Altan, A.; Maskan, M. Microwave Assisted Drying of Short-Cut (Ditalini) Macaroni: Drying Characteristics and Effect of Drying Processes on Starch Properties. *Food Research International*, **2005**, *38* (7), 787–796.
5. Anonymous. An Array of New Applications are Evolving for Radio Frequency Drying. *Food Engineering*, **1987**, *59* (5), 180–188.
6. Askari, G. R.; Emam-Djomeh, Z.; Mousavi, S. M. Heat and Mass Transfer in Apple Cubes in a Microwave-Assisted Fluidized Bed Drier. *Food and Bioproducts Processing*, **2013**, *91* (3), 207–215.
7. Aydogdu, A.; Sumnu, G.; Sahin, S. Effects of Microwave-Infrared Combination Drying on Quality of Eggplants. *Food and Bioprocess Technology*, **2015**, *8* (6), 1198–1210.
8. Basaran, P.; Akhan, Ü. Microwave Irradiation of Hazelnuts for the Control of Aflatoxin Producing *Aspergillus parasiticus*. *Innovative Food Science and Emerging Technologies*, **2010**, *11* (1), 113–117.

9. Bernussi, A. L. M.; Chang, Y. K.; Martinez-Bustos, F. Effects of Production by Microwave Heating after Conventional Baking on Moisture Gradient and Product Quality of Biscuits (Cookies). *Cereal Chemistry*, **1998**, *75* (5), 606–611.

10. Brunton, N. P.; Lyng, J. G.; Li, W.; Cronin, D. A.; Morgan, D.; McKenna, B. Effect of Radio Frequency (RF) Heating on the Texture, Color and Sensory Properties of a Comminuted Pork Meat Product. *Food Research International*, **2005**, *38* (3), 337–344.

11. Cathcart, W. H.; Parker, J. J.; Beattie, H. G. The Treatment of Packaged Bread with High Frequency Heat. *Food Technology*, **1947**, *1* (2), 174–177.

12. Chavan, R. S.; Chavan, S. R. Microwave Baking in Food Industry: A Review. *International Journal of Dairy Science*, **2010**, *5* (3), 113–127.

13. Cho, W. I.; Yi, J. Y.; Chung, M. S. Pasteurization of Fermented Red Pepper Paste by Ohmic Heating. *Innovative Food Science and Emerging Technologies*, **2016**, *34*, 180–186.

14. Cottee, C. J.; Duncan, S. R. Design of Matching Circuit Controllers for Radio-Frequency Heating. *IEEE Transactions on Control Systems Technology*, **2003**, *11* (1), 91–100.

15. Cravotto, G.; Boffa, L.; Mantegna, S.; Perego, P.; Avogadro, M.; Cintas, P. Improved Extraction of Vegetable Oils under High-Intensity Ultrasound and/or Microwaves. *Ultrasonics Sonochemistry*, **2008**, *15* (5), 898–902.

16. Demeczky, M. Continuous Pasteurization of Bottled Fruit Juices by High Frequency Energy. In: *Proceedings of IV International Congress on Food Science and Technology,* Madrid, Spain, **1974**, *4*, 11–20.

17. Desai, M.; Parikh, J.; Parikh, P. A. Extraction of Natural Products using Microwaves as a Heat Source. *Separation and Purification Reviews*, **2010**, *39* (1–2), 1–32.

18. Dev, S. R. S.; Birla, S. L.; Raghavan, G. S. V.; Subbiah, J. Microbial Decontamination of Food by Microwave (MW) and Radio Frequency (RF). In: *Microbial Decontamination in the Food Industry; Novel Methods and Applications*. Sawston: Woodhead Publishing Limited; **2012**; 274–299.

19. Dev, S. R. S.; Raghavan, G. S. V.; Gariepy, Y. Dielectric Properties of Egg Components and Microwave Heating for In-shell Pasteurization of Eggs. *Journal of Food Engineering*, **2008**, *86* (2), 207–214.

20. Duan, X.; Zhang, M.; Mujumdar, A. S.; Wang, R. Trends in Microwave Assisted Freeze Drying of Foods. *Drying Technology*, **2010**, *28* (4), 444–453.

21. Feng, H.; Tang, J.; Cavalieri, R. P. Dielectric Properties of Dehydrated Apples as Affected by Moisture and Temperature. *Transactions of the ASAE*, **2002**, *45* (1), 129–135.

22. Gao, M.; Tang, J.; Wang, Y.; Powers, J.; Wang, S. Almond Quality as Influenced by Radio Frequency Heat Treatments for Disinfestation. *Postharvest Biology and Technology*, **2010**, *58* (3), 225–231.

23. García-Mosqueda, C.; Salas-Araiza, M. D. Microwave Heating as a Post-Harvest Treatment for White Corn (Zea Mays) against *Sitotroga cerealella*. *Journal of Microwave Power and Electromagnetic Energy*, **2019**, *53* (3), 145–154.

24. Geveke, D. J.; Bigley, A. B. W.; Brunkhorst, C. D. Pasteurization of Shell Eggs Using Radio Frequency Heating. *Journal of Food Engineering*, **2017**, *193*, 53–57.

25. Geveke, D. J.; Brunkhorst, C. Radio Frequency Electric Fields Inactivation of *Escherichia Coli* in Apple Cider. *Journal of Food Engineering*, **2008**, *85* (2), 215–221.

26. Geveke, D. J.; Kozempel, M.; Scullen, O. J.; Brunkhorst, C. Radio Frequency Energy Effects on Microorganisms in Foods. *Innovative Food Science and Emerging Technologies*, **2002**, *3* (2), 133–138.

27. Giacometti, J.; Bursać Kovačević, D.; Putnik, P. Extraction of Bioactive Compounds and Essential Oils from Mediterranean Herbs by Conventional and Green Innovative Techniques: A Review. *Food Research International,* **2018,** *113,* 245–262.

28. Goksu, E. I.; Sumnu, G.; Esin, A. Effect of Microwave on Fluidized Bed Drying of Macaroni Beads. *Journal of Food Engineering,* **2005,** *66* (4), 463–468.

29. González-Arenzana, L.; Santamaría, P. Microwave Technology as a New Tool to Improve Microbiological Control of Oak Barrels: A Preliminary Study. *Food Control,* **2013,** *30* (2), 536–539.

30. Guo, W.; Wang, S.; Tiwari, G.; Johnson, J. A.; Tang, J. Temperature and Moisture Dependent Dielectric Properties of Legume Flour Associated with Dielectric Heating. *LWT—Food Science and Technology,* **2010,** *43* (2), 193–201.

31. Heinemann, R. J. B.; Fagundes, P. D. L. Comparative Study of Nutrient Composition of Commercial Brown, Parboiled and Milled Rice from Brazil. *Journal of Food Composition and Analysis,* **2005,** *18* (4), 287–296.

32. Ibrahim, M.; Rahim, R. A.; Nordin, J. M.; Nyzam, S. Z. A.; Abdullah, M. T. Thermal Distribution Analysis of Rice Weevil Disinfestation using Microwave Heating Treatment. *Indonesian Journal of Electrical Engineering and Computer Science,* **2019,** *13* (2), 759–765.

33. Izadifar, M.; Baik, O. D. Ã. Dielectric Properties of a Packed Bed of the Rhizome of *P. Peltatum* with an Ethanol/Water Solution for Radio Frequency-Assisted Extraction of Podophyllotoxin. *Biosystems Engineering,* **2008,** *100,* 376–388.

34. Jason, A. C.; Sanders, H. R. Dielectric Thawing of Fish., II: Experiments with Frozen White Fish. *Food Technology,* **1962,** *16* (6), 107.

35. Jiao, J.; Li, Z. G.; Gai, Q. Y.; Li, X. J.; Wei, F. Y.; Fu, Y. J.; Ma, W. Microwave Assisted Aqueous Enzymatic Extraction of Oil from Pumpkin Seeds and Evaluation of its Physicochemical Properties, Fatty Acid Compositions and Antioxidant Activities. *Food Chemistry,* **2014,** *147,* 17–24.

36. Jones, P. L. Dielectric Heating for Food Processing. *Nutrition and Food Science,* **1992,** *92* (2), 14–15.

37. Karabulut, O. A.; Baykal, N. Evaluation of the Use of Microwave Power for the Control of Postharvest Diseases of Peaches. *Postharvest Biology and Technology,* **2002,** *26* (2), 237–240.

38. Kim, S. Y.; Sagong, H. G.; Choi, S. H.; Ryu, S.; Kang, D. H. Radio-Frequency Heating to Inactivate *Salmonella Typhimurium* and *Escherichia Coli* O157:H7 on Black and Red Pepper Spice. *International Journal of Food Microbiology,* **2012,** *153,* (1–2), 171–175.

39. Kinn, T. P. Basic Theory and Limitations of High Frequency Heating Equipment. *Food Technology,* 1947, *1* (2), 161–173.

40. Koral, T. Radio Frequency Heating and Post-Baking. *Biscuit World Issue,* **2004,** *7* (4), 1–6.

41. Lagunas-Solar, M. C.; Zeng, N. X. Disinfection of Fishmeal with Radiofrequency Heating for Improved Quality and Energy Efficiency. *Journal of the Science of Food and Agriculture,* **2005,** *85* (13), 2273–2280.

42. Laycock, L.; Piyasena, P.; Mittal, G. S. Radio Frequency Cooking of Ground, Comminuted and Muscle Meat Products. *Meat Science,* **2003,** *65,* (3), 959–965.

43. Li, R.; Kou, X.; Cheng, T.; Zheng, A.; Wang, S. Verification of Radio Frequency Pasteurization Process for In-shell Almonds. *Journal of Food Engineering,* **2017,** *192,* 103–110.

44. Li, Z. Design of a Microcontroller Based, Power Control System for Microwave Drying. Doctorate Dissertation; McGill University, Montreal, **2004**, 157.

45. Liu, S.; Ozturk, S.; Xu, J.; Kong, F.; Gray, P.; Zhu, M. J.; Sablani, S. S.; Tang, J. Microbial Validation of Radio Frequency Pasteurization of Wheat Flour by Inoculated Pack Studies. *Journal of Food Engineering*, **2018**, *217*, 68–74.

46. Liu, S.; Tang, J.; Tadapaneni, R. K.; Yang, R.; Zhu, M. J. Exponentially Increased Thermal Resistance of *Salmonella* spp. and *Enterococcus Faecium* at Reduced Water Activity. *Applied and Environmental Microbiology*, **2018**, *84* (8), e02742–17

47. Liu, Y.; Tang, J.; Mao, Z. Analysis of Bread Dielectric Properties using Mixture Equations. *Journal of Food Engineering*, **2009**, *93*, (1), 72–79.

48. Macana, R. J.; Baik, O. D. Disinfestation of Insect Pests in Stored Agricultural Materials using Microwave and Radio Frequency Heating: A Review. *Food Reviews International*, **2018**, *34* (5), 483–510.

49. Metaxas, A. C. Foundations of Electroheat: A Unified Approach. *Fuel and Energy Abstracts*, **1996**, *37* (3), 193–201.

50. Mirade, P. S.; Daudin, J. D.; Ducept, F.; Trystram, G.; Clément, J. Characterization and CFD Modelling of Air Temperature and Velocity Profiles in an Industrial Biscuit Baking Tunnel Oven. *Food Research International*, **2004**, *37* (10), 1031–1039.

51. Monteiro, R. L.; Carciofi, B. A. M.; Marsaioli, Jr. A.; Laurindo, J. B. How to Make a Microwave Vacuum Dryer with Turntable. *Journal of Food Engineering*, **2015**, *166*, 276–284.

52. Mothibe, K. J.; Wang, C. Y.; Mujumdar, A. S.; Zhang, M. Microwave Assisted Pulse-Spouted Vacuum Drying of Apple Cubes. *Drying Technology*, **2014**, *32* (15), 1762–1768.

53. Mousa, N.; Farid, M. Microwave Vacuum Drying of Banana Slices. *Drying Technology*, **2002**, *20* (10), 2055–2066.

54. Mujumdar, A. S.; Law, C. L. Drying Technology: Trends and Applications in Postharvest Processing. *Food and Bioprocess Technology*, **2010**, *3* (6), 843–852.

55. Okyere, A. Y.; Bertoft, E.; Annor, G. A. Modification of Cereal and Tuber Waxy Starches with Radio Frequency Cold Plasma and its Effects on Waxy Starch Properties. *Carbohydrate Polymers,* **2019**, *223*, 115075.

56. Orsat, V.; Bai, L.; Raghavan, G. S. V.; Smith, J. P. Radio Frequency Heating of Ham to Enhance Shelf-Life in Vacuum Packaging. *Journal of Food Process Engineering*, **2004**, *27* (4), 267–283.

57. Orsat, V.; Gariepy, Y.; Raghavan, G. S. V.; Lyew, D. Radio Frequency Treatment for Ready to Eat Fresh Carrots. *Food Research International*, **2001**, *34* (6), 527–536.

58. Orsat, V.; Raghavan, G. S. V. Radio-Frequency Processing. Chapter 21; In: *Emerging Technologies for Food Processing*; Da-Wen Sun (Eds.); Cambridge, MA: Apple Academic Press Inc.; **2014**; pages 385–398.

59. Osepchuk, J. M.; Petersen, R. C. Historical Review of RF Exposure Standards and the International Committee on Electromagnetic Safety (ICES). *Bioelectromagnetics*, **2003**, *24* (S6), S7-S16.

60. Ozkahraman, B. C.; Sumnu, G.; Sahin, S. Effect of Different Flours on Quality of Legume Cakes to be Baked in Microwave-Infrared Combination Oven and Conventional Oven. *Journal of Food Science and Technology*, **2016**, *53* (3), 1567–1575.

61. Ozkoc, S. O.; Seyhun, N. Effect of Gum Type and Flaxseed Concentration on Quality of Gluten-Free Breads made from Frozen Dough Baked in Infrared-Microwave Combination Oven. *Food and Bioprocess Technology*, **2015**, *8* (12), 2500–2506.

62. Öztürk, S.; Özge, Ş.; Dielectric Properties and Microwave and Infrared-Microwave Combination Drying Characteristics of Banana and Kiwifruit. *Journal of Food Process Engineering,* **2017,** *40* (3), e12502.

63. Ozturk, S.; Kong, F.; Trabelsi, S.; Singh, R. K. Dielectric Properties of Dried Vegetable Powders and their Temperature Profile during Radio Frequency Heating. *Journal of Food Engineering,* **2016,** *169,* 91–100.

64. Palazouglu, T. K.; Miran, W. Computational Investigation of the Effect of Orientation and Rotation of Shell Egg on Radio Frequency Heating Rate and Uniformity. *Innovative Food Science and Emerging Technologies,* **2019,** *58,* 102238.

65. Peng, J.; Tang, J.; Jiao, Y.; Bohnet, S. G.; Barrett, D. M. Dielectric Properties of Tomatoes Assisting in the Development of microwave Pasteurization and Sterilization Processes. *LWT—Food Science and Technology,* **2013,** *54* (2), 367–376.

66. Peng, J.; Tang, J.; Luan, D.; Liu, F.; Tang, Z.; Li, F.; Zhang, W. Microwave Pasteurization of Pre-packaged Carrots. *Journal of Food Engineering,* **2017,** *202,* 56–64.

67. Périno-Issartier, S.; Ginies, C.; Cravotto, G.; Chemat, F. A Comparison of Essential Oils obtained from Lavandin via Different Extraction Processes: Ultrasound, Microwave, Turbohydrodistillation, Steam and Hydrodistillation. *Journal of Chromatography A,* **2013,** *1305,* 41–47.

68. Pettinato, M.; Casazza, A. A.; Perego, P. The Role of Heating Step in Microwave-Assisted Extraction of Polyphenols from Spent Coffee Grounds. *Food and Bioproducts Processing,* **2019,** *114,* 227–234.

69. Prabhanjan, D. G.; Ramaswamy, H. S.; Raghavan, G. S. V. Microwave-Assisted Convective Air Drying of Thin Layer Carrots. *Journal of Food Engineering,* **1995,** *25* (2), 283–293.

70. Puligundla, P.; Mok, C. Microwave and Radio-Frequency-Powered Cold Plasma Applications for Food Safety and Preservation. Chapter 11; In: *Advances in Cold Plasma Applications for Food Safety and Preservation*; Bermudez-Aguirre, D. (Eds.); Cambridge, MA: Apple Academic Press Inc.; **2020,** 309–329.

71. Purohit, P.; Jayas, D. S.; Yadav, B. K.; Chelladurai, V.; Fields, P. G.; White, N. D. G. Microwaves to Control *Callosobruchus maculatus* in Stored Mung Bean (*Vigna Radiata*). *Journal of Stored Products Research,* **2013,** *53,* 19–22.

72. Raman, G.; Gaikar, V. G. Microwave-Assisted Extraction of Piperine from *Piper Nigrum. Industrial and Engineering Chemistry Research,* **2002,** *41* (10), 2521–2528.

73. Ramaswamy, H.; Tang, J. Microwave and Radio Frequency Heating. *Food Science and Technology International,* **2008,** *14* (5), 423–427.

74. Risman, P. Terminology and Notation of Microwave Power and Electromagnetic Energy. *Journal of Microwave Power and Electromagnetic Energy,* **1991,** *26* (4), 243–250.

75. Russo, J. R. Microwave Proof Donuts. *Food Engineering,* **1971,** 55–58.

76. Sale, A. J. H. A Review of Microwaves for Food Processing. *International Journal of Food Science and Technology,* **1976,** *11* (4), 319–329.

77. Sánchez-Pardo, M. E.; Ortiz-Moreno, A. Comparison of Crumb Microstructure from Pound Cakes Baked in a Microwave or Conventional Oven. *LWT—Food Science and Technology,* **2008, 41** (4), 620–627.

78. Schiffmann, R. F. Update on the Applications of Microwave Power in the Food Industry in the United States. *Journal of Microwave Power,* **1976,** *11* (3), 222–224.

79. Sebastian, J.; Rouissi, T.; Brar, S. K.; Hegde, K.; Verma, M. Microwave-Assisted Extraction of Chitosan from *Rhizopus Oryzae* NRRL 1526 Biomass. *Carbohydrate Polymers*, **2019**, *219*, 431–440.

80. Shinde, A.; Das, S.; Datta, A. K. Quality Improvement of Orthodox and CTC Tea and Performance Enhancement by Hybrid Hot Air Radio Frequency (RF) Dryer. *Journal of Food Engineering*, **2013**, *116* (2), 444–449.

81. Shivhare, U. S.; Raghavan, G. S. V.; Bosisio, R. G. Microwave Drying of Corn II. Constant Power, Continuous Operation. *Transactions of the ASAE*, **1992**, *35* (3), 951–957.

82. Si, X.; Chen, Q.; Bi, J.; Yi, J.; Zhou, L.; Wu, X. Infrared Radiation and Microwave Vacuum Combined Drying Kinetics and Quality of Raspberry. *Journal of Food Process Engineering*, **2016**, 39 (4), 377–390.

83. Singh-Chouhan, K. B.; Tandey, R.; Sen, K. K.; Mehta, R.; Mandal, V. Critical Analysis of Microwave Hydrodiffusion and Gravity as a Green Tool for Extraction of Essential Oils: Time to Replace Traditional Distillation. *Trends in Food Science and Technology*, **2019**, *92*, 12–21.

84. Sosa-morales, M. E.; Valerio-junco, L.; López-malo, A.; García, H. S. Dielectric Properties of Foods : Reported Data in the 21st Century and their Potential Applications. *LWT—Food Science and Technology*, **2010**, *43*, 1169–1179.

85. Sumnu, G.; Sahin, S. Recent Developments in Microwave Heating. Chapter 16; In: *Emerging Technologies For Food Processing*; Da-Wen Sun, (Eds.); Cambridge, MA: Apple Academic Press Inc.; **2005**; Elsevier; pages 419–444.

86. Therdthai, N.; Zhou, W. Recent Advances in the Studies of Bread Baking Process and their Impacts on the Bread Baking Technology. *Food Science and Technology Research*, **2003**, *9* (3), 219–226.

87. Thostenson, E. T.; Chou, T-W. Microwave Processing: Fundamentals and Applications. *Composites Part A: Applied Science and Manufacturing*, **1999**, *30* (9), 1055–1071.

88. Tsen, C. C. Microwave Energy for Bread Baking and its Effect on the Nutritive Value of Bread: A Review. *Journal of Food Protection*, **1980**, *43* (8), 638–640.

89. Tulasidas, T. N.; Raghavan, G. S. V. Microwave and Convective Drying of Grapes. *Transactions of the ASAE*, **1993**, *36* (6), 1861–1865.

90. Venkatesh, M. S.; Raghavan, G. S. V. An Overview of Microwave Processing and Dielectric Properties of Agri-Food Materials. *Biosystems Engineering*, **2004**, *88* (1), 1–18.

91. Wang, J.; Luechapattanaporn, K.; Wang, Y.; Tang, J. Radio Frequency Heating of Heterogeneous Food—Meat Lasagna. *Journal of Food Engineering*, **2012**, *108* (1), 183–193.

92. Wang, S.; Monzon, M.; Johnson, J. A.; Mitcham, E. J.; Tang, J. Industrial Scale Radio Frequency Treatments for Insect Control in Walnuts, I: Heating Uniformity and Energy Efficiency. *Post-harvest Biology and Technology*, **2007**, *45*, 240–246.

93. Wang, Y.; Li, Y.; Wang, S.; Zhang, L.; Gao, M.; Tang, J. Review of Dielectric Drying of Foods and Agricultural Products. *International Journal of Agricultural and Biological Engineering*, **2011**, *4* (1), 1–19.

94. Flugstad, B. A.; Ling, Q.; Kolbe, E. R.; Wells, J. H.; Zhao, Y.; Park, J. W. *U.S. Patent No. 6,784,405*, **2004**, Washington, DC: U.S. Patent and Trademark Office, 15.

95. Yao, X. H.; Zhang, D. Y.; Luo, M.; Jin, S.; Zu, Y. G.; Efferth, T.; Fu, Y. J. Negative Pressure Cavitation-Microwave Assisted Preparation of Extract of *Pyrola Incarnata Fisch*: Rich in Hyperin, 2′-O-Galloylhyperin and Chimaphilin and Evaluation of Its Antioxidant Activity. *Food Chemistry*, **2015**, *169*, 270–276.

96. Yu, D.; Shrestha, B.; Baik, O. D. Thermal Death Kinetics of Adult Red Flour Beetle *Tribolium Castaneum* (Herbst) in Canola Seeds during Radio Frequency Heating. *International Journal of Food Properties*, **2017**, *20* (12), 3064–3075.

97. Zareifard, M. R.; Boissonneault, V.; Marcotte, M. Bakery Product Characteristics as Influenced by Convection Heat Flux. *Food Research International*, **2009**, *42* (7), 856–864.

98. Zhao, Y.; Flugstad, B.; Kolbe, E.; Park, J. W.; Wells, J. H. Using Capacitive (Radio Frequency) Dielectric Heating in Food Processing and Preservation: A Review. *Journal of Food Process Engineering*, **2000**, *23* (1), 25–55.

99. Zhou, L.; Ling, B.; Zheng, A.; Zhang, B.; Wang, S. Developing Radio Frequency Technology for Postharvest Insect Control in Milled Rice. *Journal of Stored Products Research*, **2015**, *62*, 22–31.

100. Zhou, X.; Li, R.; Lyng, J. G.; Wang, S. Dielectric Properties of Kiwifruit Associated with a Combined Radio Frequency Vacuum and Osmotic Drying. *Journal of Food Engineering*, **2018**, *239*, 72–82.

101. Zhou, X.; Ramaswamy, H.; Qu, Y.; Xu, R.; Wang, S. Combined Radio Frequency-Vacuum and Hot Air Drying of Kiwifruits: Effect on Drying Uniformity, Energy Efficiency and Product Quality. *Innovative Food Science and Emerging Technologies*, **2019**, 102182.

102. Zhou, X.; Wang, S. Recent Developments in Radio Frequency Drying of Food and Agricultural Products: A Review. *Drying Technology*, **2019**, *37* (3), 271–286.

CHAPTER 2

OHMIC HEATING IN FOOD PROCESSING: A FUTURISTIC TECHNOLOGY

SWASTIKA DAS*, ROBINA RAI, and SHYAM K. SINGH

ABSTRACT

Ohmic heating (OH) has been a prominent approach for attaining instant and consistent heating of food products. The OH has a modus operandi that provides microbiologically innocuous products with minimal damage to their quality attributes. The food products are simply subjected to a monitored flow of electricity that generates heat due to the structural resistance to the flow. This chapter discusses the basic principle and engineering aspects of the novel technique along with a detailed perusal of its impact on microbial sterility and desirability values of the food product. The chapter also enlightens the currently operating as well as futuristic potential applications of OH like sterilization, pasteurization, blanching, fermentation, etc. Considering the current marketability of various food processing techniques, a commercial and legislative comprehension has been provided for a varying range of food products that are susceptible to OH.

2.1 INTRODUCTION

India has been one of the leading countries in agricultural and food production, processing, export, and consumption. The food processing sector of India has lately emerged as a highly profitable commerce due to the emerging value addition and energy-efficient technologies. Since the beginning, pasteurization, and sterilization were an integral part of the food processing

*Corresponding author. E-mail: swastika.5396@gmail.com.

industry and provided satisfactory outlines for shelf-life and consumer safety. However, these methods involve the application of high temperatures with a longer duration of processing that instigates product deterioration and resulted in unmarketable sensorial and nutritive changes, especially in the heat-sensitive products [7].

The new generation processing technologies for food products demand maximum retention of nutrition value along with adequate microbial inactivation in the respective product. Many novel advanced nonthermal and thermal technologies that use mechanical, electrical, and electromagnetic energy have been reported to successfully deliver to this latest commercial clamor [8]. Technologies like high-pressure processing (HPP), vacuum frying, electrical impedance spectroscopy, ultrasound, irradiation, innovative transduction have been heavily researched in the past decades and adopted in the food industry to supply qualitatively rich and safe packaged food products. High hydrostatic pressure processing has been successfully applied in a wide range of food products like raw meat, low water activity foods, heat-sensitive drinks, seafood, etc., for the efficient killing of microbes without compromising the quality. Vacuum frying of the processed vegetables has been reported to better sustain the quality attributes [70].

Carrillo-Lopez et al. [14] reported that the application of ultrasound in food generates acoustic cavitation and consequently enhances the physico-chemical properties of the food. Among all the novel techniques introduced, nonthermal technologies like HPP and PEF processing have acquired much attention in the food industry, recently. However, they are only deemed effective in inactivating bacterial spores and enzymes with proper thermal assistance. Ohmic heating (OH) bears advantages over other conventional methods by facilitating rapid inactivation of microbes while sustaining the sensory and nutritive product values [25, 95].

Achieving microbial safety with a novel low energy or energy-efficient method like OH is foundational in industrial applications and continues to attract attention in the food sector [76]. Due to the excellent conducting properties of fruits and vegetables, OH has been successfully employed for the sterilization of artichoke heads [32], carrots [57], cauliflower [23], papaya pulp [30], and fruit desserts [17]. While OH has been recognized as the most efficient food processing technique, its usage is still limited to sterilization or pasteurization of particulates. Ohmic heaters are available in a wide range of electrical conductivities but the technique has not gained complete maturity for gross application and needs to be engineered according to varied potential applications.

Among all the unit operations involved in the processing of food products, thermal processing plays a key role in ensuring the apposite inactivation of microbes and food preservation. The conventional methods of heating food were based on conduction and convection mechanism, which induced heterogeneous heating and affected the textural and qualitative integrity of the respective product [40]. According to International Commission on Microbiological Specifications for Foods [86], homogenous heat treatment of food is crucial for the public health aspect and preventing foodborne diseases. Over the decades, thermal processing of food has experienced several advancements in energy-efficient and uniform heating techniques for food like ohmic and dielectric heating.

OH is not a new concept today and has been around since 1897 [52]. The fundamental law of electricity, Ohm's law was scripted by Georg Ohm in the year 1827. However, the phenomenon from which the OH method originated, that is, heating by conduction of electricity, was elucidated by James Prescott Joule in the year 1841. He illustrated the relation between heat and the resistance of a wire and the current conducted. Hence, it is also termed as Joule heating. Soon after its introduction in the heating of flowable materials, the application of electric heating in the food sector boomed in the early 1900s [5]. The United States of America extensively used this technique for the inactivation of microbes in milk replacing the retort sterilization but soon the technique faced dormancy due to the higher capital investment involved in the process [29].

Applications in other food products also experienced a backlash due to undesirable electrolysis and food contamination due to the inapt electrode substance [67]. The technique was later revived in the 1980s by APV Baker Ltd. and found application in aseptic processing of particulate foods like pasteurization of eggs, packaged soups, processing of fruit products, etc. [96]. The available conventional thermal processing techniques for solid–liquid mixtures as well as liquid products were limited by lower heat transfer rates and inefficient process outcomes. Albeit better results were obtained with the advancements in heat exchangers and adaptation of high-temperature processes, the qualitative attributes of the end product were fairly compromised.

On the other hand, OH continuously ensured rapid heat transfer rates within different phases of the material and minimal or no adverse impact on the inherent qualitative properties of the respective product. However, the contemporary electrothermal techniques of microwave, radiofrequency, and induction heating manifested nonuniform heating in the food media. Over the decades, many new manufacturers of ohmic heaters have emerged and the prices have beneficially declined. This prompted the adaptation of OH in various sectors of the food industry including milk, fruits, vegetables and their products, meat products, seafood, flours, and starches, etc [4]. According to Kumar [52], higher temperatures in particulates can be achieved than the liquid medium without burning the food product. The multifarious advantages of OH over other conventional and advanced techniques continue to attract researchers and industrial institutions in the food sector.

According to WHO (2015), almost 600 million people, that is, 1 in 10 people around the world, get affected by foodborne diseases after consumption of contaminated food, and about 420,000 deaths due to the same are reported annually. Children are the most vulnerable section toward foodborne diseases and constitute about 40% of the diseased population [52]. In order to fight this crisis, processing techniques assuring food safety with value-added benefits of nutritive fortification have become mandatory in the present times. The conventional methods of preservation that are used by the food industries like canning, freezing, and drying are recognized as the most energy-intensive processes [36]. A total of 50–100 MJ of energy is estimated to be required for the processing and packaging of 1 kg of retail food products [107].

Consequently, the current need of the hour in the food sector is a thermal processing technology that is cost-efficient, derived from inexpensive energy sources, and provides absolute assurance for qualitative integrity and consumer health. Advanced technology like OH uses electrical energy and converts it into thermal energy and is thus considered as a highly efficient method as the heat is induced inherently in the food product. It certifies as a superior energy-efficient process as 90% of the electric energy applied is converted to heat [106].

Apart from being energy efficient, the technique of OH can also be deemed as a superior processing method in expediting improved heating of the product irrespective of the structural or textural design. While there are many novel techniques available for the processing of various food products, most of them lack in facilitating uniform heating across multiple phases in a food mixture like solid in liquid mixtures. The heat generation induced by the internal resistance of the individual particulate in OH enables the uniform heating of both the solid and liquid particles concurrently. Microwave and radiofrequency techniques are also efficient methods of utilizing electric energy for processing. However, they often fail to detect cold spots and result in heating patterns due to the complex electric fields applied. Unlike conventional heating technologies, a greater range of temperatures can be achieved in OH without the risk of burning. Being an electrically induced process, there are possibilities of an anomaly in the case of solid–liquid mixtures in OH. However, they can be easily detected by proper monitoring of the energy inputs online [90]. This facilitates intelligent decision-making during the process depending on the flow and design of the product. With the recent advancements in the design and manufacturing of ohmic heaters, the technology has become more feasible for the food industry and promises a colossal development in the sector.

This chapter further delineates the role of this novel technique as a futuristic solution for the overall issues in the food processing industry, with a firm emphasis on the process mechanism, factors involved, equipment design, effect on the qualitative, and commercial attributes of the food product and potential applications.

2.2   PRINCIPLE OF OPERATION

The process of OH is performed by the fundamental concept of electrical resistance of the food products. Nearly all food products comprise a high ratio of water and dissolved salts that enable electrolytic conduction. Due to the Joule effect, electrical energy is dissipated upon the conduction of electricity in the treated food product resulting in the intended heating. Depending on the homogeneity of the objective food product, the effect of ohmic heat may vary. Homogenous products like liquid, pasty, or food with high water content will bear the advantages of volumetric heating (conjunctive thermal heating with conduction, convection, and radiation) and high-temperature short-time process which results in enhanced quality retention in the final product. However, in the case of nonhomogenous food products, the ultimate heating will be dependent on the complex relative conductivities of the particles and the solution around them. This may result in different heat generation rates in distinct localized regions.

OH applies to all food products within the electrical conductivity range of 0.1–10 S/m [45]. Electrodes are used to heat the sample product by providing a suitable electric field. They can be either in direct contact with the product surface or coupled to them through an electrically conductive medium (electrolyte) like a salt solution of sodium chloride. Following Ohm's law, the distinct arrangement of the objective product, electrodes, and the electric field induce a current flow within the food sample which potentially converts to thermal heat with high energy density and shorter heating periods. Due to the electric field applied, the electrolyte ions move toward the oppositely charged electrodes. During this process, the ions even collide with each other which instigates resistance in the medium and accelerates the kinetic energy, and consequently, the product is heated [104]. Samprovalaki et al. [92] elucidated the heat generation rate in the ohmic process with the following equations:

$$Q = \sigma E^2 \tag{2.1}$$

$$\nabla(\sigma \nabla E) = 0 \tag{2.2}$$

Equation (2.1) is similar to the I^2R equation, where Q is the rate of internal heat generation in the food sample in W/m^3, σ denotes the electrical conductivity of the medium (S/m), and E denotes the electric field strength applied in the system (V/m). Equation (2.2) depicts the voltage distribution, which discernibly depends upon the conductive properties of the medium as well as the geometry of the system.

An energy voltage range of 400–4000 V and a field strength of 20–400 V/cm are applied for optimal OH of the products with an electrode gap of 10–50 cm [45]. The electrical energy required to achieve functional heating of the food product can be derived from the mass flow, temperature rise, and properties of the food product like conductivity, specific heat, viscosity. The process can be performed in both batch and continuous methods.

2.2.1 INACTIVATION MECHANISM

Apart from the significant thermal effect of OH, it also imparts an additional nonthermal impact on the treated food product which leads to the formation of pores in the microorganism cell membrane. This results in enhanced permeability and material diffusion through the membrane by electro-osmosis which further results in disruption and death of the microbial cell [4, 50, 20, 60]. This phenomenon is known as electroporation and has a significant role in achieving the improved degree of microbial inactivation. Application of electric field in OH has both direct and indirect impact on the microbial cells that causes exudation of the organic substances from the cell-like amino acids, proteins, co-enzymes into the media [119].

The electroporation effect induced by OH at sublethal temperatures aids fermentative processes. The pore formation in the cell membrane promotes efficient intracellular nutrient transport and thereby, decreases the lag phase of fermentation. Later in the final stages of the process, electroporation allows higher rates of absorption of the metabolites from the culture medium and hence, inhibits further microbial growth [19, 61, 62]. This additional effect of electroporation plays a crucial role in providing a more effective heating process while decreasing the thermal resistance of the microbial cells and reducing the required energy input. Food products with higher quality and safety attributes can be produced by studying the parameters that enhance the electroporation effect in OH.

2.3   DESIGN OF OHMIC HEATING

The specific process and product parameters that define the effectiveness of the ohmic system have been discussed in this section. Based on the combined analysis of these parameters, the suitable processing units for the equipment are designed.

2.3.1   PRODUCT PARAMETERS

2.3.1.1   ELECTRICAL CONDUCTIVITY

Apprehension of the electrical conductivity of the objective food particle or product is imperative in performing effective ohmic processing of food. Electrical conductivity is defined as the ability of any substance to transmit electrical charge. It is presented in units of Siemens per meter (S/m) and given by the general equation below [123]:

$$\sigma = \frac{L}{A} \times \frac{I}{V} \tag{2.3}$$

where L denotes the distance between the electrodes in m, A denotes the cross-sectional area perpendicular to the direction of current flow in m^2, I is the alternating electricity in amperes, and V is the voltage applied in V.

The rate of conduction of electricity varies with the phase and concentration of the food product. At lower concentrations, the conductive properties of fluids are superior; however, at higher concentrations, the solid particles may conjure superior heating rates than the surrounding liquid [98]. The electrical conductivity of any product is mainly governed by its structural composition and may vary with the rise in temperature (heating); however, in certain food products, the effect of heat is negligible on the structure and hence, the electrical conductivity remains unaltered [83].

Products with electrical conductivity values less than 0.01 S/m and more than 10 S/m are not susceptible to OH as very high voltage or amperes of current would be required to induce the Joule effect in the product structure, rendering the process ineffective [82]. Uniform heating in particulate foods can be attained by adding salts in the medium or soaking or blanching of vegetables prior to heating, consequently increasing the electrical conductivity of the food products and enhancing the process [95].

Several studies in OH over the decades have also suggested that the conductive properties of vegetables can also be enhanced by infiltration of salt solution [35, 78, 116]. Electrical conductivity values also differ with the orientation of the product structure. Structural alignment of the vascular bundles as well as the geometry of the parenchyma cells may alter the electrical conductivity of plant perishables. Apart from the chemical composition of the product treated and the methodology used, the orientation of the product during measurement also leads to varying results in conductivity [115]. The electrical conductivity of the product often exhibits a slight variation during OH, often leading to a reduction in the system impedance [54, 83].

2.3.1.2   SIZE AND SHAPE

The rate of OH is largely affected by the physical shape and size of the food product to be treated, even in the case of isotropic foods. Alwis and Fryer [3] developed a computer program that illustrates the complicated relationship between the size and geometry of the food product and the heating rates. The electrical conductivity of especially solid food products is widely affected by the particle size and shape. A few studies have described the influence of the variation in particle size of the product on the conductive properties. It has been reported that an increase in the particle size of the solid food products or dispersed solids in fluid products results in an enhanced resistance of ionic movement and thereby, reduction in the electrical conductivity [16, 99, 123]. Decreased levels of electrical conductivity will hinder the effectiveness of the heating process.

2.3.1.3   TEXTURE AND VISCOSITY

Physical properties of any food product, especially agricultural products like vegetables, meat, and dairy, are mostly dependent on the aging period and the processing temperature. With the aging of agricultural products during the postharvest storage, the texture and viscosity of the perishables exhibit changes and consequently affect the thermal properties. The freshness and the structural composition of the plant perishables are highly foundational in determining the impedance generation. Hence, the changes in textural properties due to postharvest aging shall affect the impedance of the products and influence the OH process [101]. Soft textured fruits like strawberries and peach bear superior conductive properties than hard textured fruits like apples and pineapples [94]. Due to the substandard rates of convective heat transfer through the interphase of liquid food products like fruit juices with higher viscosity, they exhibit a higher rate of OH than the ones with lower viscosity [106].

2.3.1.4 MOISTURE CONTENT

The moisture content and the concentration of the product play a significant role in determining the efficacy of OH of fluid food products. Since the movement of free ions is responsible for establishing electrical conductivity in liquids, moisture content of the product holds a crucial role in determining the rate of conduction [102]. The conductivity of the product is increased during OH at 50 °C due to the evaporation of a certain amount of water [51]. The concentration of the products has an inversely proportional relationship with the electrical conductivity of liquid juices with insoluble solids which may be due to the hindrance in the mobility of ions with increasing concentration [71, 41].

2.3.2 PROCESS PARAMETERS

2.3.2.1 FREQUENCY AND WAVEFORM

Frequencies as high as 100 Hz are adopted for OH of food. Usually, the application of power with the lower frequencies often engender electrolysis effects and dissolution of metallic electrodes [84, 85]. The frequency of the alternating current used in the process is known to influence the measured electrical conductivity of the products. The products with a cellular structure like vegetables and fruits exhibit different heating rates with varying frequencies. However, the heating rate of the food products with no cellular structure can be simply increased by elevating the frequency [101].

It was reported by Lakkakula et al. [53] that OH performed at lower frequencies of alternating current resulted in a higher yield during oil extraction. Lower frequencies of alternating current also reportedly enhance enzyme stabilization and other concurrent chemical processes [60]. Apart from the frequency, the waveform also has a significant impact on the electrical conductivity and efficiency of the OH of food. It has been found that square waveform manifests poor performance for certain food products like soymilk and turnip vegetable as compared to the sine and sawtooth waveform and often results in the dissolution of chemical constituents as well [48, 59].

2.3.2.2   ELECTRODES

The choice of electrodes, in combination with the frequency of the power input, is an essential aspect in designing a perfect OH system for food. It is the foundational component of the OH system. Figure 2.1 illustrates the design of a basic OH system. Since the electrodes are the primary factor for heat loss, they have a significant impact on the performance of the OH system [3]. The frequency and density of the applied current and the electrode material have a direct influence on the occurrence of electrochemical reactions at the electrode-product junction. The reactions can potentially result in the electrolysis of the biological components in the food product as well as corrode the electrodes, decreasing its functional life. Presently, current densities of 0.5–20 A/cm^2 are applied at the electrode tip. The application of higher frequencies of alternating current in the heating system reduces the oxidation reaction of the metallic component due to the reversed field effect and inhibition of Faraday reactions [45]. It was reported by Zell et al. [124] that the rate of temperature rise through the electrode interface is a function of its thickness as heat transfer and electrical resistance are dependent on mass. Because of the higher resistance to oxidation, stainless steel is usually adopted for the electrode material.

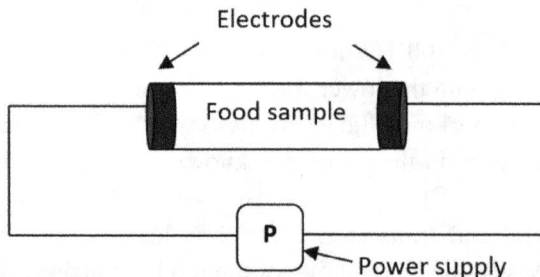

FIGURE 2.1   Basic OH system.

2.3.2.3   TEMPERATURE

The electrical conductivity of any food item is a thermophysical property and inherently a function of temperature. Considering the several past studies on the impact of temperature on the electrical conductivity and heating of food products, it can be firmly stated that the conductive properties of nearly all food products exhibit a linear increase with temperature rise unless they undergo a phase transition during the process, such as starch gelatinization or melting of fat [35, 87, 102, 117, 119, 123].

For efficient heating of the product, it is essential to supervise the temperature distribution throughout the structure of the food product. Often, in the conventional process of OH, nonuniformity is observed in the treated product that results in hot spots (the region with high temperatures) and cold spots (areas with low temperatures) [32]. Knowledge of the factors which influence the heating rate in the objective food product like thermal diffusivity, structural and chemical composition, etc., is primarily important to detect the order of temperature distribution in the product that will further assist in assuring adequate food quality and safety [45, 111].

2.3.2.4   PROCESSING TIME

The applied temperature and the time of processing are the crucial defining parameters in any food processing technique. An ideal selection of the temperature and time combination should be such that the qualitative integrity of the end product is best maintained, and the process ensures complete safety against the hazardous microorganisms and enzymes. The use of pretreatment processes like osmotic dehydration is known to significantly reduce the processing time by assisting in the extraction of water content in the food. A higher amount of moisture loss has been recorded in the samples of apples treated at 50 °C with a combined treatment of osmotic dehydration and OH in a shorter period of processing time [71].

2.3.2.5   PROCESS EFFICIENCY

Unlike the conventional heating processes, OH provides an efficient method for processing a wide range of food products. Providing volumetric and direct-resistance heating of food without any medium for conduction, the OH technique reduces the processing time with a minimal range of processing temperatures. The accelerated thermal kinetics and uniform heating with considerable electric input facilitate a faster heating process while avoiding the degradation of thermosensitive constituents in the food product. As discussed in previous sections, the application of alternating current at lower frequencies results in corrosion of the electrodes and lower conductive properties; thereby, it is usually suggested that high-frequency current must be applied for OH. While, at the industrial level, OH with higher frequencies seems to have economically limited aspects, but, OH is centralized at achieving thermal efficiency rather than the input power.

Many studies have reported that the thermal efficiency for OH irrespective of the opted system imparts a thermal efficiency close to 100% [28, 30, 58]. Ghnimi et al. [30] reported that the OH system can manifest high energy efficiencies of 70%–90% by adjusting the pulse parameters like delay time and duty cycle of the power input.

2.3.3   SYSTEM COMPONENTS

2.3.3.1   ELECTRICAL CONDUCTIVITY MEASURING TUBE

It is a simple cylindrical tube made of glass, polycarbonate, or Teflon material with electrodes attached at its ends. The food product to be treated is held in the tube in continuous contact with the electrodes. The temperature in the tube can be monitored by inserting thermocouples from the top. For heterogeneous liquid products, the unit can be set up on a shaker to enable proper mixing of the components [117]. The system unit can also be jacketed to maintain a constant temperature in the cell and prevent heat loss [16]. A more advanced measuring unit was introduced by Tulsiyan et al. [112] that is functional for a wide range of food products and can carry 10 electrodes at a time. Several samples of the food products can be simultaneously placed in the electrical conductivity measuring unit by sandwiching them between the electrodes. The units are made strong enough to be sealed and heated up to the sterilizing temperatures of the respective products.

2.3.3.2   BATCH AND CONTINUOUS HEATERS

Batch heaters like the rectangular trough heaters and T-tube heater were used at a laboratory scale. The rectangular trough heater, consisting of a transparent box covered with foam, mounted on a magnetic stirrer, and flat electrodes with rectangular cross-section are placed at the opposite sides of the box [34, 115]. The foam prevents heat loss to the surrounding. The glass T-tube heater has cylindrical electrodes attached on the tube ends fixed with plastic spacers and multiple temperature measuring thermocouples on the top. The cell can be pressurized to attain sterilization temperatures [90, 91, 95]. The continuous type heater has four ends perpendicular to each other. Two cylindrical electrodes are attached to one pair of opposite ends with plastic spacers and the other ends are connected to either plastic or nonconducting tubes. The sample material is continuously pumped in and out of the system and the temperatures at the respective spots are constantly monitored.

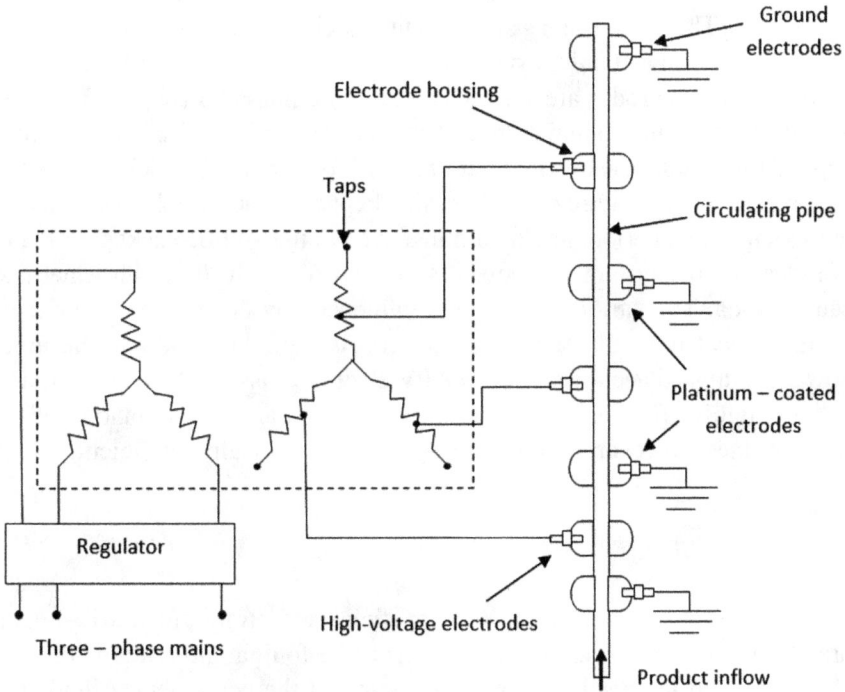

FIGURE 2.2   Schematic diagram of APV baker OH unit.

2.3.3.3   REACTOR VESSELS

For specific microlevel studies such as growth kinetics of lactic acid bacteria [63, 19] and degradation kinetics of ascorbic acid [6], experimental fermentor or reaction vessels were developed. Both the fermentor and the reaction vessel shared a similar design of a glass vessel with a fluid jacket for maintaining the system temperature and a lid with an opening for thermocouples and electrodes. The fermenter lid also had openings for pH probe tube, inoculation, and fraction collector. The magnetic stirrer can be used for mixing the medium. Samaranayake and Sastry [92] also used a similar vessel for observing the electrochemical reaction in the ohmic processing of food.

2.3.3.4   APV BAKER UNIT

The basic idea of the APV baker unit design for OH is represented in Figure 2.2. The system consists of multiple electrode housings (minimum four) enclosed in a stainless steel vertical tube with an upward flow of feed product. The electrodes are connected to a three-phase 50 Hz power supply, and the product tube is maintained such that uniform electrical impedance is applied throughout the heating section. The food product is quickly heated to the objective temperature of sterilization, kept at the achieved conditions for the specific time period, and then transferred to the aseptic packaging plant. For cleaning the system, the power is turned off and flushed with water and caustic soda. The various parameters influenced by the temperature rise or heating effect through the process are constantly monitored and the input power is adjusted accordingly. The APV electrodes are made up of platinum and are highly efficient in preventing polarization and contamination of the food product; however, it makes the system economically inefficient.

2.3.3.5   FLUID JET UNIT

While the tubular baker unit was beneficial and straightforward, several drawbacks were associated with a system like fouling along the walls and electrodes and the need for regular cleaning of the equipment. Alleviating these drawbacks, Emmepiamme, Italy, proposed a fluid jet system for the OH of especially, viscous food products. It is based on the idea of direct exposure of the jet of fluid flowing between the two electrodes with the alternating current. The schematic representation of the continuous OH system using fluid jets is illustrated in Figure 2.3. The system consists of a glass tube holding the electrodes and a high-pressure nozzle tip attached to the electrodes. The treated liquid is ejected from the system through the conical receptacle attached to a ground electrode. The length of the glass tube is a critical process parameter in the fluid jet system. The system design allows for the adjustment of the length of the glass tube for different food products with varying electrical conductivities. The required power for the process is continuously monitored based on the feedback mechanism to prevent any damage. Product deposits were observed in the case of dairy. However, the deposit had no significant effect on the flow parameters and could be cleaned by flushing with an alkaline solution.

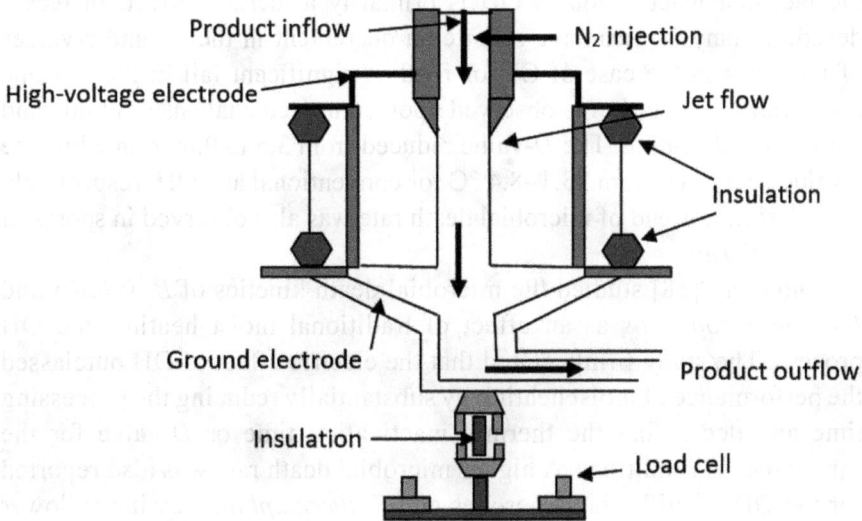

FIGURE 2.3 Schematic diagram of continuous fluid jet OH system.

2.4 EFFECT ON MICROBIAL DEATH RATE

Microbial inactivation in OH of food products is mainly a function of the thermal effect. However, the application of the electric field may induce the subjunctive nonthermal effect of electroporation which reportedly elevates the inactivation process in food. Only a handful of studies are available describing the impact of OH on microbial inactivation in food. No microorganism or pathogen strain has been reported so far as the reference index for microbial safety of food products subjected to ohmic cooking. The processes are performed by considering the most thermal resistant microorganism as the reference, like the conventional methods. Hot and cold spots developed in the product during the heating process due to nonuniform heating can also affect the inactivation rate and lead to a food safety compromise [45].

For an easy understanding of the degree of inactivation in food products, knowledge of the D- and z-values of the microbes is essential because of their eminent influence on the processing of microbes and determination of kinetic parameters [50, 74]. Initial experimental reports [77] claimed that there was no statistically significant impact of OH on the microbial death rate of yeast (*Zygosaccharomyces bailii*) as compared to the conventional heating methods under similar process conditions. This advocates the fact that inactivation in OH is primarily a thermal effect. In recent decades, many studies have reported a decrement in the D- and z-values of microbes in the case of OH of food. A significant fall in the D-value and z-value of *E. coli* was observed upon compared analysis of ohmic and conventional heating. The D-value reduced from 3.5 to 0.86 min while the z-value decreased from 23.1–8.4 °C for conventional and OH, respectively [80]. A similar trend of microbial death rate was also observed in spores of *B. licheniformis*.

Cho et al. [18] studied the microbial death kinetics of *B. subtilis* and *Bacillus atrophaeus* as an effect of traditional moist heating and OH process. The study firmly stated that the electric effect of OH outclassed the performance of moist heating by substantially reducing the processing time and decreasing the thermal inactivation time or D-value for the same process conditions. A higher microbial death rate was also reported for the OH of milk-viable aerobes and *S. thermophillus* owing to lower D-values [110].

The influence of the electrical effect during OH was further investigated and highlighted by several researchers in recent studies. A temperature-independent relationship was discovered between the microbial death rate and the applied voltage for *Alicyclobacillus acidoterrestris* spores [10]. Parallel correlations were also observed for other electric parameters like field strength [89] and frequency [53] of the flowing current in OH. Effect of the chemical attributes of the food product like the pH value also influences the inactivation of microbes (*Bacillus and Geobacillus endospores*) during OH of the food [76].

Elevated inactivation rates and reduced kinetic values (D- and z-values) were observed in the case of several microbial spores during OH that aids in a process with lower thermal intensity and reduced degradation of nutritional compounds while assuring microbial stability of the product. Disruption of the cell wall and increased permeability as an effect of the electric field potentially facilitates the incremented rates of inactivation; however, the underlying mechanism still needs to be expounded.

2.5 EFFECT ON FOOD PRODUCT

Like any other heating or processing technique in food, OH may engender some changes in the physicochemical properties of the treated food product like texture, color, and taste. As the food product is in direct contact with the metallic electrodes, there is a fair probability of transfusion of unwanted ions in the food product which may tether the desirable characteristics in the food and corrode the electrode as well [38]. A comprehended effect of the OH process on various sensorial and compositional attributes of the food product is explained in this section.

2.5.1 COLOR

Scientifically, the change in color of the food products is determined using the CIE parameters; L (representing lightness or darkness), a (representing greenness or redness), and b (representing blueness or yellowness). These parameters can be further processed in specific equations to express the complex features of chroma, hue, and total color difference. Reports suggest that fresh-like lightness and color ($a*$ and $b*$) values are better retained in OH assisted with osmotic dehydration than vacuum impregnation [69]. Change in color of certain food products is often an effect of the degradation of particular enzymes like PPO in sugarcane juice causing nonenzymatic browning in the product with the increase in temperature. However, the degree of change in color decreased with increasing electric field strength. Improved lightness values were also observed in the sugarcane juice because of the chlorophyll degradation [100].

2.5.2 TEXTURE

The texture is one of the crucial marketable parameters of food products that defines the quality of the product. Textural or structural change in the cells of the plant perishables is inevitable during the heating process. Continuous and prolonged heating will adversely affect the cell structure and thereby, the electrical conductivity of the product. Food samples subjected to the pretreatment of osmotic dehydration prior to OH exhibited a decrease in firmness values. The most significant decrement was observed at atmospheric pressure and temperature of 40 °C. The application of vacuum during the heating process manifested improved firmness values with a more compact and less deformed tissue structure [69]. OH of surimi (crab meat imitation) resulted in a viscosity higher than the conventional methods. An enhanced network structure due to the reduced decomposition of myosin and actin resulted in improved gel quality [75, 83]. The application of high-frequency current reportedly reduces the degree of texture degradation in delicate fruits like peaches [104]. Cell disruption was also observed in products with harder textures like potatoes and increased with temperature rise and moderate electric field [54].

2.5.3   ENZYMES

The biological components in food products, especially enzymes, are highly susceptible to the applied process parameters during heating operations. Heat-sensitive enzymes like PPO have reportedly exhibited an undesirable rise in activity during the ohmic processing of agricultural products like grapes and different fruit juices. The enzyme activity rose until a specific critical temperature and then showed a gradual decrease [43]. Increased activity of PPO leads to a change in the color and texture of the perishable food products and the production of toxins in some instances. These changes induce adverse effects on the acceptability and desirability of the products in the market.

Few contradictory reports of Leizerson and Shimoni [57], Demidroven and Baysal [22] claim that OH has a positive impact on the enzyme activity of orange juice by inactivating pectin esterase to a greater extent (by 90%–98%) as compared to the conventional techniques. The activity of peroxidases in pea puree [44] and carrot pieces [58] was successfully restricted during OH in comparison to the conventional technique by tethering the process parameters in a shorter processing time.

2.5.4   OTHER ESSENTIAL COMPONENTS

Food products are complex biological structures made of organic compounds that are highly vulnerable to heat and temperature. Ideal thermal processing methods must ensure adequate retention of the nutritional and value-added components in the food product. In comparison to the traditional methods, minimal decomposition of the flavoring substances like limonene, pinene, and myrcene was reported during OH [57]. While a similar trend in the degradation kinetics of anthocyanins in acerola puree was observed in both traditional and OH processes under identical process conditions by Mercali et al. [64]. A strong correlation between the degradation of anthocyanins and vitamin C and the voltage of the applied current was claimed by many researchers [65, 96]. The degradation rates exhibited an increasing trend with a rise in the current and voltage. Sterilization of vegetable and meat purees intended for baby food using OH exhibits better performance than conventional retort heating by limiting the production of furan and other processing contaminants during the sterilization and storage period [39].

2.6 APPLICATIONS

2.6.1 STERILIZATION AND PASTEURIZATION

Since its introduction, thermal processing of food products by pasteurization or sterilization has been recognized as the standard technical operations for the preservation of perishables in the food industry. The idea is to heat the products such that optimum microbial inactivation is achieved in the product, thereby increasing the shelf life. Traditionally, retort heating is adopted for sterilization in which the heat is conveyed from the source to the sample via conduction. The phenomenon of heat generation directly within the product structure lays superior advantages for OH over the traditionally opted methods. The lethal effect of the electric field on the microbial population during OH of fruit juices aids for a shelf life twice as long as that attained in conventional methods [57].

On the other hand, a study on ohmic pasteurization of strawberry by Castro et al. [16] suggested that there was no imminent influence of the electric field on the decomposition of ascorbic acid in the fruit. Elzubier et al. [25] also reported the successful attempt of sterilization of guava juice using the OH method. Nutritional compounds like vitamins and carotenoids are better retained in ohmic sterilization [1, 113]. Amino acid content remained nearly intact during ohmic sterilization of tomato puree while it showcased a significant decrease when subjected to retort sterilization [66].

2.6.2 BLANCHING AND THAWING

Blanching and thawing are pre- and post-thermal treatment processes for food. Blanching of food products prior to any rigorous heat treatment helps in retaining the quality attributes of the product. Conventionally, blanching is performed by immersing the vegetables in hot water for a short period. Blanching with an application of OH potentially reduces the undesirable leaching of solutes encountered during the conventional method, irrespective of the shape and size of the vegetable [68]. Due to the applied electric field, the rate of moisture loss is accelerated in ohmic-assisted blanching [118]. Thawing is the process of slowly and mildly heating the frozen products to bring them to room temperature without affecting the textural properties. It is an essential operation in meat processing. OH of frozen products by the application of the electric field is the perfect technology for thawing. Enhanced microbial stability and superior product quality procured by ohmic-assisted thawing as compared to the conventional methods, advocates for the greater potential of the technique [42]. Favorable thawing rates, energy utilization ratio, and minimum weight loss can be achieved by ohmic thawing of frozen meat like beef cut [12, 23]

2.6.3   FERMENTATION

The assisting nonthermal effect of the cell disruption by the electric field in OH facilitates easy and rapid transfusion of metabolic compounds and thereby, decreases the lag period of fermenting bacteria. Hence, fermentation in the production of yogurt, beer, cheese, or wine can be achieved efficiently with minimal processing time [19]. Reports on the impact of electric field on fermenting yeast like *Saccharomyces cerevisiae* have shown that there is a faster release of the nutrient compounds in the microbial cells in OH than in the case of traditional heating techniques [120].

2.6.4   EXTRACTION

Many valuable organic compounds are lost during the industrial processing of food products either due to decomposition by the process conditions or discarded as wastes. Most of these compounds like lipids, polyphenols, lignin, cellulose, pigments, etc., can be extracted from the products and utilized as high value-added components for various purposes like energy generation, pharmaceutical products, cosmetics, and other commercial applications. With the growing interest in the recovery of such compounds, several extraction techniques were developed, focusing on energy-efficient techniques with reduced usage of chemicals and organic solvents and higher degrees of extraction. It has been reported that the utilization of OH processes for the extraction of anthocyanins and phenolic compounds from colored potatoes delivers a greater yield of extraction without any use of organic solvents in a shorter processing period [81]. A higher rate of recovery was also reported for the extraction of lipids from rice bran using OH [53]. Successful extraction of sucrose from sugar beets [46] and soymilk [48] was also reported using OH.

2.7 RANGE OF POTENTIAL PRODUCTS

2.7.1 DAIRY

OH was originally proposed as a processing technique for the dairy industry by Anderson and Finkelstein [3]. Appropriate treatment of milk is essential before processing the dairy products for effective inactivation of the pathogens that contaminate the raw milk, like *E. coli, Salmonella spores*, and *L. monocytogenes* [72]. OH for low fat (0%–3%) dairy products exhibit superior performance in sterilization with higher microbial death rates. However, milk rich in fats impart reduced microbial stability. Fat being an insulator, milk with a higher amount of fat content hinders the electrical conductance through the fluid resulting in cold spots that consequently decrease the rate of microbial inactivation [49]. Despite having certain anomalies, OH has illustrated superior performances in dairy processing as compared to the conventional methods. Several studies have claimed it to be the best method for milk pasteurization with minimum fouling [9, 109].

2.7.2 FRUITS AND VEGETABLES

After the introduction of the technique of OH in food, several researchers have reported its influence on the structural, thermal, and compositional properties of fruits, vegetables, and their processed products like extracted juices and concentrates. OH of the plant perishables has been known to increase the shelf life with better retained nutritive and sensorial values. An apprehended delineation of the various findings in some of the past reports of OH of vegetables and fruits is presented in Table 2.1.

2.7.3 MEAT AND SEAFOODS

The advantage of inherent structural heating in OH, by the action of an electric field, makes it the perfect technology for meat cooking and thawing.

The complex organic composition of meat and seafood products makes it difficult to achieve uniform heating in the product without compromising the qualitative attributes of the product.

According to Sarang et al. [94] and Icier et al. [42], OH is the best alternative technique for the cooking and thawing of meat. Application of OH in turkey meat reportedly yielded superior quality products with minimal processing time [124]. Better yields of seafood were also reported by Yongsawatdigul et al. [119] and Pongviratchai and Park [83], upon application of OH. A few examples are illustrated in Table 2.1 justifying the effectiveness of the technique in this sector.

TABLE 2.1 Ohmic Heating on Various Perishable Fruits and Vegetables

Product	Effect of Ohmic Heating	Reference
Apricot pieces in syrup	Shelf life prolonged to one year at a storage temperature of 25 °C with greater retention of the sensory and nutritive values	[79]
Goat milk	No change in FFA after OH, hence, no effect on the product quality	[13]
Grape juice	Optimum inactivation of PPO	[43]
Milk	No effect on the chemical composition and pH of pasteurized milk, minimum processing time	[103]
Orange juice	Reduction of pectin methyl esterase activity by 98% and vitamin C decomposition by 15%	[47, 57]
Pacific whiting surimi	Twofold rise in shear stress and shear strain of the gel as compared to conventional technique, higher viscosity, shorter processing time	[83, 119]
Pea puree	Compared to conventional heating, faster inactivation of peroxidise enzyme, better color retention	[44, 47]
Pears	Permeability of the cells increased with vacuum impregnation	[70]
Red beets, carrots	Quick softening of the texture	[27]
Spinach puree	Better retention of chlorophyll content, carotene content, and color attributes as compared to conventional heating	[122]
Sugarcane juice	Enhanced inactivation of PPO (98%), shorter processing time, prominent change in color	[100]
Sweet potato cubes	Reduced processing time for vacuum frying	[125]
Tomato Paste	Better retention of color and gloss, decrease in pH values	[11]
Turkey meat	8–15-fold reduction in processing time	[124]

2.8   CONSUMER AND COMMERCIAL ACCEPTABILITY

Acceptability and marketability of any food product in the present millennial age largely depend upon the compositional richness of the product along with the aesthetic values. The evolved health concerned generation today, demands for more fresh-like attributes and value-added organic compounds in the product. On the other hand, for successful industrial adaptation, the processing technique should bear higher product yield and energy efficiency. As advocated by several reports in the past decades the novel technique of OH has a strong potential to deliver for every aspect of the current market demands [16, 21].

The heating of the substance at relatively lower temperatures in the presence of an electric field constitutes a safe processing atmosphere for food products in OH. It facilitates a minimally processed food product with assured microbial stabilization and greater retention of the sensory (color, texture) and nutritive (vitamins, proteins, carotenoids) attributes. The technique of OH of milk has largely evolved in the past years and has shown impressive results like less aggregation and major retention of protein, vitamin C, and color during sterilization [81, 88]. Qualitative attributes of fish like color, hardness, temperature, water activity, and organoleptic score were evaluated after applying an OH treatment, and based on the performance values obtained for the optimal process conditions, satisfactory overall acceptability of 7.81 was obtained for the treated fish sample [75].

In the world of rapidly expanding industrial pollution, OH provides an environment-friendly green processing method for food products that promote its acceptability on ethical grounds. Inspired by the growing consumer interests, currently, eight companies and 18 plants are working with different kinds of OH units for pasteurization and sterilization of fluidic products processed from fruits and vegetables, milk, whey, egg, ice cream mixture, soup mixture, etc. [15, 40, 121].

A comparative analysis of the energy utilization conducted by De Halleux et al. [36] claimed that application of ohmic cooking at an industrial scale would allow energy efficiency as high as 90% and a huge reduction (82%–97%) in the units of energy consumption as compared to the traditional smokehouse cooking. OH is the perfect food processing technique that delivers microbial stability with improved qualitative properties and shorter processing time that helps in avoiding economic losses, proving to be highly marketable [45, 47, 97].

2.9   SUMMARY

A detailed study on the physical, thermal, chemical, and structural attributes of different kinds of food products with variable data must be performed for a serviceable evolution of the technique. Processing of food products with nonconductive properties such as fat globules deems peremptory for an improved commercial adaptation of the process. In order to prevent the undesirable effects in the ohmic heated product, it is essential to monitor the process-induced chemical changes like allergenicity at molecular levels, overall pH, density, softening, and electrochemical reactions. Adequate indicators must be developed to detect these unwanted changes and allow for potential process windows. Designing efficient simulation models for the evaluation of the process, product, and system parameters with precision in microbial stabilization demands optimal attention. A combination of other processing techniques such as osmotic dehydration, vacuum heating, and radio-frequency heating needs to be explored for elevating the potential of this novel processing technology.

KEYWORDS

- **consumer acceptability**
- **design parameters**
- **electrical conductivity**
- **food quality**
- **microbial**
- **death kinetics**
- **ohmic heating**

REFERENCES

1. Achir, N.; Dhuique-Mayer, C.; Hadjal, T.; Madani, K.; Pain, J.P; Dornier, M. Pasteurization of Citrus Juices with Ohmic Heating to Preserve the Carotenoid Profile. *Innovative Food Science and Emerging Technologies,* **2016,** *33,* 397–404.
2. Akdemir, E.G.; Baysal, T.; Icier, F.; Yildiz, H.; Demirdoven, A.; Bozkurt, H. Processing of Fruits and Fruit Juices by Novel Electrotechnologies. *Food Engineering Reviews,* **2012,** *4,* 68–87.

3. de Alwis, A.A.P.; Fryer, P.J. A Finite Element Analysis of Heat Generation and Transfer During Ohmic Heating of Foods. *Chemical Engineering Science,* **1990,** *45* (6), 1547–1559.

4. An, H.J; King, J.M.. Thermal Characteristics of Ohmically Heated Rice Starch and Rice Flours. *Journal of Food Science,* **2007,** *72* (1), 84–88.

5. Anderson, A.K.; Finkelsten, R. A Study of the Electro-pure process of treating milk. *Journal of Dairy Science,* **1919,** *2,* 374–406.

6. Assiry, A.M.; Sastry, S.K.; Samaranayake, C. Influence of Temperature, Electrical Conductivity, Power and pH on Ascorbic Acid Degradation Kinetics During Ohmic Heating with Stainless Steel Electrodes. *Bioelectrochemistry,* **2006,** *68,* 7–13.

7. Awuah, G.B.; Ramaswamy, H.S.; Economides, A. Thermal Processing and Quality: Principles and Overview. *Chemical Engineering and Processing,* **2007,** *46,* 584–602.

8. Bari, M.L.; Alexandru G.; Ukuku, D.O.; Dey, G.; Miyaji, T. New Food Processing Technologies and Food Safety. *Hindawi Journal of Food Quality,* **2017,** *2017,* 2.

9. Bansal, B.; Chen, X.D. Critical Review of Milk Fouling in Heat Exchangers. *Comprehensive Reviews in Food Science and Food Safety,* **2006,** *5,* 27–33.

10. Baysal, A.H.; Icier, F. Inactivation Kinetics of *Alicyclobacillus acidoterrestris* Spores in Orange Juice by Ohmic Heating: Effects of Voltage Gradient and Temperature on Inactivation. *Journal of Food Protection,* **2010,** *73,* 299–304.

11. Boldaji, M.T.; Borghei, A.M.; Beheshti, B.; Hosseini, S.E. The Process of Producing Tomato Paste by Ohmic Heating Method. *Journal of Food Science and Technology,* **2015,** *52* (6), 3598–3606.

12. Bozkurt, H.; Icier, F. Ohmic Thawing of Frozen Beef Cuts. *Journal of Food Process Engineering,* **2012,** *35,* 16–36.

13. Cappato, L.P.; Ferreira, M.V.S. Ohmic Heating in Dairy Processing: Relevant Aspects for Safety and Quality. *Trends in Food Science & Technology,* **2017,** *62,* 104–112.

14. Carrillo, L.; Alma, A.R.; Rodriguez, L.L.; Villagrana, R.A.R. Modification of Food Systems by Ultrasound. *Journal of Food Quality,* **2017,** *3,* 1–12.

15. Castro, I. *Ohmic Heating as an Alternative to Conventional Thermal Treatment.* PhD Dissertation, Universidade do Minho, Braga, Portugal, **2007,** 236.

16. Castro, I.; Teixeira, J.A.; Salengke, S.; Sastry, S.K.; Vicente, A.A. Ohmic Heating of Strawberry Products: Electrical Conductivity Measurements and Ascorbic Acid Degradation Kinetics. *Innovative Food Science and Emerging Technologies,* **2004,** *5,* 27–36.

17. Castro, I.; Oliveira, C.; Domingues, L.; Teixeira, J.A.; Vicente, A.A. The Effect of the Electric Field on Lag Phase, β-Galactosidase Production and Plasmid Stability of a Recombinant *Saccharomyces cerevisiae* Strain Growing on Lactose. *Food and Bioprocess Technology,* **2012,** *8* (5), 3014–3020.

18. Cho, H.Y.; Yousef, A.E.; Sastry, S.K. Kinetics of Inactivation of *Bacillus subtilis* Spores by Continuous or Intermittent Ohmic and Conventional Heating. *Biotechnology and Bioengineer,* **1999,** *62* (3), 368–372.

19. Cho, H.Y.; Yousef, A.E.; Sastry, S.K. Growth Kinetics of *Lactobacillus acidophilus* under Ohmic Heating. *Biotechnology and Bioengineering,* **1996,** *49* (3), 334–340.

20. Coster, H.G.; Zimmermann, U. The Mechanism of Electric Breakdown in Membranes of *Valoniautricularis. Journal of Membrane Biology,* **1975,** *22,* 73–90.

21. Darvishi, H.; Khoshtaghaza, M.H.; Najafi, G. Ohmic Heating of Pomegranate Juice: Electrical Conductivity and pH Change. *Journal of the Saudi Society of Agricultural Sciences,* **2013,** *12,* 101–108.

22. Demirdoven, A.; Baysal, T. Optimization of Ohmic Heating Applications for Pectin Methylesterase Inactivation in Orange Juice. *Journal of Food Science and Technology,* **2014,** *51,* 1817–1826.

23. Duygu, B.; Umit, G. Application of Ohmic Heating System in Meat Thawing. *Procedia Social and Behavioural Sciences,* **2015,** *195,* 2822–2828.

24. Eliot-Godéreaux, S.C.; Zuber, F. Processing and Stabilisation of Cauliflower by Ohmic Heating Technology. *Innovative Food Science & Emerging Technologies,* **2001,** *2,* 279–287.

25. Elzubier, A.S.; Thomas, C.S.Y.; Sergie, S.Y. The Effect of Buoyancy Force in Computational Fluid Dynamics Simulation of a Two-Dimensional Continuous Ohmic Heating process. *American Journal of Applied Science,* **2009,** *6* (11), 1902–1908.

26. Evrendilek, G.A.; Baysal, T.; Icier, F.; Yildiz, H., Demidroven, A.; Bozkurt, H. Processing of Fruits and Fruit Juices by Novel Electrotechnologies. *Food Engineering Reviews,* **2012,** *4,* 68–87.

27. Farahnaky, A.; Azizi, R.; Gavahian, M. Accelerated Texture Softening of Some Root Vegetables by Ohmic Heating. *Journal of Food Engineering,* **2012,** *113,* 275–280.

28. Fillaudeau, L.; Winterton, P.; Leuliet, J.C. Heat Treatment of Whole Milk by Direct Joule Effect—Experimental and Numerical Approaches to Fouling Phenomena. *Journal of Dairy Science,* **2006,** *89,* 4475–4489.

29. Fryer, P.J.; Li, Z. Electrical Resistance Heating of Foods. *Trends in* Food *Science and Technology,* **1993,** *4,* 364–369.

30. Ghnimi, S.; Zaîd, I.; Maingonnat, J.F.; Delaplace, G. Axial Temperature Profile of Ohmically Heated Fluid Jet: Analytical Model and Experimentation Validation. *Chemical Engineering Science,* **2009,** *64,* 3188–3196.

31. Gomathy, K.; Thangavel, K.; Balakrishnan, M.; Kasthuri, R. Effect of Ohmic Heating on the Electrical Conductivity, Biochemical and Rheological Properties of Papaya Pulp. *Journal of Food Process Engineering,* **2015,** *38,* 405–413.

32. Goullieux, A.; Pain, J.P. (Eds.) Emerging Technologies for Food Processing. Academic Press , London; **2005,** 469–505.

33. Guida, V.; Ferrari, G.; Pataro, G.; Chambery, A.; DiMaro, A.; Parente, A. The Effects of Ohmic and Conventional Blanching on the Nutritional, Bioactive Compounds and Quality Parameters of Artichoke Heads. *LWT—Food Science and Technology,* **2013,** *53,* 569–579.

34. Gupta, S.N. Application of Ohmic Heating in Prevention of Fouling of Concentrated Milk and Lye Peeling of Pears. M.Sc. Thesis; The Ohio State University, Columbus, OH, **2010,** 201.

35. Halden, K.; De Awis, A.A.P.; Fryer, P.J. Changes in the Electrical Conductivity of Foods During Ohmic Heating. *International Journal of Food Science and Technology,* **1990,** *25,* 9–25.

36. de Halleux, D.; Chiu, L.; Raymond, Y.; Ramaswamy, H.S.; Ohmic Heating of Processed Meat and its Effects on Product Quality. *Journal of Food Science,* **2004,** *69* (2), 71–77.

37. Hendrickson, J. Energy Use in the U.S. Food System: A Summary of Existing Research and Analysis. University of Wisconsin, Center for Integrated Agricultural Systems, Madison, WI, **1996,** 21.

38. Herting, G.; Wallinder, I.O.; Leygraf, C. Corrosion-Induced Release of Chromium and Iron from Ferritic Stainless Steel Grade AISI 430 in Simulated Food Contact. *Journal of Food Engineering,* **2008,** *87,* 291–300.

39. Hradecky, J.; Kludska, E.; Belkova, B.; Wagner, M.; Hajslova J. Ohmic Heating: A Promising Technology to Reduce Furan Formation in Sterilized Vegetable and Vegetable/Meat Baby foods. *Innovative Food Science and Emerging Technologies,* **2017,** *43,* 1–6.

40. Icier F. Ohmic Heating of Fluid Foods. In: *Novel Thermal and Non-Thermal Technologies for Fluid Foods*; Cullen, P.J., Tiwari, B.K., and Valdramidis, V.P. (Eds), Academic Press, Cambridge, MA, **2012,** 305–368.

41. Icier, F.; Ilicali, C. Electrical Conductivity of Apple and Sourcherry Juice Concentrates During Ohmic Heating. *Journal of Food Process Engineering,* **2004,** *27,* 159–180.

42. Icier, F.; Izzetoglu, G.T; Bozkurt, H.; Ober, A. Effects of Ohmic Thawing on Histological and Textural Properties of Beef Cuts. *Journal of Food Engineering,* **2010,** *99,* 360–365.

43. Icier, F.; Yildiz, H.; Baysal, T. Polyphenoloxidase Deactivation Kinetics During Ohmic Heating of Grape Juice. *Journal of Food Engineering,* **2008,** *85,* 410–417.

44. Icier, F.; Yildiz, H.; Baysal, T. Peroxidase Inactivation and Colour Changes During Ohmic Blanching of Pea Puree. *Journal of Food Engineering,* **2006,** *74,* 424–429.

45. Jaeger, H.; Roth, A.; Toepfl, S.; Holzhauser, T. Opinion on the Use of Ohmic Heating for the Treatment of Foods. *Trends in Food Science & Technology,* **2016,** *55,* 84–97.

46. Katrokha, I.; Matvienko, A.; Vorona, L.; Kupchik, M.; Zaets, V. Intensification of Sugar Extraction from Sweet Sugar Beet Cossettes in an Electric Field. *Sakharnaya Promyshlennost,* **1984,** *7,* 28–31.

47. Kaur, R.G.K.; Singh, A.K. Nutritional Impact of Ohmic Heating on Fruits and Vegetables—A review. *Cogent Food & Agriculture,* **2016,** *2* (1), 11–59.

48. Kim, J.; Pyun, Y. Extraction of Soy Milk Using Ohmic Heating. *Abstract, 9th Congress on Food Science and Technology*, Budapest, **1995.**

49. Kim, S.S.; Kang, D.H. Comparative Effects of Ohmic and Conventional Heating for Inactivation of *Escherichia coli O157:H7, Salmonella enterica Serovartyphimurium*, and *Listeria monocytogenes* in Skim Milk and Cream. *Journal of Food Protection,* **2015,** *78,* 1208–1214.

50. Knirsch, M.C.; Alves, S.C.; Vicente, A.A.M.O.S.; Penna, T.C.V. Ohmic Heating—A review. *Trends in Food Science & Technology,* **2010,** *21,* 436–441.

51. Kong, C.S.; Ogawa, H.; Iso, N. Rheological Analysis of the Effect of Gelatinization of Starch Added to Fish-Meat Gel Based on Volume Changes. *Fish Science,* **1999,** *65* (6), 930–6.

52. Kumar, T. A Review on Ohmic Heating Technology: Principle, Applications and Scope. *International Journal of Agriculture, Environment and Biotechnology*, **2018,** *11* (4), 679–687.

53. Lakkakula, R.N; Lima, M.; Walker, T. Rice Bran Stabilization and Rice Bran Oil Extraction Using Ohmic Heating. *Bioresource Technology,* **2004,** *92,* 157–161.

54. Lebovka, N.I.; Praporscic, I.; Ghnimi, S.; Vorobiev, E. Does Electroporation Occur During the Ohmic Heating of Food? *Journal of Food Science,* **2005,** *70* (5), 308–311.

55. Lee, S.Y.; Sagong, H.G.; Ryu, S.; Kang, D.H. Effect of Continuous Ohmic Heating to Inactivate *Escherichia coli O157:H7*, Salmonella Typhimurium and Listeria monocytogenes in orange juice and tomato juice. *Journal of Applied Microbiology,* **2012,** *112,* 723–731

56. Legrand, A.; Leuliet, J.C.; Duquesne, S.; Kesteloot, R. Physical, Mechanical, Thermal and Electrical Properties of Cooked Red Bean (*Phaseolus vulgaris L.*) for Continuous Ohmic Heating Process. *Journal of Food Engineering,* **2007**, *81*, 447–458.

57. Leizerson, S.; Shimoni, E. Effect of Ultra-high Temperature Continuous Ohmic Heating Treatment on Fresh Orange Juice. *Journal of Agricultural and Food Chemistry,* **2005**, *53*, 3519–3524.

58. Lemmens, L.; Tiback, E.; Svelander, C. Thermal Pretreatments of Carrot Pieces Using Different Heating Techniques: Effect on Quality Related Aspects. *Innovative Food Science & Emerging Technologies,* **2009**, *10*, 522–529.

59. Lima, M.; Heskitt, B.F.; Burianek, L.L.; Nokes, S.E.; Sastry, S.K. Ascorbic Acid Degradation Kinetics During Conventional and Ohmic Heating. *Journal of Food Preservation,* **1999**, *23*, 421–434.

60. Lima, M.; Sastry, S.K. The Effects of Ohmic Heating Frequency on Hot-Air Drying Rate and Juice yield. *Journal of Food Engineering,* **1999**, *41*, 115–119.

61. Loghavi, L.; Sastry, S.K.; Yousef, A.E. Effects of Moderate Electric Field Frequency and Growth Stage on the Cell Membrane Permeability of *Lactobacillus acidophilus*. *Biotechnology Progress,* **2009**, *25* (1), 85–94.

62. Loghavi, L.; Sastry, S.K.; Yousef, A.E. Effect of Moderate Electric Field Frequency on Growth Kinetics and Metabolic Activity of *Lactobacillus acidophilus*. *Biotechnology Progress,* **2008**, *24*, 148–153.

63. Loghavi, L.; Sastry, S.K.; Yousef, A.E. Effect of Moderate Electric Field on the Metabolic Activity and Growth Kinetics of *Lactobacillus acidophilus*. *Biotechnology and Bioengineering,* **2007**, *98*(4), 872–881.

64. Mercali, G.D.; Jaeschke, D.P.; Tessaro, I.C.; Marczak, L.D.F. Degradation Kinetics of Anthocyanins in Acerola Pulp: Comparison Between Ohmic and Conventional Heat Treatment. *Food Chemistry,* **2013**, *136*, 853–857.

65. Mercali, G.D.; Jaeschke, D.P.; Tessaro, I.C.; Marczak, L.D.F. Study of Vitamin C Degradation in Acerola Pulp During Ohmic and Conventional Heat Treatment. *LWT— Food Science and Technology,* **2012**, *47*, 91–95.

66. Mesías, M.; Wagner, M.; George, S.; Morales, F.J. Impact of Conventional Sterilization and Ohmic Heating on the Amino Acid Profile in Vegetable Baby Foods. *Innovative Food Science and Emerging Technologies,* **2016**, *34*, 24–28

67. Mizrahi, S.; Kopelman, I.; Perlaman, J. Blanching by Electroconductive Heating. *Journal of Food Technology,* **1975**, *10*, 281–288.

68. Mizrahi, S. Leaching of Soluble Solids During Blanching of Vegetables by Ohmic Heating. *Journal of Food Engineering,* **1996**, *29*, 153–166.

69. Moreno, J.; Bugueño, G.; Velasco, V.; Petzold G.; Tabilo-Munizaga, G. Osmotic Dehydration and Vacuum Impregnation on Physicochemical Properties of Chilean Papaya (*Caricacandamarcensis*). *Journal of Food Science,* **2004**, *69* (3), 102–106.

70. Moreno, J.; Simpson, R.; Sayas, M.; Segura, I.; Aldana, O.; Almonacid, S. Influence of Ohmic Heating and Vacuum Impregnation on the Osmotic Dehydration Kinetics and Microstructure of Pears (*cv. Packham's Triumph*). *Journal of Food Engineering,* **2011**, *104*, 621–627.

71. Moura, S.C.S.R.; Vitali, A.A.; Hubinger, M.D. A Study of Water Activity and Electrical Conductivity in Fruit juices: Influence of Temperature and Concentration. *Brazilian Journal of Food Technology,* **1999**, *2* (12), 31–38.

72. Murinda, S.; Nguyen, L.; Nam, H.; Almeida, R.; Headrick, S.; Oliver, S. Detection of Sorbitol-Negative and Sorbitol-Positive Shiga Toxin-Producing *Escherichia coli, Listeria monocytogenes, Campylobacter jejuni,* and *Salmonella spp.* in Dairy Farm Environmental Samples. *Food borne Pathogens & Disease*, **2004**, *1*, 97–104.

73. Oyedeji, A.B.; Sobukola, O.P. Effect of Frying Treatments on Texture and Colour Parameters of Deep Fat Fried Yellow Fleshed Cassava Chips. *Journal of Food Quality*, **2017**, Special Issue, 10.

74. Park, I.K.; Kang, D.H. Effect of Electropermeabilization by Ohmic Heating for Inactivation of *Escherichia coli O157:H7, Salmonella enterica serovar typhimurium,* and *Listeria monocytogenes* in Buffered Peptone Water and Apple Juice. *Applied and Environmental Microbiology*, **2013**, *79*, 7122–7129.

75. Park, J.W.; Yongsawatdigul, J.; Kolbe, E. Proteolysis and Gelation of Fish Proteins Under Ohmic Heating. *Process-Induced Chemical Changes in Food*, **1998**, *434*, 25–34.

76. Park, S.H.; Balasubramaniam, V.M.; Sastry, S.K.; Lee, J.Y. Pressure-Ohmic Thermal Sterilization: A Feasible Approach for the Inactivation of *Bacillus amyloliquefaciens* and *Geobacillus stearothermophilus* Spores. *Innovative Food Science & Emerging Technologies*, **2013**, *19*, 115–123.

77. Palaniappan, S.; Sastry, S.K. Effects of Electroconductive Heat Treatment and Electrical Pretreatment on Thermal Death Kinetics of Selected Microorganisms. *Biotechnology and Bioengineering*, **1992**, *39*, 225–232.

78. Palaniappan, S.; Sastry, S.K. Electrical Conductivities of Selected Solid Foods During Ohmic Heating. *Journal of Food Process Engineering*, **1991**, *14*, 221–236.

79. Pataro, G.; Donsi, G.; Ferrari, G. Aseptic Processing of Apricots in Syrup by Means of a Continuous Pilot Scale Ohmic Unit. *LWT—Food Science and Technology*, **2011**, *44*, 1546–1554.

80. Pereira, R.; Martins, R.C.; Vicente, A. Goat Milk Free Fatty Acid Characterization During Conventional and Ohmic Heating Pasteurization. *Journal of Dairy Science*, **2008**, *91*, 2925–2937.

81. Pereira, R.N.; Rodrigues, R.M. Effects of Ohmic Heating on Extraction of Food-Grade Phytochemicals from of Colored Potato. *LWT—Food Science and Technology*, **2016**, *74*, 493–503.

82. Piette, G.; Buteau, M.L.; de Halleux, D. Ohmic Cooking of Processed Meats and its Effects on Product Quality. *Journal of Food Science*, **2006**, *69* (2), 71–78.

83. Pongviratchai, P.; Park, J.W. Electrical Conductivity and Physical Properties of Surimi-Potato Starch Under Ohmic Heating. *Journal of Food Science*, **2007**, *72* (9), 503–507.

84. Remik, D. Apparatus and Method for Electrical Heating of Food Products. Patent No. 4, 739, 140, USA; **1988**, 21.

85. Reznick, D. Ohmic Heating of Fluid Foods. *Food Technology*, **1996**, *50* (5), 250–251.

86. Roberts, J.S.; Balaban, M.O.; Zimmerman, R.; Luzuriaga, D. Design and Testing of a Prototype Ohmic Thawing Unit. *Computers and Electronics in Agriculture*, **1998**, *19*, 211–222.

87. Roberts, T.A.; Pitt, J.I.; Farkas, J.; Grau, F.H. (Eds.) *Microorganisms in Foods*. Blackie Academic and Professional. London. *International Commission on Microbiological Specifications for Foods*, **1998**, 119.

88. Roux, S.; Courel, M.; Birlouez-Aragon, I.; Municino, F.; Massa, M.; Pain, J.P. Comparison of Continuous Ohmic Heating and Steam Injection for the Sterilization of

Infant Formula. *Conference on Novel Technologies for Sustainable Dairy Products*. IDF World Dairy Summit, Parma, **2011.**

89. Sagong, H.G.; Park, S.H.; Choi, Y.J.; Ryu, S.; Kang, D.H. Inactivation of *Escherichia coli* O157:H7, Salmonella Typhimurium, and Listeria Monocytogenes in Orange and Tomato Juice Using Ohmic Heating. *Journal of Food Protection*, **2011**, *74*, 899–904.

90. Salengke, S.; Sastry, S.K. Models for Ohmic Heating of Solid–Liquid Mixtures Under Worst-Case Heating Scenarios. *Journal of Food Engineering*, **2007**, *83*, 337–355.

91. Salengke, S.; Sastry, S.K. Experimental Investigation of Ohmic Heating of Solid–Liquid Mixtures Under Worst-Case Heating Scenarios. *Journal of Food Engineering*, **2007**, *83*, 324–336.

92. Samaranayake, C.P.; Sastry, S.K. Electrode and pH Effects on Electrochemical Reactions During Ohmic Heating. *Journal of Electroanalytical Chemistry*, **2005**, *557*, 125–135.

93. Samprovalaki, K.; Bakalis, S.; Fryer, P.J. Ohmic Heating: Models and Measurements. In: *Heat Transfer in Food Processing*; Yanniotis, S. and Sunden, B. (Eds); WIT press, Southampton, NY, **2007**, 159–186.

94. Sarang, S.; Sastry, S.K.; Knipe, L. Electrical Conductivity of Fruits and Meats During Ohmic Heating. *Journal of Food Engineering*, **2008**, *87*, 351–356.

95. Sarang, S.; Sastry, S.K.; Gaines, J.; Yang, T.C.S.; Dunne, P. Product Formulation for Ohmic Heating: Blanching as a Pretreatment Method to Improve Uniformity in Heating of Solid-Liquid Food Mixtures. *Journal of Food Science: Food Engineering and Physical Properties*, **2007**, *72* (5), 227–234.

96. Sarkis, J.R.; Jaeschke, D.P.; Tessaro, I.C.; Marczak, L.D.P. Effects of Ohmic and Conventional Heating on Anthocyanin Degradation During the Processing of Blueberry Pulp. *LWT—Food Science and Technology*, **2013**, *51*, 79–85.

97. Sastry, S.K. Overview of Ohmic Heating. Chapter 2; In: *Ohmic Heating in Food*; Ramaswamy, H.S., Marcotte, M., Sastry, S., and Abdelrahim (Eds.); CRC Press, Boca Raton, FL, **2014**, 3–6.

98. Sastry, S.K.; Heskitt, B.F.; Sarang, S.S.; Somavat, R.; Ayotte, K. Why Ohmic Heating? Advantages, Applications, Technology, and Limitations. Chapter 2; In: *Ohmic Heating in Food*; Ramaswamy, H.S., Marcotte, M., Sastry, S. and Abdelrahim (Eds.), CRC Press, Boca Raton, **2014**, 7–14.

99. Sastry, S.K.; Palaniappan. S. Ohmic Heating of Liquid Particle Mixtures. *Food Technology*, **1991**, *45*(12), 64–67.

100. Saxena, J.; Makroo, H.A.; Srivastava, B. Effect of Ohmic Heating on Polyphenol Oxidase (PPO) Inactivation and Color Change in Sugarcane Juice. *Journal of Food Processing Engineering*, **2016**, *40* (3), 12485.

101. Sensoy, L.; Sastry, S.K. Ohmic Blanching of Mushrooms. *Journal of Food Process Engineering*, **2004**, *27*, 1–15.

102. Shirsat, N.; Brunton, N.P.; Lyng, J.G.; McKenna, B. Water Holding Capacity, Dielectric Properties and Light Microscopy of Conventionally and Ohmically Cooked Meat Emulsion Batter. *European Food Research and Technology*, **2004**, *219*, 1–5.

103. Shivmurti, S.; Harshit, P.; Rinkita, P.; Smit, P. Comparison of Chemical Properties of Milk when Conventionally and Ohmically Heated. *International Food Research Journal*, **2014**, *21* (4), 1425–1428.

104. Shynkaryk, M.V.; Ji, T.; Alvarez, V.B.; Sastry, S.K. Ohmic Heating of Peaches in the Wide Range of Frequencies (50 Hz to 1 MHz). *Journal of Food Science*, **2010**, *75*, 493–500.

105. Singh, R.P.; Heldman, D.R. (Eds). *Introduction to Food Engineering*, 5th Edition; Academic Press Incorporation, Orlando, FL. **2014.**

106. Singh, S.P.; Tarsikka, P.S.; Singh, H. Study on Viscosity and Electrical Conductivity of Fruit Juices. *Journal of Food Science and Technology,* **2008,** *45* (4), 371–372.

107. Skudder, P.J. Ohmic Heating: New Alternative of Aseptic Processing of Viscous Foods. *Food Engineering,* **1988,** *60,* 99–101.

108. Smil V. (Ed.). Energy in Nature and Society: General Energetics of Complex Systems. MIT Press, Cambridge, MA, **2008,** 219.

109. Stancl, J.; Zitny, R. Milk Fouling at Direct Ohmic Heating. *Journal of Food Engineering,* **2010,** *99,* 437–444.

110. Sun, H.X.; Kawamura, S.; Himoto, J.I.; Itoh, K.; Wada, T.; Kimura, T. Effects of Ohmic Heating on Microbial Counts and Denaturation of Proteins in Milk. *Food Science and Technology Research,* **2008,** *14,* 117–123.

111. Tucker, G.S. Food Waste Management and Value-Added Products: Using the Process to Add Value to Heat-Treated Products. *Journal of Food Science,* **2004,** *69* (3), 102–104.

112. Tulsiyan, P.; Sarang, S.; Sastry, S.K. Electrical Conductivity of Multicomponent Systems During Ohmic Heating. *International Journal of Food Properties,* **2008,** *11*(1), 233–241.

113. Vikram, V.B.; Ramesh, M.N.; Prapulla, S.G. Thermal Degradation Kinetics of Nutrients in Orange Juice Heated by Electromagnetic and Conventional Methods. *Journal of Food Engineering,* **2005,** *69* (1), 31–40.

114. Wang, C.S.; Kuo, S.Z.; Kuo-Huang, L.L.; Wu, J.S.B. Effect of Tissue Infrastructure on Electric Conductance of Vegetable Stems. *Journal of Food Science: Food Engineering and Physical Properties,* **2001,** *66* (2), 284–288.

115. Wang, L.; Li, D.; Tatsumi, E.; Liu, Z.; Chen, X.; Li, L. Application of Two-Stage Ohmic Heating to Tofu Processing. *Chemical Engineering and Processing,* **2007,** *46* (5), 486–490.

116. Wang, W.; Sastry, S.K. Salt Diffusion into Vegetables Tissue as a Pretreatment for Ohmic Heating: Electrical Conductivity Profiles and Vacuum Infusion Studies. *Journal of Food Engineering,* **1993,** *20,* 299–309.

117. Wang, W.; Sastry, S.K. Changes in Electrical Conductivity of Selected Vegetables During Multiple Thermal Treatments. *Journal of Food Process Engineering,* **1997,** *20,* 499–516.

118. Wigerstrom, K. Passing an Electric Current of 50–60 Hz Through Potato Pieces During Blanching. US Patent Number 3,997,678, USA, **1976.**

119. Yongsawatdigul, J.; Park, J.W.; Kolbe, E. Electrical Conductivity of Pacific Whiting Surimi Paste During Ohmic Heating. *Journal of Food Science,* **1995,** *60* (5), 922–935.

120. Yoon, S.W.; Lee, C.Y.J.; Kim, K.M.; Lee, C.H. Leakage of Cellular Material from *Saccharomyces cerevisiae* by Ohmic Heating. *Journal of Microbiology and Biotechnology,* **2002,** *12,* 183–188.

121. Yildiz, H.; Guven, E. Industrial Application and Potential Use of Ohmic Heating for Fluid Foods. *Bulgarian Chemical Communications,* **2014,** *46,* 98–102.

122. Yildiz, H.; Icier, F.; Baysal, T. Changes in Beta-Carotene, Chlorophyll and Color of Spinach Puree during Ohmic Heating. *Journal of Food Process Engineering,* **2010,** *33,* 763–779.

123. Zareifard, M.R.; Ramaswamy, H.S.; Trigui, M.; Marcotte, M. Ohmic Heating Behavior and Electrical Conductivity of Two-phase Food Systems. *Innovative Food Science and Emerging Technologies,* **2003,** *4,* 45–55.

124. Zell, M.; Lyng, J.G.; Morgan, D.J.; Cronin, D.A. Minimizing Heat Losses During Batch Ohmic Heating of Solid Food. *Food Bioproduct Processes,* **2011,** *89,* 128–134.

125. Zhong, T.; Lima, M. The Effect of Ohmic Heating on Vacuum Drying Rate of Sweet Potato Tissue. *Bioresource Technology,* **2003,** *87,* 215–220.

CHAPTER 3

POWER ULTRASOUND: A GREEN TECHNOLOGY FOR PROCESSING OF FOOD

YASHINI MUTHUKRISHNAN, CHIKKABALLPUR K. SUNIAL, and ASHISH RAWSON*

ABSTRACT

The green approach of ultrasound processing is to protect the environment and health of consumers by eradicating the usage of toxic prone solvents and release of hazardous substances; reducing the consumption of energy and water; recycling of wastes; and ensuring safety and quality of food. This chapter focuses on the application of ultrasound in extraction, food preservation, food modification, size reduction, membrane separation, and delivery of food. It will detail mechanisms and novel methods of ultrasound processing in food processing. Further, the application of hazard analysis and critical control point in ultrasound food processing will be discussed.

3.1 INTRODUCTION

Consumers' awareness regarding the quality, safety, and eco-friendly aspect of processed foods is increasing day by day. This growth of "Green consumerism" has led to the innovation of green technologies in the food processing sector. The introduction of green technologies could lead to a reduction of the processing time and energy requirement or an improvement in operating conditions thereby ultimately wrapping up the entire process more economically. Both financial and environmental expenses can be brought down substantially by adopting the innovative thermal and nonthermal processing techniques [58]. But the nonthermal processing more effectively

*Corresponding author. E-mail: ashishrawson@gmail.com.

preserves the natural characteristics of food in addition to the reduction of environmental footprints than thermal processing. This has promoted the wide application of nonthermal technology in food industries. Ultrasound is a nonthermal technology, which is also known as clean, cheap, rapid, ease of operation, energy-saving, nondestructive, and eco-friendly technology. Several research works and reviews ascertained that ultrasound to be a potential tool for green food processing [14].

Food processing can be practiced using green technologies by integrating green chemistry and green engineering, which basically relies on the development of existing methods or design of new processing methods to improve food production; to reduce the consumption of energy, water, solvent, and environmental impact; to use nontoxic solvent; to intensify biowaste valorization; and to ensure safety and quality of food. The opportunities of power ultrasound in aspects of green chemistry have been found in sonocatalysis and organic sonochemistry, sonochemical preparation of materials, polymer chemistry, lignocellulosic biomass conversion, extraction, combination of technologies, and industrial applications [13].

Ultrasound waves are acoustic waves that are usually in the frequency range from 16 to 20 kHz that are normally higher than the human hearing limit. The ultrasound waves change their properties such as velocity, attenuation, and frequency spectrum when traveling through a medium. Based on the power and frequency the ultrasound waves are further classified as low power ultrasound and power ultrasound. Low power ultrasound waves have a power range that is usually within 10 W. These low power waves are usually transmitted at a very high frequency and these waves would not disturb the transmitting medium. Such low power ultrasound could very well be applied for food analysis and controlling food quality [5].

Low-frequency ultrasound or power ultrasound are those in the frequency range from 20 to 100 kHz. They possess energy levels in the range of few tens of Watts and hence impact the medium through which they propagate. Further, they can result in the formation of cavitation or acoustic streaming by causing the disturbance in a fluid. This resulting phenomenon exhibits a strong macroscopic effects which could be applied for enhancing the heat transfer process. This type of ultrasound is found to be effective in food processing, food preservation, food modification, and food safety due to its effect on the change of physical, chemical, and biological properties of food [6].

This chapter discusses the application of power ultrasound as a green technology in extraction, food preservation, food modification, size reduction, membrane separation, and delivery of food.

3.2 APPLICATION OF ULTRASOUND AS A GREEN FOOD PROCESSING

In existing conventional food processing technology, there are many short-comings, as they consume higher energy, emit higher levels of CO_2, and utilize many harmful chemicals in their processing techniques, thereby driving the food industries toward adopting a greener approach. The emergence of power ultrasound as a green alternative had attracted the attention of the food industry which had begun to take its step toward maintaining a positive environmental effect. Power ultrasound has its wide application in food processing includes extraction, inactivation of microbes and enzymes, drying, encapsulation, emulsification, homogenization, size reduction, membrane filtration, food modification, fermentation, and crystallization. The ultrasound had attracted the industries which in turn could avail the ultrasound equipment in a more reliable and practically feasible manner. The green approaches of power ultrasound application in food processing had been depicted in Figure 3.1.

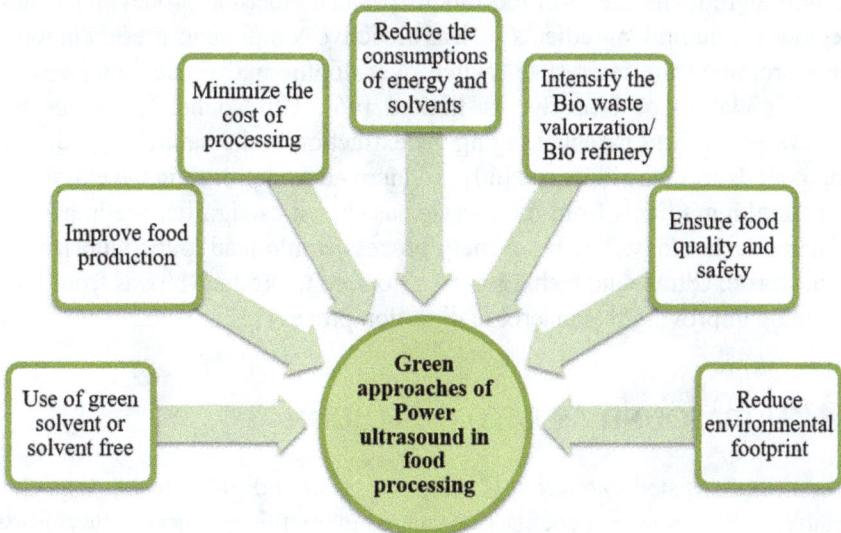

FIGURE 3.1   Green approaches of power ultrasound in food processing.

In ultrasound processing, the use of green solvents mitigates adverse effects on the consumer and the environment. This helps to reduce the usage of organic solvents that are prone to hazards due to their toxicity. The numbers of green solvents used in the extraction or separation process are water, gas-expanded solvents, supercritical fluids, deep eutectic mixtures, switchable solvents, ionic liquids, supramolecular solvents, and bio-based molecular solvents [57]. Ultrasound has shown promising results in micro-bial inactivation leading to a pasteurized or sterilized food product. It ensures food safety along with the use of a green solvent.

The intensities resulting from acoustic and hydrodynamic cavitation can heavily impact the food system's constituents like carbohydrates, protein, lipids, etc., in the form of both chemical and physical modifications. The quality of food can be preserved and altered to meet the consumer require-ment, whereas the conventional method shows a negative effect in quality compared to the ultrasound processing. Cavitation offers its benefits in food processing by effectively reducing the reaction time, substantially increasing the yield, and employing low temperature and pressure when compared to other conventional techniques.

As the consumer demand has been increasing toward natural ingredients instead of synthetic ingredients, the thermal and nonthermal extraction of natural ingredients raised in food industries. The thermal processing would degrade the natural ingredients such as bioactive components present in food. This promoted the interest to switch over nonthermal processing because the degradation of bioactive is limited [66]. Ultrasound intensifies the biorefinery process by encouraging the extraction of the natural ingredients/compounds from the biowaste [60]. An innovative approach in the extraction of natural ingredients from food waste has shown its significance in aspects of green technology. The biorefinery process would lead to the valorization of biowaste. Ultrasound technology is also used to produce biogas from food waste by improving the anaerobic digestion process [32].

3.3 ULTRASOUND-ASSISTED EXTRACTION

Ultrasound-assisted extraction (UAE) is a clean and green technology that relatively shows good benefits in terms of extraction efficiency, flexibility, low investment, easiness of operation to other existing, and emerging green technology, such as maceration, subcritical, and supercritical fluid extrac-tion, pulsed electric field, high-pressure processing, microwave-assisted extraction, and accelerated solvent extraction. UAE techniques integrate exceptionally with the existing processes with minimal modification result in high yield and hence gained interest in food industries. The UAE techniques can replace the usual organic solvents with generally recognized as safe solvents (GRAS) and reduce the solvent utilization or make the entire extraction of the solvent-free method. They are also helpful in efficiently extracting the components that are sensitive to heat and shorten the entire extraction time effectively. The extraction of costly raw materials using the ultrasound is highly economical when compared to the other existing conventional processes thereby fulfilling the demand of the industry [82].

3.3.1   EXTRACTION MECHANISMS

When the ultrasound treatment is implemented in the pretreatment of extraction process, they substantially increase the extraction of many components including polyphenolics, anthocyanins, aromatic compounds, antioxidants, polysaccharides, oils, and other functional compounds to a much higher level.

When a sound wave is subjected to an elastic medium, the particles on its path get displaced longitudinally. The source from which the sound waves emit behaves like a piston on the medium's surface leading to compression and rarefaction phases taking place successively into the medium. Compression in the medium occurs when the source (piston) is held in its open state and the rarefaction results when the source (piston) is held in its contracted position. Cavitation bubbles that had resulted from the void formed into the medium respond to the ultrasonic effect. These cavitation bubbles expand (grow) during the rarefaction phases and contract in size during compression phases. Upon reaching maximum size, the cavitation bubbles implode during the compression cycle resulting in the release of large quantities of energy. During implosion, the temperature and the pressure had been assessed to be up to 5000 K and 2000 atm, respectively, in an ultrasonic bath observed at room temperature. This results in the creation of hot spots, accelerating an intense chemical reaction into the medium.

Implosion of the bubbles onto the surface of the solid material releases high pressure and temperature leading to the formation of microjets that get directed toward the plant matrix. The cell walls of the plant matrix get destroyed due to the high pressure and temperature and ejects out the contents from the matrix into the medium. This entire procedure can be applied effectively for extracting the ingredients from the natural products. The factors affecting the UAE are ultrasound intensity, ultrasonic frequency, extraction temperature and time, and properties of solvent and food [82].

3.3.2 NOVEL METHODS OF UAE

3.3.2.1 UAE USING CLEAN AND GREEN SOLVENT

Solvent is an important variable in the UAE method. The solvent is chosen based on its physical characteristics like viscosity, surface tension, and vapor pressure that play a major role in influencing the solubility of the target compounds against solvent the extraction process. Acoustic cavitation is affected by all these physical properties. If the viscosity or surface tension increases, then it significantly raises the cavitation threshold. In this case, the amplitude should be increased. Low vapor pressure solvent is employed mostly due to its positive impact in the intensification of the collapse of the bubble where vapor pressure mainly relies on the temperature of the medium [14].

Ultrasound-assisted green extraction involves the use of green solvent (GRAS) and less consumption of solvent. The use of green solvents promotes the safety and sustainability aspects. Krishnan et al. [38] indicated the extraction of oil from rice bran with the assistance of ultrasound using ethanol that eventually replaced the conventional method of extraction using hexane due to the higher unsaturated free fatty acids obtained in ultrasound extraction than conventional extraction. To enhance the efficiency of the UAE, the researchers started to study the potential of a mixture of solvent in the extraction process. β-cyclodextrin-enhanced UAE of bioactive compounds using a mixture of 80% ethanol and 20% water solvent, reported to increase extraction efficiency and stability of bioactive compounds in red beetroot extract [84]. Philippi et al. [61] experimented on the extraction kinetics of polyphenol from eggplant peel using two mixtures of green solvent, that is, water–ethanol and water–glycerol combinations, and suggested glycerol to replace ethanol in extraction due to the high boiling point, inflammability, nontoxicity, and low cost involved in the use of glycerol but there was no change in composition of polyphenols in the use of two mixtures.

Solvent-free extraction is one of the emerging technologies in point of green extraction. In this type of extraction, the vegetable oils and other co-solvents are used in the UAE [82]. Boukroufa et al. [9] had stated a green extraction process that was free from solvents to extract oils, polyphenols, and pectin from orange peels waste adopting green technologies, such as ultrasound and microwave. They used water instead of solvent and their observation supported as a green technology in the approach of reduction of energy, time, and wastewater.

Goula et al. [25] used vegetable oil to extract carotenoids from pomegranate waste and it was reported to yield carotenoids with extraction time of 30 min. Ionic solvent was used in the ultrasound-assisted completely solvent-free method to extract the carotenoids instead of organic solvent [49].

Deep eutectic solvents (DESs) are a green solvent alternative for conventional solvent which have been used in the extraction and separation process. It is a mixture of a halide salt or other hydrogen bond acceptors and one or two hydrogen bond donors. There are many types of DESs, but most of the DES is prepared using choline–chloride because it has nontoxic quaternary ammonium salt with hydrogen bond donors.

Bosiljkov et al. [8] proposed a green UAE technique using DES solvent to extract wine lees anthocyanins and indicated the optimum condition of extraction, that is, 30.6 min, 341.5 W, natural DESs (choline–chloride-based DES with malic acid + 35.4% water) to increase the extraction level. Furthermore, the comparison of DES with acidified ethanol reported that the use of DES to be a promising option in anthocyanin extraction. Another study was carried out using DES (choline–chloride-based DES with triethylene glycol + 20% water) in UAE to extract flavonoids from common buckwheat sprouts [47]. They observed more than 97% recovery yield of extracted flavonoids of UAE using C18 solid-phase extraction. Hence DES had been highly recommended for the extraction of bioactive compounds from plants for its greener approach. Some of the recent application of ultrasound-assisted green extraction is listed in Table 3.1.

3.3.2.2   UAE IN COMBINATION WITH OTHER TECHNOLOGIES

The coupling of ultrasound with other technologies takes advantage to demonstrate as a green technology. This results in an improvement of yield and quality; and reduction of extraction time, energy, and temperature. There are several reviews and research works that have been published on the combination of ultrasound with other technologies [48]. Some of the most common combination technologies of ultrasound in the extraction processes are ultrasound-assisted supercritical fluid extraction, ultrasound-assisted microwave extraction, ultrasound-assisted maceration extraction, and ultrasound-assisted high-pressure extraction.

TABLE 3.1 Some Recent Application of Ultrasound-assisted Green Extraction in Food Processing

Food Material	Target Compound	Treatment Conditions	Benefits/Outcomes	Reference
Pomegranate peel	Phenolic compound	Solvent: water and ethanol + water, temperature: 70 °C, power: 480 W	• Water-based extraction yields phenolic compounds higher than ethanol • Replaces organic solvent, prevents from clogging, reduces extraction time, and adds value to waste into other form	[76]
Eggplant peel	Polyphenols	Solvent: glycerol + water and absolute ethanol + water, power: 140 W, frequency: 37 kHz, time: 90 min.	• Waste valorization concept • Glycerol has been recommended to replace ethanol due to high polyphenol yield	[61]
Buckwheat sprouts	Flavonoids	Solvent: deep eutectic solvent, power: 520 and 700 W, frequency: 40 kHz, temperature: 40 °C, Time: 40 min	• 80% (Choline chloride + triethylene glycol+20% water) deep eutectic solvent shows higher yield • Recovery yield >97%	[47]
Tremella fuciformis (snow fungus)	Polysaccharides	Ultrasound-assisted extraction: solvent–polyethylene glycol Ultrafiltration: temperature- 20 °C, transmembrane pressure—0.1 MPa, permeate flux—3.97×10^{-4} m s^{-1}	• Increases the recovery and purity of polysaccharides • Reduces the operation time, consumption of solvent	[88]

TABLE 3.1 (Continued)

Artichoke solid wastes	Phenolic compound	Ultrasound-assisted extraction solvent: 50% ethanol, frequency: 20 kHz, power: 240W Nanofiltration–pressure: 20 bar temperature: 25 °C, Effective membrane area: 7.7×10^{-3} m^2	• Prevents from fouling • High retention of phenolic compounds	[64]
Seaweed	Pigments such as chlorophyll and carotenoid	Ultrasound-assisted extraction solvent: Deionized water, power: 100–300 W, temperature: 40–60 °C, frequency–20 kHz, liquid/solid ratio–30/1 (v/w) Ultrafiltration: 10 kDa membrane, transmembrane pressure: 0.2, 0.3, and 0.4 MPa, effective membrane area –4.1 $\times 10^{-4}$ m^2	• Enhances the extraction efficiency of pigments • reduces the chance of fouling	[90]

Thermosonication is one of the alternatives to thermal processing. Rawson et al. [67] observed higher retention of ascorbic acid, total phenol, and lycopene content of watermelon juice for low treatment time, whereas there was a decrease in all these properties for high treatment time and amplitude. This method of processing influences the quality of food. Sengar et al. [73] suggested ultrasound-assisted microwave extraction as an efficient green extraction method in the pectin extraction from the tomato processing waste compared to UAE, microwave-assisted extraction, ohmic-heating-assisted extraction, and ultrasound-assisted ohmic heating extraction.

Turrini et al. [83] adapted the green extraction of bioactive compounds mainly total phenols and ellagitannins from pomegranate peel using traditional drying, microwave drying, and microwave hydrodiffusion and gravity coupled with ultrasound. They reported that the efficiency of extraction in the relationship between the time and extracted bioactive compounds was high in microwave coupled with ultrasound than the traditional drying coupled with ultrasound. Furthermore, it was observed that time got reduced with an increase in yield than traditional. Ultrasound-assisted pressurized liquid extraction exhibited as a clean and green extraction of phenol from pomegranate peel and the optimum extraction condition established with water as solvent, the temperature at 70 °C, power at 480 W, and 3 cycles [76].

The ultrasonic combination technologies are also used to determine trace elements in food as a sample preparation method for instrumentation [3, 46]. Ultrasound-assisted dispersive liquid–liquid microextraction, ultrasound-assisted emulsification microextraction, and ultrasound-assisted magnetic solid-phase extraction are the sample preparation methods in combination with ultrasound [3].

Ultrasound-assisted emulsification microextraction is an efficient, eco-friendly, inexpensive technique, and its application is mostly on the sample preparation process. It also can reduce the limit of detection in the determination methods [59]. Ultrasound-assisted cloud point extraction, a novel green extraction technique had been used for the preconcentration and determination of trace elements, that is, manganese, zinc, tin, and antimony in food samples using flame atomic absorption spectrometry. Here, the aqueous solution of nonionic surfactants would separate into the surfactant-rich phase and an aqueous phase in heating at temperature above a certain temperature. Therefore, the preconcentration of trace elements can be achieved. The benefits of this preconcentration technique over other analytical instrumentation show high recovery, reasonable precision, inexpensive, rapid, minimum organic solvent consumption and safety [4].

Another green extraction technique, that is, ultrasound-assisted-enzyme based hydrolytic microextraction method has been used to detect the trace elements such as manganese present in food, in which enzyme catalyzes the breakdown of the bonds in protein, starch, and lipid molecules but the use of pepsin enzyme recovered 95% manganese [87].

3.4   FOOD PRESERVATION

Food preservation is a promising method in shelf life improvement of the product. Food preservation can be carried out as an individual or as a hurdle technology. The need for ultrasound in food preservation is to overcome the shortcoming like loss of essential nutrients, enzymatic browning, long exposing time, and spoilage in conventional preservation. This section discusses the green application of ultrasound in drying, high-temperature treatment, and freezing process.

3.4.1   DRYING

Drying or dehydration is a major process involved in the food processing. Food products can be effectively preserved by implementing dehydration techniques. Food materials can be subjected to dehydration using high-intensity ultrasonic waves. Drying food products with the help of ultrasound-assisted techniques depends on the food composition, power of ultrasound, frequency of ultrasound, and treatment time. The advantage of this technology is an effective drying rate and the fast removal of bound water from the product. In ultrasonic drying, heat and mass transfer occurs by cavitation which generates micro streaming channels on the food products.

Applying the airborne ultrasonic energy for drying materials had been tried since several decades. Dehydration of the food products can be carried out with the help of power ultrasonic generators by implementing two experimental procedures: process of forced-air drying that could be assisted by air-borne ultrasound and the process of ultrasonic dehydration where the ultrasound is brought in direct contact with the food material. Ultrasound-assisted convective drying of rough rice at 35 °C, 0.8 m/s velocity, and 90 W power increases bed evaporation up to 38.93%, reduces drying time up to 27.92%, and energy consumption up to 25.98% [18].

Magalhães et al. [45] intensified the drying process of apple cubes by coupling ultrasound pretreatment and ultrasound-assisted air drying and reported to decrease 58% drying time, increase 30% external mass transfer and increase in water diffusivity to 93% than conventional drying. Ultrasound-assisted vacuum drying increases the heat and mass transfer and results in shortening the time taken to drying by 25% and increasing the moisture diffusivity to 89% and prevents the bioactive degradation than conventional drying of red pepper [80].

The synergistic mechanism of ultrasound and other technologies have been used in the drying of foods to improve quality, shorten time, and increase production yield. To improve the convective drying of pumpkin, the combination of ultrasound and ethanol pretreatment has been employed by Rojas et al. [70]. It was indicated that combined effects reduce 59% drying time and 44% energy consumption, improves rehydration properties, and preserves carotenoid during drying.

Sunil et al. [78] observed the significant reduction in microwave drying time and high rehydration ratio in ultrasound pretreatment of okra and also reported to retain the color and textural properties of okra. In ultrasound-assisted osmotic dehydration, the lower solution temperature is employed to achieve higher water loss and solute gain. But the low solution temperature and less treatment time retain the quality of food. Ultrasound-assisted Osmo-dehydrated kiwifruit (35 kHz for 10, 20, and 30 min.; 61.5% sucrose solution at 25 °C) showed the reduction in metabolic heat without affecting respiratory pathway; partial loss of cell viability and structural damage with the increase in treatment time [52].

Rojas et al. [69] indicated the ultrasound pretreatment in the incorporation of nutrients into the food matrix and they performed the incorporation of iron and carotenoids microcapsules in apple and pumpkin by immersing the food material into microcapsules dispersed solvents in ultrasound chamber.

3.4.2   HIGH-TEMPERATURE PRESERVATION

Ultrasound has an interesting role in the inactivation of microorganism which is also known as nonthermal sterilization. The effects of ultrasound on the microbial inactivation occur in a combination of the creation of the compression, rarefaction, and shearing effect, generation of cavitation and free radicals (H^+ and OH^-), and influence of the temperature and pressure of the food products. The formation of free radicals in an aqueous medium affects the structure of the cell wall of the microorganism and leads to the disintegration of the cell wall. It had been observed to show no negative impact on the human consumption of food because the radicals are short lifetimes. Ultrasound treatment exhibits DNA breakdown and enzyme inactivation in microorganisms present in the medium, the cell wall disruption is carried out injected free radicals into cells, not a mechanical effect responsible for it [41].

Li et al. [40] investigated the impact resulting from ultrasound on *Escherichia coli* and *Staphylococcus aureus* and had reported that the sterilization process shows impairment of cell membrane, enzymatic activity inactivation, and inhibition of metabolic performance. Manothermosonication (MTS) treatment is mostly employed to inactivate the microorganism, such as *E. coli* and *S. aureus* [34, 12].

The combination of pressure and heat promotes the cavitation effects. It has the potential to achieve a minimum of 5 log reduction in ultrasound treatment to meet FDA standard faster than ultrasound treatment alone. MTS treatment inactivated the *E. coli* present in apple–carrot juice at 60 °C for 30 s to achieve 5 log reduction [34].

Inactivation of the enzyme can be carried out using ultrasound-assisted thermal processing of food. The choice of ultrasound inactivation mechanisms varies according to the specific type of enzymes depending upon their amino acid composition and the structural conformation of the enzymes. The enzyme inactivation carried out using the sonication technique is a result of the physical and chemical effects of cavitation. The shear force generated by the power ultrasound results in the breaking down of hydrogen bonding and van der Waals interaction in the polypeptide chain thereby modifying the protein's secondary and tertiary structure. This ultimately leads to the inactivation of enzymes. Other mechanism which is responsible for the inactivation of enzymes is by the free radicals that are produced during sonolysis of water molecules. The hydroxyl and hydrogen-free radicals that are produced react with the amino acid residues resulting in stabilizing the enzyme, substrate binding, or in the catalytic activity of the enzyme, consequently changing the activity of the enzyme. There are also a few enzymes like catalase, yeast invertase, or pepsin that are unresponsive to ultrasound [72]. Ultrasound pretreatment has been proven to sensitize the high heat resistance enzyme in green coconut water [71]. This would reduce the negative effects of the use of high temperature and long time processing.

The combination of high-pressure carbon dioxide and power ultrasound is a novel technology that has been used to pasteurize the fresh-cut carrot. Ultrasound enhances the contact between the microorganism and CO_2 which accelerates the diffusion through the membrane; thereby changes occur in pH, these entire phenomena help to inactivate the microorganism. Ferrentino et al. [21] indicated that the combination of high-pressure carbon dioxide and power ultrasound results in a faster reduction to 8 log reductions of *E. coli* spiked on the fresh carrot for 3 min at 10 MPa and 10 W delivered every 2 min of treatment as a function of temperature 35 °C than traditional processing.

3.4.3   FREEZING AND THAWING OR DEFROSTING

Ultrasonic-assisted freezing is developing in the field of food processing. The main features of ultrasonic-assisted freezing are the initiation of nucleation, control of ice crystals size, acceleration of freezing rate, and improvement of frozen food quality. Hickling and Molecular segregation theories explain the mechanism of nucleation. Hickling theory is mostly used in food freezing and states that the nucleation begins immediately after the cavitation bubbles collapse violently where localized pressure reaches up to 5 GPa in a very short time and increases the degree of supercooling. Molecular segregation theory states that the pressure gradient around the cavitation bubbles results in a pressure-controlled diffusion of particles and induces the nucleation process. As the potential of power ultrasound in the facilitation of nucleation at high temperature could be harnessed to control the size of the crystal and its distribution in food [15].

The main mechanism of power ultrasound-assisted crystallization is small bubbles act as primary nuclei in cavitation, the fragments act as primary nuclei, and mass and heat transfer act as a driving force for crystal growth. Ice crystal nucleation has been enhanced by the infusion of CO_2 in the solid matrix followed by power ultrasound application (20 kHz, 0.17 W/cm^2 and the duty cycle of 10 s on/10 s off). This had resulted in the reduction of the freezing time and the crystal size [85].

The success of the freezing process depends upon optimum thawing conditions for enhancing the shelf life of food. As frozen foods are easily susceptible to physicochemical and biological changes in the thawing process due to the suboptimal processing condition and also it consumes more time and expensive, therefore ultrasound-assisted thawing technology has been introduced in food processing. Ultrasound-assisted thawing at 25 °C retains the nutritional and sensory properties, improves thawing efficiency, and reduces thawing time by 51%–73% than normal thawing at 4 °C of mango pulp [44]. The combination of infusion of CO_2 and power ultrasound treated freeze sample undergoes thawing at a constant temperature (20 ± 0.5 °C) and relative humidity (70 ± 5% RH) which significantly results in low water losses, high gel strength, decrease the α-helix, and increase the β-sheet and random coil [85].

3.5 FOOD MODIFICATION

Food modification in the application of ultrasound exhibits beneficial changes in the quality of food. In this process, the physicochemical, biological, sensory, and functional properties of the food are altered to get the desired quality. The occurrence of cavitation in food changes the physicochemical, structural, and functional properties of food. During cavitation, the water molecules break up leading to the formation of free radicals. Such formation results in chemical reactions that force the process of cross-linking of protein molecules in an aqueous medium [11]. Due to the creation of local turbulence and microcirculation of liquid (acoustic streaming), the mass transport reactions get highly increased [24].

With the increase in temperature and pressure generated due to the collapsing of gas bubbles in the cavitation process (sonolysis) [68], dissociation of water molecules in aqueous solutions takes place leading to the formation of hydroxide radicals (OH–) and hydrogen atoms. The formation of free radicals in the cavitation process can also be witnessed in the nonaqueous solutions and polymers. Nevertheless, the cavitation resulting from an aqueous medium is best suited than the ones resulting from organic media. The point to be noted is that the formation of free radicals from ultrasound may or may not be helpful [29]. The reason cited for ignoring the free radicals is that they may lead to potential oxidative damage. Also, it is considered as a disadvantage associated with the preservation of phenols but on the contrary, it increases the antioxidant efficiency in the case of flavonoids [5]. The resulting oxidation may turn to be advantageous for a few food items like for instance chocolates but could be disadvantageous for other food products. Consider, for instance, that in the case of chocolates, it supports the development of flavor but in the case of milk it could bring unpleasantness to its taste. The applications that get adversely affected by free radicals could prefer adopting high frequencies which would deteriorate the count of free radicals as well as the bubbles. The other option is to make use of radical quenchers like ethanol or ascorbic acid. Some of the most common food modifications are done in protein, starch, fat, enzyme activity, tenderization, brining, hydration, germination, and dielectric property. In which the process shortens the time, consumes less energy, uses green solvent, improves food quality and properties, ensures food safety, and increases the production output. Recent ultrasound-assisted modification of food is outlined in Table 3.2.

The ultrasound-treated protein shows the improvement in functional properties, reduction in bulk viscosity, increment in hydrophobicity, and enhancement in emulsion-related property [55, 26]. Janghu et al. [30] maximized the diastase activity and minimized the hydroxymethyl furfural content in honey at 8 min, 60% amplitude, and 60 mL volume. In application of power ultrasound to improve the hydration process, it increases the size of starch granules, improves the water holding capacity, requires short time, and maintains the quality of food [7].

Ran et al. [65] indicated the combination of microwave blanching with ultrasound in salt solutions enhances the dielectric properties of carrot and shortens time in microwave vacuum frying. The synergistic effects of ultrasound (0, 300, and 600 W) and microwave (600, 800, and 1000 W) on the vacuum frying of pumpkin slices were reported to reduce the moisture content and an oil uptake; enhance the crispness and retain the structure and color [27].

Brining is a two-way mass transfer process as the water migrates from the meat to the brine and solute from the brine to the meat. Ultrasound reduced the salting time with no significant negative effects in quality of the meat. It also helped in achieving uniform salting of meat product in addition by improving the texture at high ultrasound intensity which involved in developing the meat products with reduced or low salt [53]. Inguglia et al. [28] studied the effect of geometric parameter that includes the probe size and the distance between the probe and the sample on the efficiency of salt uptake in meat and observed the decrease in the salt enhancement with the increase in the distance between the probe and sample. Based on consumer likings, tenderness remains the most favored delicacy and the same could be considered as the quality factor for ascertaining the quality of meat. Ultrasound treatment improves the tenderness of meat and fragmentation of meat in a short time [56]. This is caused by the release of cathepsin and calcium. Zou et al. [93] presented the close relationship between the tenderization of meat with dissociation of actomyosin in the ultrasound-treated sample.

TABLE 3.2 Some Recent Advances in Ultrasonic-assisted Modification of Foods

Food Material	Treatment Conditions	Salient Findings	Reference
Malt barley	Ultrasound: water, 35 and 40 °C, 37 kHz, 154 W	Approximately 33% hydration time reduced than thermal treatment	[7]
Maize starch	Ozone: 15 min, 25 °C, 1 L/min Ultrasound: water, 8 h, 24–26 °C, 25 kHz, 72 W/L	Combined treatment increases the paste clarity, apparent amylase, carbonyl, and carboxyl groups but decreases the pH and molecular size distribution	[10]
Wine	Ultrasound: water, 30 min, 30 °C and 150 W	40.44% reduction of alcohol content	[89]
Oats	Ultrasound-assisted Germination: water, 25 kHz, 16 W/L, treatment time: 5 min, soaking time: 96 h, 24 ± 2 °C, RH: 95 ± 3%	Improves the GABA, free sugars, and antioxidant ability	[19]
Egg white—ovalbumin	Ultrasound-assisted glycation: ovalbumin + xylose (2:1), 55 °C, 20 kHz, 300 W	Improves the solubility and foaming properties due to the increase in the glycation and the alteration in molecular structure	[22]
Cholesterol-reduced lard emulsion	Ultrasound: emulsion + 4% whey protein isolate, 35 °C, 475 W, 5 min	Enhances interfacial protein adsorption and fat crystallization	[31]
Chicken liver-water-soluble protein	Ultrasound: 4 min, 20 kHz, amplitude 50%, 100 W, 11.5 W/cm²	Conformation change in protein and increases the surface hydrophobicity, sulfhydryl groups and emulsifying properties	[91]
Chicken breast	Ultrasound: 20 kHz, 350 W, 5 min, 0.2 M sodium bicarbonate solution	Conformation change in actomyosin, increases myofibril fragmentation index and decreases the filtering residues, and cooking loss	[92]

3.6 SIZE REDUCTION

3.6.1 CUTTING

Ultrasonic cutting or ultrasound-assisted cutting (UAC) is a size-reduction unit operation utilizing the vibration energy of ultrasound that superposes with a conventional blade movement to improve cutting quality. Generally, the UAC setup consists of transducer, power supply unit, cutting knife, and booster/horn. The electrical energy from the power supply is converted into mechanical vibration using a transducer that is amplified by a booster and then transmitted to the cutting blade. During cutting, the energy of the ultrasound also heats and even partially melts the material. The quality of the food cutting by an ultrasonic cutter is affected by the geometry of the blade, the direction of vibration of the knife relative to the movement of the food, and the frequency and amplitude of the ultrasound.

Liu et al. [42] explained the advantages of UAC of bakery products such as breads and cakes through the experiment. It was reported to enhance cut surface quality and minimize crumbling, squeezing, debris, and smearing. And it showed the ability to cut into thin food slices than conventional cutting. One of the recent applications of UAC was on the cheese types namely, cheddar, mozzarella, and swiss [86]. UAC of cheeses showed a smooth and shining cut surface appearance than conventional cutting, which was obtained at 30%, 40%, and 50% amplitude levels. The quality attributes of stored cheese cuts exhibited low peroxide values, less browning, decrease in pH, and retains sensory attributes up to 21 days compared to conventional cutting.

Another characteristic of this technique is that ultrasonic vibrations generated during the process enable the "auto-cleaning" of the blade. This is mainly due to the prevention of adherence to product and microorganism's growth on the blade surface.

3.6.2 MILLING

Milling is a unit operation in food processing that has been used to break down the whole materials into smaller particles. Dehulling is one of the pulses milling operations in the removal of the hull from cotyledon. The ultrasound pretreatment on black gram enhances dehulling yield due to the change in the microstructure and denaturation of protein and gums in between the hull and cotyledon, thereby it loosens the hulls [77].

Ultrasound-assisted turbine bead milling is a novel method in cell disintegration of microalgae [43]. As microalgae are abundant in nutrients such as proteins, lipids, and polysaccharides, this technique of microalgae cell disintegration would help to produce biofuels and biochemicals. The setup consists of a turbine bead mill, cylindrical hopper with ultrasonic generator, and pump with driving motor. This technology exhibits a decrement in suspension viscosity and an increment in disintegration efficiency. The optimum processing condition, that is, solvent—water, power—100 W, time interval—5 s, disintegration time—30 min was found for *Nannochloropsis oculata* cells disintegration of 97.4% efficiency.

3.6.3   EMULSIFICATION AND HOMOGENIZATION

Emulsification and homogenization are the size reduction processes in liquids. Emulsification is the process employed in emulsion formation. Whereas in homogenization, one immiscible liquid undergoes size reduction and dispersed uniformly into another immiscible liquid to generate stable emulsion. The basic hydrodynamics phenomenon in the application of ultrasound in a liquid or dispersed phase in the continuous liquid phase is acoustic cavitation and acoustic streaming.

Krasulya et al. [37] described two approaches of preparation of food emulsions using ultrasound to study the impact of acoustic cavitation. The first approach is a direct ultrasonication of a medium containing both continuous and dispersed phases, for example, sunflower oil in water, and the impact of cavitation was determined by diameter drop with time. The second approach is a creation of cavitation only into a continuous phase (brine, syrup, etc.), for example, brine treatment for meat sausages, and the impact of cavitation has resulted in structure formation and relaxation. The ultrasonic emulsification process can be carried out in batch or continuous processing methods depending upon the volume of liquids.

The time unit of continuous processing method is in milliseconds, whereas in batch processing, it is in seconds. The desired submicron emulsion is obtained in both cases, due to the generated acoustic energy and its effective utilization [54]. The coupling of ultrasound and other techniques saves the processing time and energy in emulsion preparation and enhances the stability of emulsion [75].

The application of power ultrasound in dairy processing such as yogurt and ice cream production gained benefits due to the improvement in the homogenization and emulsification process assisted by ultrasound [2]. da Silva et al. [17] studied the colloidal stability of cupuaçu juice using ultrasound-assisted homogenization and natural hydrocolloids (i.e., gum Arabic). They had carried out to overcome the rapid separation of phase by reducing particle size and increasing viscosity.

McCarthy et al. [50] observed the functional properties of pea protein isolate powder by dispersing in water using homogenization, microfluidization, and ultrasonication. It was concluded that ultrasonication creates uniform small particle size emulsion, whereas the homogenization and microfluidation generate high energy cold-set gels. Taha et al. [79] investigated the effect of types of oils and ultrasound emulsification time on the physicochemical properties of emulsion stabilized by soy protein and reported to change physicochemical properties of emulsions concerning ultrasonic emulsification time and type of oils.

Gavahian et al. [23] proposed the novel in-pack ultrasonic emulsification process without using special packaging material. In their study, the emulsion was prepared using a mechanical homogenizer, power ultrasound, and in-pack sonication, with and without emulsifier. In-pack sonication technique was reported to improve emulsion stability using a natural emulsifier and to save 73% of energy in comparison to mechanical homogenization. It also addresses the shortcomings of conventional power ultrasound processing by increasing the life-span of the ultrasound probe and eradicating the movement of metal ions from horn directly into food emulsion. Therefore, ultrasonic emulsification is a green processing method in comparison with other techniques such as high-pressure homogenization, high-speed mixing, and microfluidization due to the evidence of ultrasonication easiness in cleaning and maintenance, less energy consumption, and maintain sanitary environment over other techniques.

3.7   MEMBRANE SEPARATION

Membrane separation is an important process where a solid product free from liquid or a solid product that is separated from its mother liquor in the food industry, the separation of solids from liquids is produced. Membrane fouling is a common problem that occurs during a membrane separation process due to the deposition of solid materials on the membrane surface. It was addressed by the application of ultrasound on the membrane which in turn reduces time, energy, maintenance cost, and increases the life span of membrane; permeate flux; and membrane productivity [63].

Two basic concepts in the prevention of membrane fouling using ultrasound are cavitational and acoustic effects which stop the particle accumulation on the membrane by reducing the cake resistance and concentration polarization. Ultrasound-assisted filtration setup has the membrane filter immersed in the ultrasonic bath. Dead end and crossflow modes are the flow configurations that influence the ultrasound-assisted membrane ultrafiltration process [51]. Ultrasound-assisted ultrafiltration is widely used in dairy industries to recover the whey proteins and lactose [62, 35].

Prabhuzanty et al. [62] enhanced the recovery of whey proteins in the application of ultrasound on the ultrafiltration unit and the result intensified the benefits of ultrasound on the membrane separation and spray drying of food by reducing the fouling and increasing permeate flux. Ultrasound-assisted ultrafiltration and nanofiltration process are used in the filtration and purification of polyphenols and polysaccharides [64, 88, 21].

dos Santos Sousa et al. [21] indicated that the ultrasound-assisted ultra-filtration process yields high purity catechins, permeate flux, and permeate stability by permeation of phenolic compounds through the 20 kDa membrane. In juice manufacturing industries, ultrasound-assisted filtration is employed to clarify the juice. Aghdam et al. [1] investigated the ultrasound-assisted clarification (28–32 kHz, 80–150 W and pore size 0.45 μm, area 78×10^{-4} m^2) of pomegranate juice and reported to decrease the cake thickness and intensity of cake formation.

3.8   DELIVERY OF FOOD

Ultrasonic encapsulation has a vital role in the targeted delivery of compounds by generating it with specific physical and functional properties and further protecting it from the environment. The particle size of internal material being treated with ultrasound ranges from 100 nm to 20 μm. The main factor influencing the particle size is ultrasound processing conditions such as frequency, power, time, and design of setup employed. The major applications of ultrasonic encapsulation are the formation of food emulsions, polymeric particles, and protein-coated microspheres. Ultrasounds could be applied for emulsions using cavitation and shear force principle for dispersing the different fluids. This gives attributes such as droplet size and polydispersity which manages the functionality and stability of emulsions. The self-regulated and externally regulated polymeric particles formed during ultrasound treatment only exhibit controlled drug release in a biological system. The protein-coated microspheres are formed in the combination of emulsification and polymerization [39].

Ultrasound-assisted microencapsulation has its significance in compound embedding and loading ratios and preservation of flavor in spices for a long period [81].

Kaderides et al. [33] had applied the UAE for valorizing the pomegranate peels into a food ingredient and encapsulating them using spray drying. UAE of phenolic compounds using green solvent, that is, water was encapsulated using the wall material maltodextrin/whey protein isolate (50:50) with an efficiency of 99.8%. They had applied the encapsulated phenolics extract on hazelnut paste to investigate the shelf life and the result reported that the encapsulated extract increased the shelf life of hazelnut paste.

de Barros Fernandes et al. [17] studied the stabilization kinetics of ultrasound-assisted emulsion of ginger essential oil using a wall material blend of gum Arabic, maltodextrin, and inulin, which was injected into a spray dryer to produce microencapsulates. This method restores the oil quality from the influence of environmental factors and improves the bioavailability in consumption.

The drawbacks of developing green nanosized structures are the raising risks in fabrication and consumption. It has been addressed using ultrasound technology. High-intensity ultrasound alone or in combination with other technologies has the potential to produce encapsulates in the nanometer scale which is also known as edible carrier nanostructures with a low polydispersity and adequate stability. These structures are derived from lipids, surfactants, and polymers. Nanostructured delivery system improves the bioavailability and bioaccessibility of bioactive ingredients [36].

3.9   HACCP IN ULTRASOUND FOOD PROCESSING

Hazard analysis and critical control point (HACCP) is a preventive program implemented in the food industries to ensure food safety by identifying and controlling hazards from raw material procurement to the finished product before distribution. This program manages the risks associated with the physical, chemical, and biological hazards without affecting the food production, workers, consumers, and environment. It mainly relies on the following basic seven principles in setting up of HACCP plan.

1. Conduct a hazard analysis from the raw material procurement to the end of process.
2. Determine the critical control points (CCPs) for process steps that are easily susceptible to hazards.

3. Determine the limits for CCPs.
4. Set up monitoring system required for the CCPs.
5. Implement corrective action for CCPs that solve the exceeding limits.
6. Establish verification procedures to check whether the CCPs and HACCP are monitored and operated effectively.
7. Establish documentation procedures for record and documents effectively.

HACCP plan for ultrasound food processing has been needed in the industries to ensure safety [74]. HACCP plan is designed by the quality team using seven principles. For example, consider the ultrasonic extraction and filtration of juice. The stepwise process involved is procurement of raw material and its subsequent storage; ultrasound processing; filtration; pasteurization; packaging; and storage. Hazard analysis is carried out for stepwise process design, the physical, chemical, and biological hazards may occur due to the processing condition, such as temperature, flow rate, power, time, frequency, and other factors include the storage time, storage temperature, chemical contamination, microbial contamination, foreign materials, and properties of product. After the hazard analysis, the CCP should be identified, established, monitored, corrected, verified, and documented in HACCP design. The HACCP plan of the UAE and filtration is shown in Table 3.3.

3.10 SUMMARY

Ultrasound technology is the best alternative for conventional food processing in terms of quality, production, and shelf life of food. Owing to the strong potential of ultrasound processing in aspects of green technology, there is a good scope of their utilization in the future as it can mitigate the use of toxic solvents, the release of hazardous substances, and the energy or resource consumption but can also enhance the food waste valorization, food quality, and food safety. Thereby, it maintains a clean, green, and sustainable environment. However, there is a gap between the research conducted in the laboratory and the requirements in the food industry. To overcome this challenge, pilot plant studies have to be carried out, and to date, only limited studies have been reported on industrial scale ultrasound processors.

TABLE 3.3 HACCP Plan for Ultrasound-assisted Extraction and Filtration of Juice (B—biological Hazards, P—physical Hazards, and C—chemical Hazards)

Process Step	Potential Hazard	CCP	Control Measures	Monitoring	Correction
Raw material procurement and storage	Microbial spoilage (B), foreign material (P), chemical contamination (C)	CCP	Visual check during procurement from supplier	Visual inspection	Reject any defects found and return to supplier
Cleaning	Foreign material (P), contamination of water (C, B)	CCP	Material should be free of dirt and foreign matters	Check the quality of water before the use	Again repeat cleaning process
Ultrasonic extraction and filtration	Microbial spoilage (B), chemical reaction (C), metal (P)	CCP	Control the processing conditions such as flow rate, temperature, power, frequency	Check the process parameter in an interval of time and maintain daily records	Adjust the processing parameter; Stop production and clean properly
Packaging and storage	Microbial and pathogen (B), metal, packaging material (P)	CCP	Remove the packed goods which do not meet the requirements	Visual inspection and quality check	Rejection and discard in extreme condition or reprocess it

KEYWORDS

- **biowaste valorization**
- **cavitation**
- **emulsification**
- **encapsulation**
- **green solvent**
- **green technology**
- **power ultrasound**
- **ultrasonication**

REFERENCES

1. Aghdam, M.A.; Mirsaeedghazi, H.; Aboonajmi, M.; Kianmehr, M.H. Effect of Ultrasound on Different Mechanisms of Fouling During Membrane Clarification of Pomegranate Juice. *Innovative Food Science and Emerging Technologies*, **2015**, *30*, 127–131.
2. Akdeniz, V.; Akalın, A.S. New Approach for Yoghurt and Ice cream Production: High-Intensity Ultrasound. *Trends in Food Science and Technology*, **2019**, *2019*, 110–119.
3. Albero, B.; Tadeo, J.L.; Pérez, R.A. Ultrasound-Assisted Extraction of Organic Contaminants. *TrAC Trends in Analytical Chemistry*, **2019**, *2019*, 156–165.
4. Altunay, N.; Gürkan, R.; Yıldırım, E. A New Ultrasound Assisted-Cloud Point Extraction Method for the Determination of Trace Levels of Tin and Antimony in Food and Beverages by Flame Atomic Absorption Spectrometry. *Food Analytical Methods*, **2016**, *9* (10), 2960–2971.
5. Ashokkumar, M.; Sunartio, D. Modification of Food Ingredients by Ultrasound to Improve Functionality: A Preliminary Study on a Model System. *Innovative Food Science and Emerging Technologies*, **2008**, *9* (2), 155–160.
6. Awad, T.S.; Moharram, H.A.; Shaltout, O.E.; Asker, D.; Youssef, M.M. Applications of Ultrasound in Analysis, Processing and Quality Control of Food: A Review. *Food Research International*, **2012**, *48* (2), 410–427.
7. Borsato, V.M.; Jorge, L.M.; Mathias, A.L.; Jorge, R.M. Ultrasound Assisted Hydration Improves the Quality of the Malt Barley. *Journal of Food Process Engineering*, **2019**, *42*, e13208.
8. Bosiljkov, T.; Dujmić, F.; Bubalo, M.C. Natural Deep Eutectic Solvents and Ultrasound-Assisted Extraction: Green Approaches for Extraction of Wine Lees Anthocyanins. *Food and Bioproducts Processing*, **2017**, *102*, 195–203.
9. Boukroufa, M.; Boutekedjiret, C.; Petigny, L.; Rakotomanomana, N.; Chemat, F. Bio Refinery of Orange Peels Waste: A New Concept Based on Integrated Green and Solvent Free Extraction Processes Using Ultrasound and Microwave Techniques to Obtain Essential Oil, Polyphenols and Pectin. *Ultrasonics Sonochemistry*, **2015**, *24*, 72–79.

10. Castanha, N.; Lima, D.C.; Junior, M.D.M. Combining Ozone and Ultrasound Technologies to Modify Maize Starch. *International Journal of Biological Macromolecules*, **2019**, *139*, 63–74.

11. Cavalieri, F.; Ashokkumar, M.; Grieser, F.; Caruso, F. Ultrasonic Synthesis of Stable, Functional Lysozyme Microbubbles. *Langmuir*, **2008**, *24* (18), 10078–10083

12. Chantapakul, T.; Lv, R.; Wang, W.; Chummalee, W.; Ding, T.; Liu, D. Manothermosonication: Inactivation of *Escherichia coli* and *Staphylococcus aureus*. *Journal of Food Engineering*, **2019**, *246*, 16–24.

13. Chatel, G. How Sonochemistry Contributes to Green Chemistry? *Ultrasonics Sonochemistry*, **2018**, *40*, 117–122.

14. Chemat, F.; Rombaut, N.; Meullemiestre, A. Review of Green Food Processing Techniques: Preservation, Transformation, and Extraction. *Innovative Food Science and Emerging Technologies*, **2017**, *41*, 357–377.

15. Cheng, X.; Zhang, M.; Xu, B.; Adhikari, B.; Sun, J. The Principles of Ultrasound and its Application in Freezing Related Processes of Food Materials: A Review. *Ultrasonics Sonochemistry*, **2015**, *27*, 576–585.

16. de Barros Fernandes, R.V.; Borges, S.V.; Silva, E.K. Study of Ultrasound-Assisted Emulsions on Microencapsulation of Ginger Essential Oil by Spray Drying. *Industrial Crops and Products*, **2016**, *94*, 413–423.

17. da Silva, L.F.R.; Gomes, A.D.S.; Castro, D.R.G. Ultrasound-Assisted Homogenization and Gum Arabic Combined to Physicochemical Quality of Cupuaçu Juice. *Journal of Food Processing and Preservation*, **2019**, 2019, 14072.

18. Dibagar, N.; Chayjan, R.A.; Kowalski, S.J.; Peyman, S.H. Deep Bed Rough Rice Air-Drying Assisted with Airborne Ultrasound Set at 21 kHz Frequency: A Physicochemical Investigation and Optimization. *Ultrasonics Sonochemistry*, **2019**, *53*, 25–43.

19. Ding, J.; Johnson, J.; Chu, Y.F.; Feng, H. Enhancement of γ-aminobutyric Acid, Avenanthramides, and Other Health-Promoting Metabolites in Germinating Oats (*Avena sativa L.*) Treated With and Without Power Ultrasound. *Food Chemistry*, **2019**, *283*, 239–247.

20. dos Santos Sousa, L.; Cabral, B.V.; Madrona, G.S.; Cardoso, V.L.; Reis, M.H.M. Purification of Polyphenols from Green Tea Leaves by Ultrasound Assisted Ultrafiltration Process. *Separation and Purification Technology*, **2016**, *168*, 188–198.

21. Ferrentino, G.; Spilimbergo, S. High Pressure Carbon Dioxide Combined with High Power Ultrasound Pasteurization of Fresh Cut Carrot. *The Journal of Supercritical Fluids*, **2015**, *105*, 170–178.

22. Fu, X.; Liu, Q.; Tang, C.; Luo, J.; Wu, X.; Lu, L.; Cai, Z. Study on Structural, Rheological and Foaming Properties of Ovalbumin by Ultrasound-Assisted Glycation with Xylose. *Ultrasonics Sonochemistry*, **2019**, 104644.

23. Gavahian, M.; Chen, Y.M.; Khaneghah, A.M.; Barba, F.J.; Yang, B.B. In-pack Sonication Technique for Edible Emulsions: Understanding the Impact of Acacia Gum and Lecithin Emulsifiers and Ultrasound Homogenization on Salad Dressing Emulsions Stability. *Food Hydrocolloids*, **2018**, *83*, 79–87.

24. Gogate, P.R.; Pandit, A.B. Sonocrystallization and its Application in Food and Bioprocessing; In: *Ultrasound Technologies for Food and Bioprocessing*; Feng, H., Barbosa-Canovas, G. and Weiss J. (Eds.) Springer, New York, NY, **2011**, 467–493.

25. Goula, A.M.; Ververi, M.; Adamopoulou, A.; Kaderides, K. Green Ultrasound-Assisted Extraction of Carotenoids from Pomegranate Wastes Using Vegetable Oils. *Ultrasonics Sonochemistry*, **2017**, *34*, 821–830.

26. Higuera-Barraza, O.A.; Del Toro-Sanchez, C.L.; Ruiz-Cruz, S.; Márquez-Ríos, E. Effects of High-Energy Ultrasound on the Functional Properties of Proteins. *Ultrasonics Sonochemistry*, **2016**, *31*, 558–562.

27. Huang, M.S.; Zhang, M.; Bhandari, B. Synergistic Effects of Ultrasound and Microwave on the Pumpkin Slices Qualities During Ultrasound-Assisted Microwave Vacuum Frying. *Journal of Food Process Engineering*, **2018**, *41* (6), 12835.

28. Inguglia, E.S.; Zhang, Z.; Burgess, C.; Kerry, J.P.; Tiwari, B.K. Influence of Extrinsic Operational Parameters on Salt Diffusion During Ultrasound Assisted Meat Curing. *Ultrasonics*, **2018**, *83*, 164–170.

29. Jambrak, A.R.; Mason, T.J.; Paniwnyk, L.; Lelas, V. Accelerated Drying of Button Mushrooms, Brussels Sprouts and Cauliflower by Applying Power Ultrasound and its Rehydration Properties. *Journal of Food Engineering*, **2007**, *81* (1), 88–97.

30. Janghu, S.; Bera, M.B.; Nanda, V.; Rawson, A. Study on Power Ultrasound Optimization and its Comparison with Conventional Thermal Processing for Treatment of Raw Honey. *Food Technology and Biotechnology*, **2017**, *55* (4), 570.

31. Jiang, J.; Song, Z.; Wang, Q.; Xu, X.; Liu, Y.; Xiong, Y.L. Ultrasound-Mediated Interfacial Protein Adsorption and Fat Crystallization in Cholesterol-Reduced Lard Emulsion. *Ultrasonics Sonochemistry*, **2019**, *59*, 104641.

32. Joshi, S.M.; Gogate, P.R. Intensifying the Biogas Production from Food Waste Using Ultrasound: Understanding into Effect of Operating Parameters. *Ultrasonics Sonochemistry*, **2019**, *59*, 104755.

33. Kaderides, K.; Goula, A.M.; Adamopoulos, K.G. A Process for Turning Pomegranate Peels into a Valuable Food Ingredient Using Ultrasound-Assisted Extraction and Encapsulation. *Innovative Food Science and Emerging Technologies*, **2015**, *31*, 204–215.

34. Kahraman, O.; Lee, H.; Zhang, W.; Feng, H. Manothermosonication (MTS) Treatment of Apple-Carrot Juice Blend for Inactivation of *Escherichia coli* 0157: H7. *Ultrasonics Sonochemistry*, **2017**, *38*, 820–828.

35. Khaire, R.A.; Sunny, A.A.; Gogate, P.R. Ultrasound Assisted Ultrafiltration of Whey Using Dual Frequency Ultrasound for Intensified Recovery of Lactose. *Chemical Engineering and Processing—Process Intensification*, **2019**, *142*, 107581.

36. Koshani, R.; Jafari, S.M. Ultrasound-Assisted Preparation of Different Nanocarriers Loaded with Food Bioactive Ingredients. *Advances in Colloid and Interface Science*, **2019**, *270*, 123–146.

37. Krasulya, O.; Bogush, V.; Trishina, V. Impact of Acoustic Cavitation on Food Emulsions. *Ultrasonics Sonochemistry*, **2016**, *30*, 98–102.

38. Krishnan, V.C.A.; Kuriakose, S.; Rawson, A. Ultrasound Assisted Extraction of Oil from Rice Bran: A Response Surface Methodology Approach. *Journal of Food Processing and Technology*, **2015**, *6* (454), 2–19.

39. Leong, T.S.; Martin, G.J.; Ashokkumar, M. Ultrasonic Encapsulation—A Review. *Ultrasonics Sonochemistry*, **2017**, *35*, 605–614.

40. Li, J.; Ahn, J.; Liu, D.; Chen, S.; Ye, X.; Ding, T. Evaluation of Ultrasound-Induced Damage to *Escherichia coli* and *Staphylococcus aureus* by Flow Cytometry and Transmission Electron Microscopy. *Applied and Environmental Microbiology*, **2016**, *82* (6), 1828–1837.

41. Liao, X.; Li, J.; Suo, Y.; Chen, S.; Ye, X.; Liu, D.; Ding, T. Multiple Action Sites of Ultrasound on *Escherichia coli* and *Staphylococcus aureus*. *Food Science and Human Wellness*, **2018**, *7* (1), 102–109.

42. Liu, L.; Jia, W.; Xu, D.; Li, R. Applications of Ultrasonic Cutting in Food Processing. *Journal of Food Processing and Preservation*, **2015**, *39* (6), 1762–1769.

43. Liu, X.; Pan, Z.; Wang, Y. Ultrasound-Assisted Turbine Bead Milling for Disintegration of *Nannochloropsis oculata* cells. *Journal of Applied Phycology*, **2019**, *31* (3), 1651–1659.

44. Liu, Y.; Chen, S.; Pu, Y.; Muhammad, A.I.; Hang, M.; Liu, D.; Ye, T. Ultrasound-Assisted Thawing of Mango Pulp: Effect on Thawing Rate, Sensory, and Nutritional Properties. *Food Chemistry*, **2019**, *286*, 576–583.

45. Magalhães, M.L.; Cartaxo, S.J. Drying Intensification Combining Ultrasound Pre-Treatment and Ultrasound-Assisted Air Drying. *Journal of Food Engineering*, **2017**, *215*, 72–77.

46. Manimekalai, M.; Sheshadri, A.; Kumar, K.S.; Rawson, A. Ultrasound Assisted Method Development for the Determination of Selected Sulfonamides in Honey Using Liquid Chromatography-Tandem Mass Spectrometry. *Biosciences Biotechnology Research Asia*, **2019**, *16* (2), 289.

47. Mansur, A.R.; Song, N.E.; Jang, H.W.; Lim, T.G.; Yoo, M.; Nam, T.G. Optimizing the Ultrasound-Assisted Deep Eutectic Solvent Extraction of Flavonoids in Common Buckwheat Sprouts. *Food Chemistry*, **2019**, *293*, 438–445.

48. Marić, M.; Grassino, A.N.; Zhu, Z.; Barba, F.J.; Brnčić, M.; Brnčić, S.R. An Overview of the Traditional and Innovative Approaches for Pectin Extraction from Plant Food Wastes and By-products: Ultrasound-, Microwaves-, and Enzyme-Assisted Extraction. *Trends in Food Science and Technology*, **2018**, *76*, 28–37.

49. Martins, P.L.G.; de Rosso, V.V. Thermal and Light Stabilities and Antioxidant Activity of Carotenoids from Tomatoes Extracted Using an Ultrasound-Assisted Completely Solvent-Free Method. *Food Research International*, **2016**, *82*, 156–164.

50. McCarthy, N.A.; Kennedy, D.; Hogan, S.A. Emulsification Properties of Pea Protein Isolate Using Homogenization, Microfluidization and Ultrasonication. *Food Research International*, **2016**, *89*, 415–421.

51. Naddeo, V.; Belgiorno, V.; Borea, L.; Secondes, M.F.N.; Ballesteros, F. Control of Fouling Formation in Membrane Ultrafiltration by Ultrasound Irradiation. *Environmental Technology*, **2015**, *36* (10), 1299–1307.

52. Nowacka, M.; Tappi, S.; Tylewicz, U.; Luo, W. Metabolic and Sensory Evaluation of Ultrasound-Assisted Osmo-Dehydrated Kiwifruit. *Innovative Food Science and Emerging Technologies*, **2018**, *50*, 26–33.

53. Ojha, K.S.; Keenan, D.F.; Bright, A.; Kerry, J.P.; Tiwari, B.K. Ultrasound Assisted Diffusion of Sodium Salt Replacer and Effect on Physicochemical Properties of Pork Meat. *International Journal of Food Science and Technology*, **2016**, *51* (1), 37–45.

54. O'Sullivan, J.; Murray, B.; Flynn, C.; Norton, I. Comparison of Batch and Continuous Ultrasonic Emulsification Processes. *Journal of Food Engineering*, **2015**, *167*, 114–121.

55. O'Sullivan, J.J.; Park, M.; Beevers, J.; Greenwood, R.W.; Norton, I.T. Applications of Ultrasound for the Functional Modification of Proteins and Nanoemulsion Formation: A Review. *Food Hydrocolloids*, **2017**, *71*, 299–310.

56. Peña-Gonzalez, E.; Alarcon-Rojo, A.D. Ultrasound as a Potential Process to Tenderize Beef: Sensory and Technological Parameters. *Ultrasonics Sonochemistry*, **2019**, *53*, 134–141.

57. Pena-Pereira, F.; Tobiszewski, M. (Eds). *The Application of Green Solvents in Separation Processes*. Elsevier, Amsterdam, **2017**, 7–20.

58. Pereira, R.N.; Vicente, A.A. Environmental Impact of Novel Thermal and Non-Thermal Technologies in Food Processing. *Food Research International,* **2010,** *43* (7), 1936–1943.

59. Pérez-Outeiral, J.; Millán, E.; Garcia-Arrona, R. Ultrasound-Assisted Emulsification Microextraction Coupled with High-Performance Liquid Chromatography for the Simultaneous Determination of Fragrance Allergens in Cosmetics and Water. *Journal of Separation Science,* **2015,** *38* (9), 1561–1569.

60. Perino, S.; Chemat, F. Green Process Intensification Techniques for Bio-Refinery. *Current Opinion in Food Science,* **2019,** *2019,* 109–119.

61. Philippi, K.; Tsamandouras, N.; Grigorakis, S.; Makris, D.P. Ultrasound-Assisted Green Extraction of Eggplant Peel (*Solanum melongena*) Polyphenols Using Aqueous Mixtures of Glycerol and Ethanol: Optimization and Kinetics. *Environmental Processes,* **2016,** *3* (2), 369–386.

62. Prabhuzantye, T.; Khaire, R.A.; Gogate, P.R. Enhancing the Recovery of Whey Proteins Based on Application of Ultrasound in Ultrafiltration and Spray Drying. *Ultrasonics Sonochemistry,* **2019,** *55,* 125–134.

63. Qasim, M.; Darwish, N.N.; Mhiyo, S.; Darwish, N.A.; Hilal, N. The Use of Ultrasound to Mitigate Membrane Fouling in Desalination and Water Treatment. *Desalination,* **2018,** *443,* 143–164.

64. Rabelo, R.S.; Machado, M.T.; Martínez, J.; Hubinger, M.D. Ultrasound Assisted Extraction and Nanofiltration of Phenolic Compounds from Artichoke Solid Wastes. *Journal of Food Engineering,* **2016,** *178,* 170–180.

65. Ran, X.L.; Zhang, M.; Wang, Y.; Bhandari, B. Dielectric Properties of Carrots Affected by Ultrasound Treatment in Water and Oil Medium Simulated Systems. *Ultrasonics Sonochemistry,* **2019,** *56,* 150–159.

66. Rawson, A.; Patras, A.; Tiwari, B.K.; Noci, F.; Koutchma, T.; Brunton, N. Effect of Thermal and Non-thermal Processing Technologies on the Bioactive Content of Exotic Fruits and their Products: Review of Recent Advances. *Food Research International,* **2011,** *44* (7), 1875–1887.

67. Rawson, A.; Tiwari, B.K.; Patras, A.; Brunton, N. Effect of Thermosonication on Bioactive Compounds in Watermelon Juice. *Food Research International,* **2011,** *44* (5), 1168–1173.

68. Rawson, A.; Tiwari, B.K.; Tuohy, M.G.; O' Donnell, C.P.; Brunton, N. Effect of Ultrasound and Blanching Pretreatments on Polyacetylene and Carotenoid Content of Hot Air and Freeze Dried Carrot Discs. *Ultrasonics Sonochemistry,* **2011,** 18, 1172–1179.

69. Rojas, M.L.; Alvim, I.D.; Augusto, P.E.D. Incorporation of Microencapsulated Hydrophilic and Lipophilic Nutrients into Foods by Using Ultrasound as a Pre-Treatment for Drying: A Prospective Study. *Ultrasonics Sonochemistry,* **2019,** *54,* 153–161.

70. Rojas, M.L.; Silveira, I.; Augusto, P.E.D. Ultrasound and Ethanol Pre-Treatments to Improve Convective Drying: Drying, Rehydration and Carotenoid Content of Pumpkin. *Food and Bioproducts Processing,* **2019,** *119,* 20–30.

71. Rojas, M.L.; Trevilin, J.H.; dos Santos Funcia, E.; Gut, J.A.W.; Augusto, P.E.D. Using Ultrasound Technology for the Inactivation and Thermal Sensitization of Peroxidase in Green Coconut Water. *Ultrasonics Sonochemistry,* **2017,** *36,* 173–181.

72. Sala, F.J.; Burgos, J.; Condon, P.; Lopez, P.; Raso, J. Effect of Heat and Ultrasound on Microorganisms and Enzymes; In: *New Methods of Food Preservation;* G. W Gould (Eds.); Springer, Boston, MA, **1995**, 176–204.

73. Sengar, A.S.; Rawson, A.; Muthiah, M.; Kalakandan, S.K. Comparison of Different Ultrasound Assisted Extraction Techniques for Pectin from Tomato Processing Waste. *Ultrasonics Sonochemistry,* **2020**, *61,* 104812.

74. Sicaire, A.G.; Fine, F.; Vian, M.; Chemat, F. HACCP and HAZOP in Ultrasound Food Processing. In: *Handbook of Ultrasonics and Sonochemistry,* Springer, Singapore, **2015**, 1–19.

75. Silva, E.K.; Costa, A.L.R.; Gomes, A.; Bargas, M.A.; Cunha, R.L.; Meireles, M.A.A. Coupling of High-Intensity Ultrasound and Mechanical Stirring for Producing Food Emulsions at Low-Energy Densities. *Ultrasonics Sonochemistry,* **2018**, *47,* 114–121.

76. Sumere, B.R.; de Souza, M.C.; dos Santos, M.P. Combining Pressurized Liquids with Ultrasound to Improve the Extraction of Phenolic Compounds from Pomegranate Peel (*Punica granatum L.*). *Ultrasonics Sonochemistry,* **2018**, *48,* 151–162.

77. Sunil, C.K.; Chidanand, D.V.; Manoj, D.; Choudhary, P.; Rawson, A. Effect of Ultrasound Treatment on Dehulling Efficiency of Black gram. *Journal of Food Science and Technology,* **2018**, *55*(7), 2504–2513.

78. Sunil, C.K.; Kamalapreetha, B.; Sharathchandra, J.; Aravind, K.S.; Rawson, A. Effect of Ultrasound Pre-Treatment on Microwave Drying of Okra. *Journal of Applied Horticulture,* **2017**, *9,* 58–62.

79. Taha, A.; Hu, T.; Zhang, Z.; Bakry, A.M.; Khalifa, I.; Pan, S.; Hu, H. Effect of Different Oils and Ultrasound Emulsification Conditions on the Physicochemical Properties of Emulsions Stabilized by Soy Protein Isolate. *Ultrasonics Sonochemistry,* **2018**, *49,* 283–293.

80. Tekin, Z.H.; Baslar, M. The Effect of Ultrasound-Assisted Vacuum Drying on the Drying Rate and Quality of Red Peppers. *Journal of Thermal Analysis and Calorimetry,* **2018**, *132*(2), 1131–1143.

81. Teng, X.; Zhang, M.; Devahastin, S. New Developments on Ultrasound-Assisted Processing and Flavor Detection of Spices: A Review. *Ultrasonics Sonochemistry,* **2019**, *55,* 297–307.

82. Tiwari, B.K. Ultrasound: A Clean, Green Extraction Technology. *TrAC Trends in Analytical Chemistry,* **2015**, *71,* 100–109.

83. Turrini, F.; Zunin, P.; Catena, S.; Villa, C.; Alfei, S.; Boggia, R. Traditional or Hydro-Diffusion and Gravity Microwave Coupled with Ultrasound as Green Technologies for the Valorization of Pomegranate External Peels. *Food and Bioproducts Processing,* **2019**, *117,* 30–37.

84. Tutunchi, P.; Roufegarinejad, L.; Hamishehkar, H.; Alizadeh, A. Extraction of Red Beet Extract with β-cyclodextrin-enhanced Ultrasound Assisted Extraction: A Strategy for Enhancing the Extraction Efficacy of Bioactive Compounds and their Stability in Food Models. *Food Chemistry,* **2019**, *297,* 124994.

85. Xu, B.G.; Zhang, M.; Bhandari, B.; Sun, J.; Gao, Z. Infusion of CO_2 in a Solid Food: A Novel Method to Enhance the Low-Frequency Ultrasound Effect on Immersion Freezing Process. *Innovative Food Science and Emerging Technologies,* **2016**, *35,* 194–203.

86. Yildiz, G.; Rababah, T.M.; Feng, H. Ultrasound-Assisted Cutting of Cheddar, Mozzarella and Swiss Cheeses—Effects on Quality Attributes During Storage. *Innovative Food Science and Emerging Technologies,* **2016**, *37,* 1–9.

87. Yilmaz, E.; Soylak, M. Innovative, Simple and Green Ultrasound Assisted-Enzyme Based Hydrolytic Microextraction Method for Manganese at Trace Levels in Food Samples. *Talanta*, **2017**, *174*, 605–609.
88. Zhang, L.; Wang, M. Polyethylene Glycol-Based Ultrasound-Assisted Extraction and Ultrafiltration Separation of Polysaccharides from *Tremella fuciformis* (snow fungus). *Food and Bioproducts Processing*, **2016**, *100*, 464–468.
89. Zhang, Q.A.; Xu, B.W.; Chen, B.Y.; Zhao, W.Q.; Xue, C.H. Ultrasound as an Effective Technique to Reduce Higher Alcohols of Wines and its Influencing Mechanism Investigation by Employing a Model Wine. *Ultrasonics Sonochemistry*, **2020**, *61*, 104813.
90. Zhu, Z.; Wu, Q.; Di, X.; Li, S.; Barba, F.J. Multistage Recovery Process of Seaweed Pigments: Investigation of Ultrasound Assisted Extraction and Ultra-Filtration Performances. *Food and Bioproducts Processing*, **2017**, *104*, 40–47.
91. Zou, Y.; Shi, H.; Chen, X.; Xu, P.; Jiang, D.; Xu, W.; Wang, D. Modifying the Structure, Emulsifying and Rheological Properties of Water-Soluble Protein from Chicken Liver by Low-Frequency Ultrasound Treatment. *International Journal of Biological Macromolecules*, **2019**, *139*, 810–817.
92. Zou, Y.; Shi, H.; Xu, P.; Jiang, D.; Zhang, X.; Xu, W.; Wang, D. Combined Effect of Ultrasound and Sodium bicarbonate Marination on Chicken Breast Tenderness and its Molecular Mechanism. *Ultrasonics Sonochemistry*, **2019**, *59*, 104735
93. Zou, Y.; Zhang, K.; Bian, H.; Zhang, M.; Sun, C.; Xu, W.; Wang, D. Rapid Tenderizing of Goose Breast Muscle Based on Actomyosin Dissociation by Low-Frequency Ultrasonication. *Process Biochemistry*, **2018**, *65*, 115–122.

CHAPTER 4

THREE-DIMENSIONAL (3D) PRINTING OF FOODS

ASWIN S. WARRIER

ABSTRACT

Three-dimensional (3D) printing is a booming technology that enables the creation of a 3D object from a design in a computer, by adding layer over layer of molten material. Recently, the application of this technology in food processing has gained immense interest due to its multitudinous advantages over conventional processing techniques. This chapter intends to describe the basics of 3D food printing, covering areas like the historical development of this technology, technology behind 3D printing—its methods and equipment, applications in the food industry, advantages and limitations, etc.

4.1 INTRODUCTION

Manufacturing involves the conversion of raw materials to final products, manually or using machines, and can include production, assembling, processing, or fabrication. Traditionally, the terms production and processing were used while referring to food manufacturing, while the concept of food fabrication or creating a food product as per our needs and design was unthinkable about half a century back. Though the term "fabricated foods" is being used from the early 1970s [13] referring to convenience foods, fortified foods, textured proteins, etc., food fabrication was never a very common term; and though three-dimensional (3D) printing is not the only method of fabricating foods, the advancement in 3D printing (3DP) technology has indeed expanded the scope of food fabrication, and thereby made the term slightly more popular. 3DP makes it possible to create a tangible object from

*Corresponding author. E-mail: aswinswarrier@kvasu.ac.in

a design on a computer. Many a time, the term "additive manufacturing" (AM) is also used instead of 3DP, which indicates how a 3D printer adds one layer over another to "print" something based on a design, in contrast to the traditional manufacture by machining, which may be called "subtractive manufacturing." Though the terms AM and 3DP are often used interchangeably, AM is a wider term consisting of several technologies. According to Zelinski [44], Editor-in-chief of AM Magazine, 3DP and AM are two different things, though 3DP is an integral part of AM.

This chapter intends to describe the basics of 3D food printing, covering areas like the historical development of this technology, technology behind 3DP—its methods and equipment, applications in the food industry, advantages and limitations, etc.

4.2 HISTORY

3DP is a relatively young field of study considering its history of only about half a century. In "Star Trek: The Original Series" which was aired in the 1960s, there is a reference to a machine called "Food Synthesizer" which is capable of manufacturing foods on command, which may be considered as an indicative approximation toward food printing [37].

Ellam [10] observed that that the first idea of 3DP was proposed by David E.H. Jones in 1974 through his weekly column in "New Scientist" magazine. Initially, 3DP was developed as a technique for rapid prototyping. Horvath [18], who authored an interesting book on 3DP, believes that the concept of 3DP is nothing new, as nature has been practicing this technology in the formation of seashells and certain rock formations.

The initial attempts to create 3D models with the help of lasers started in the 1960s, though the results were not very encouraging. Hideo Kodama of Japan invented a method of AM using photo hardening of polymers, in 1981. The three techniques he proposed are considered as the first working AM techniques. A few years later, Charles Hull developed a functional 3D printer working on stereolithography (SL). These two inventions are considered as the pioneering works in 3DP. Hull's company 3D Systems launched their first commercial 3D printer SLA-1 in 1988. In the same period, 3D printers based on selective laser sintering (SLS) and fused deposition modeling (FDM) were developed by Carl Deckard and Scott Crump respectively. The beginning of the next decade saw developments in several other techniques like direct metal laser sintering, ballistic particle manufacturing, etc.

The 21st century started with several impressive developments like 3D inkjet printers and desktop 3D printers. A progressive event occurred in 2005 when certain patents related to FDM expired. A revolutionary researcher Adrian Bowyer took advantage of this situation, by starting the RepRap movement, to make and propagate affordable 3D printers. This made the designs and technology open to a larger population. Fab@Home was another similar initiative that shared inexpensive designs with the common public [15, 36, 40]. The initial application of this technology in food manufacture was in 2007 when Periard et al. [32] from Cornell University used a Fab@ Home rapid prototyping machine to print 3D food items using cake frosting, cheese, chocolate, and peanut butter. Thereafter, this technology garnered enormous interest among food scientists.

4.3 TECHNOLOGY

3DP encompasses several techniques that can be used to physically create 3D structures from its computer-aided design (CAD) file. The essential steps involved in this include creation of a 3D model with the help of CAD, saving the model in STL file format, dividing the model into thin sections in the file, creating a layer and depositing further layers over it, and postprocessing to give the finishing touch [22]. Due to its versatility and convenience, 3D printers have evolved from sophisticated machines used in research laboratories and working on proprietary software to small-scale units sold online at reasonable rates or even do-it-yourself type assemblies that use open-source software. In general, any 3D food printer consists of three main parts, namely, a three-axis computer-controlled Cartesian coordinate system, a user interface, and a layer addition mechanism that operates based on extrusion, binding, or sintering [33].

The US-based standards institution, ASTM International has constituted a committee F42 on AM in 2009, which has classified AM into seven process categories namely, material jetting, binder jetting, material extrusion, vat photopolymerization, powder bed fusion, directed energy deposition, and sheet lamination [4]. The major difference between these processes lies in how layer deposition is done and the materials utilized. Out of these seven, only a few process categories are used for food printing. These are binder jetting that uses 3DP and inkjet printing technology, material extrusion that uses FDM, vat photopolymerization that uses SL, and powder bed fusion that uses SLS technology [33].

Though the oldest amongst these technologies, SL is no more popular among food scientists, though it has been reported to have been used to print cooked egg white [25]. SL is a photochemical process in which a computer-controlled laser beam is used to solidify a liquid resin layer [28]. FDM is currently the most commonly used AM technique, thanks to its simplicity and low cost [35].

FDM involves the extrusion of a heated material (heated above its melting point) through a nozzle to form a layer. Layers get attached on cooling. FDM is commonly used in 3D chocolate printing. Another extrusion technique called robocasting that uses a robotic arm to move the extrusion nozzle along a platform surface is also being used in food printing. In SLS, a powerful laser is used to sinter powder particles together, to form a 3D structure, wherein layers bind with each other due to the melting of fats and sugars [30]. SLS is usually used with sugar-rich foods. In binder jetting technologies like 3DP (not to be confused with the generalized term) developed by Sachs et al. [34], use a layer of powder bound together by a liquid binding material sprayed evenly over it. 3D inkjet printing is very much similar to the conventional two-dimensional (2D) printers, where edible materials are used instead of printing ink [20]. The printer may either be continuous or drop-on-demand and the printer head may either use a piezoelectric or a thermal mechanism [30]. This technique, being the most developed, is suitable for complex foods [25].

4.4   APPLICATIONS

3DP has brought immense progress in the field of manufacturing. This technology is being used in several areas like medical, pharmaceutical, food, industrial, fashion, defence, space, etc. 3DP is capable of bringing about revolutionary changes in food processing. It is competent enough for enhancing the designs, flavors, and textures of food; developing entirely new food items; or even in making use of under-utilized edible items to take the world closer to food security. Furthermore, a 3D digital recipe can be shared over the internet making it possible to print a specific food item anywhere in the world. 3DP makes it possible to create a food that is tailor-made suiting the preferences of the consumer, considering his/her age group, choices in taste, nutrition, color, etc. [30]. Nowadays, the concept of "digital gastronomy," that combines the traditional culinary arts with modern computer-based technology, depends heavily on 3DP.

As already mentioned under history, the earliest research works in 3DP of food involved the use of cake frosting, chocolate, processed cheese, and peanut butter, to make custom-shaped food items having complex combinations of colors or materials [32]. Sun et al. [37] have classified the materials that can be used for 3DP of edible items into three groups: natively printable materials (like cheese, pasta dough, chocolate, or hummus) which can easily be extruded through a nozzle; nonprintable traditional food materials (like cereals, fruits, vegetables, or meat) which can be made printable with the help of additives like hydrocolloids and; alternative ingredients (like insect powder) which are not commonly consumed as food but are capable of contributing to the reduction in food shortage and increased nutrition without compromising with the appeal and aesthetics.

Many of the food items like chocolate can be readily consumed after printing, while some others require some postprocessing steps like frying or baking. 3DP was identified as a feasible approach for making ice cream when a group of students from Massachusetts Institute of Technology hooked up a 3D printer with a soft-serve ice cream machine with a liquid nitrogen cooling system as part of their graduate project [12]. Within the last two decades, several food items like carbohydrate, protein, and meat puree concentrate, pasta, cookies, cakes, pizza, meat, sugar confectionaries, snacks, infant foods, mashed potato, fruit and vegetable blend, fish surimi, lemon juice gel, milk protein gel, etc., have been printed by the researchers across the globe [30].

Zhang et al. [45] used FDM to print a cereal-based food-containing wheat dough, calcium caseinate, and water, along with probiotics, thereby underlining the suitability of this technology for functional foods. Liu et al. [26] printed a milk protein-based food stimulant using varying combinations of milk protein concentrate and whey protein isolate, with an extrusion-based printer. Azam et al. [5] used an extrusion printer to create a food item from the orange concentrate-wheat starch mixture. Liu et al. [27] successfully printed a complex formulation consisting of egg white protein, gelatin, corn starch, and sucrose. An et al. [2] investigated the 3DP characteristics of an edible algae *Nostoc* and observed that printability can be improved by the addition of potato starch to the algal biomass. Huang et al. [19] studied the printability and quality of brown rice with varying operational parameters, like nozzle size and perimeter.

Hertafeld et al. [17] integrated an infrared heating mechanism with a 3D printer to process complex foods like a combination of dough and chicken that require cooking as a postprinting step. Cotabarren et al. [8] investigated the extrusion printing of a nutraceutical formulation containing monoglyceride oleogels and phytosterol mixtures. Anukiruthika et al. [3] studied the printing characteristics of egg yolk and egg white in combination with rice flour with varying parameters.

Park et al. [31] developed a callus-based ink for 3DP by combining callus with alginate. They opined that foods thus printed can have textures comparable to that of real foods.

3D food printing has been investigated not only for regular applications. In the field of geriatric nutrition, a European Union-funded project named "Development of personalized food using rapid manufacturing for the nutrition of elderly consumers" (PERFORMANCE) studied the suitability of 3DP in five European countries between 2012 and 2015 [7]. The main intention behind this multimillion-euro project was to develop personalized food for the weak and elderly population who has difficulties in chewing and swallowing food, with the help of 3DP technology. 3DP can also be advantageous to defense personnel deployed in remote areas, and those who are traveling and staying in isolated far-off places for a long time. Furthermore, long shelf-life foods can be made available in regions facing food shortages with the help of this technology. The idea of using 3DP to satisfy the taste and nutritional needs of space travelers were proposed in detail by Terfansky and Thangavelu [38].

A company named "Made in Space," operating in association with National Aeronautics and Space Administration (NASA) has developed a 3D printer capable of working in zero-gravity situations, which was launched for the voyage in 2014 [23]. NASA has taken up a project that studies the applicability of 3DP in feeding astronauts [43]. They suggest enhancement of shelf life, reduced weight, and reduced need for postprocessing to be the advantages of this application in space. Chinese scientists successfully printed moon cakes in the Green Space Interstellar Project using a 3D space food printer in 2016 [23]. Another fascinating application of 3DP is in the custom production of packaging materials.

The application of 3DP in tissue engineering is also opening up promising avenues in food manufacturing. The technology called bioprinting use biomaterials, which may be further cultured to form food items similar to traditional foods. Meat cultured in a lab can be printed to desired shapes, which may replace actual meat in the future [35]. This can not only reduce costs and the difficulties associated with rearing animals but also save the lives of numerous animals. Such developments will affect the entire farming system and even the ethical, religious, and societal concepts associated with nonvegetarianism. Vancauwenberghe et al. [41] used bioprinting technology using pectin gel, live lettuce cells, and bovine serum albumin as bioink. They demonstrated good accuracy and reproducibility for the experiment, while the viability of plant cells reduced with increasing pectin content.

4.5   ADVANTAGES AND LIMITATIONS

4.5.1   ADVANTAGES [20, 22, 24, 25, 30, 33, 35]

1. Ability to create new food designs.
2. Ability to integrate with other advanced technologies like the Internet of Things.
3. Ability to meet the personalized needs of the consumer, taking into consideration his health condition, dietary requirement, allergies, etc.
4. Ability to share digital recipes and print similar food at different places, as per requirement. This can reduce the cost of logistics.
5. Aiding in progressing toward food security.
6. Customization with respect to shape, color, texture, and nutrient content.
7. Encourages creativity in gastronomy.
8. Possibility of complete automation of production line.
9. Production of complex foods can be made easy.
10. The supply chain can be simplified.
11. Useful in some specific areas like space, defense, etc.
12. Utilization of underutilized edible items.
13. Versatility of production.
14. Waste reduction.

4.5.2   LIMITATIONS

1. Currently, the cost involved for large-scale production, speed of production, the shelf life of raw materials needed for printing, need for technical know-how, etc., are also limiting factors.

2. Hacking, security issues affecting food safety, piracy, and other intellectual property-related issues, etc., are threats that need to be addressed.

3. Some modifications may be needed with the adapted taste of the consumers.

4. Some steps in food processing like incubation, aging, leavening, etc., are difficult to simulate.

5. The popularity or awareness about this technology among regular consumers is limited.

6. The technology is still developing and is not yet fully matured.

7. With the currently available technology, it is not possible to print all food items.

4.6 ONE DIMENSION AHEAD: 4D PRINTING

In the same way, as two-dimensional printing paved the way to 3DP, the latter is leading the way to four-dimensional (4D) printing, with time considered as the fourth dimension. 4D printing is the process of additively manufacturing a 3D object, that can respond to certain environmental factors like heat, light, moisture, etc., and alter its form or function subsequently over time. The major difference between 3DP and 4D printing lies in designing and in the materials used. While the design must take into consideration, what stimuli will trigger the change, the intensity of the stimulus required, and the extent of change induced, the materials used shall be both printable as well as responsive to the specified stimuli [9]. Ali et al. [1] have written a detailed review of the technology involved, materials used, and applications of 4D printing, which may be referred for a detailed reading.

The concept of 4D printing was proposed initially by American scientist Skylar Tibbits through a TED talk he delivered in 2013 [39]. The area that is gaining most from this advanced technology is the biomedical and tissue engineering, as organs in the human body are dynamic [6]. Though numerous studies on the biomedical applications of 4D printing [14, 21, 29, 46] have been reported, investigations on the application of this technology in food processing are in the budding stage and reported research works are limited. However, 4D printing is a promising technology that is capable of reforming food manufacture in the future. Although the term "4D printing" was used nowhere in their article, Wang et al. [42] have used the concept of

4D printing in developing an edible film that transforms into a 3D structure of the desired shape during cooking in presence of water.

Ghazal et al. [11] printed a food product consisting of an anthocyanin-potato starch gel and lemon juice gel, which changes its color in response to the pH conditions. This finding can not only help in developing foods in different colors but also can help in ensuring food safety by indicating pH variations. In a similar study by He et al. [16], colorful ready-to-eat food items were developed by extruding purple sweet potato puree and mashed potatoes, which responded to pH by changing color.

4.7 SUMMARY

Within a short span of just over a decade, 3DP of food has transformed from an idea to a feasible reality. The potential of this technology is evident from the sheer volume of the reported research. 3DP may offer some solution in feeding the ever-growing population of this world. Most of the researchers who reviewed 3D food printing have proposed varying ideas about future applications possible, each being unique and creative in their own right. The advantages of this technology are very encouraging, while its limitations are only a matter of time, which will be overcome not so far in future. Not much time is required for 3D printers to become household appliances and to restructure the food habits of mankind.

KEYWORDS

- **3D printing**
- **additive manufacturing**
- **food fabrication**
- **food printing**

REFERENCES

1. Ali, M.H.; Abilgaziyev, A.; Adair, D. 4-D Printing: Critical Review of Current Developments, and Future Prospects. *The International Journal of Advanced Manufacturing Technology*, **2019**, *105* (1–4), 701–717.

2. An, Y.J.; Guo, C.F.; Zhang, M.; Zhong, Z.P. Investigation on Characteristics of 3D Printing Using *Nostoc sphaeroides* Biomass. *Journal of the Science of Food and Agriculture*, **2019**, *99* (2), 639–646.

3. Anukiruthika, T.; Moses, J.A.; Anandharamakrishnan, C. 3D Printing of Egg Yolk and White with Rice Flour Blends. *Journal of Food Engineering*, **2020**, *265*, 109691.

4. ASTM. AM Standards Structure and Primer. **2009**; https://www.astm.org/COMMIT/ F42_AMStandardsStructureAndPrimer.pdf.; Accessed on December 25, 2020.

5. Azam, S.R.; Zhang, M.; Mujumdar, A.S.; Yang, C. Study on 3D Printing of Orange Concentrate and Material Characteristics. *Journal of Food Process Engineering*, **2018**, *41* (5), article ID: 12689.

6. Choi, J.; Kwon, O.C.; Jo, W.; Lee, H.J.; Moon, M.W. 4D Printing Technology: A Review. *3D Printing and Additive Manufacturing*, **2015**, *2* (4), 159–167.

7. CORDIS. Development of Personalized Food Using Rapid Manufacturing for the Nutrition of Elderly Consumers. **2016**; https://cordis.europa.eu/project/id/312092.; Accessed on January 5, 2020.

8. Cotabarren, I.M.; Cruces, S.; Palla, C.A. Extrusion 3D Printing of Nutraceutical Oral Dosage Forms Formulated with Monoglycerides Oleogels and Phytosterols Mixtures. *Food Research International*, **2019**, *126*, 108676.

9. Ding, H.; Zhang, X.; Liu, Y.; Ramakrishna, S. Review of Mechanisms and Deformation Behaviors in 4D Printing. *The International Journal of Advanced Manufacturing Technology*, **2019**, *105* (11), 4633–4649.

10. Ellam, R. Editor's Pick: 3D Printing: You Read It Here First. *NewScientist,* **2016**, *232* (3099), 52–59.

11. Ghazal, A.F.; Zhang, M.; Liu, Z. Spontaneous Color Change of 3D Printed Healthy Food Product Over Time After Printing as a Novel Application for 4D Food Printing. *Food and Bioprocess Technology*, **2019**, *12* (10), 1627–1645.

12. Gibbs, S. MIT Students Make 3D-Printed Ice Cream. The Guardian. **2014**; https://www. theguardian.com/technology/2014/jul/17/mit-students-3d-printing-ice-cream.; Accessed on July 12, 2020.

13. Glicksman, M.; Sand, R.E. Fabricated Foods. *Critical Reviews in Food Science & Nutrition,* **1971**, *2* (1), 21–43.

14. González-Henríquez, C.M.; Sarabia-Vallejos, M.A.; Rodriguez-Hernandez, J. Polymers for Additive Manufacturing and 4D-Printing: Materials, Methodologies, and Biomedical Applications. *Progress in Polymer Science*, **2019**, *94*, 57–116.

15. Gross, B.C.; Erkal, J.L.; Lockwood, S.Y.; Chen, C.; Spence, D.M. Evaluation of 3D Printing and its Potential Impact on Biotechnology and the Chemical Sciences. *Analytical Chemistry*, **2014**, *86* (7), 3240–3253.

16. He, C.; Zhang, M.; Guo, C. 4D Printing of Mashed Potato/Purple Sweet Potato Puree with Spontaneous Color Change. *Innovative Food Science & Emerging Technologies*, **2020**, *59*, 102250.

17. Hertafeld, E.; Zhang, C.; Jin, Z.; Jakub, A. Multi-Material Three-Dimensional Food Printing with Simultaneous Infrared Cooking. *3D Printing and Additive Manufacturing*, **2019**, *6* (1), 13–19.

18. Horvath, J. *Mastering 3D Printing*; Berkeley, CA: Apress; **2014**, 224.

19. Huang, M.S.; Zhang, M.; Bhandari, B. Assessing the 3D Printing Precision and Texture Properties of Brown Rice Induced by Infill Levels and Printing Variables. *Food and Bioprocess Technology*, **2019**, *12* (7), 1185–1196.

20. Izdebska, J.; Zolek-Tryznowska, Z. 3D Food Printing—Facts and Future. *Agro Food Industry Hi Tech*, **2016**, *27* (2), 33–37.

21. Javaid, M.; Haleem, A. 4D Printing Applications in Medical Field: A Brief Review. *Clinical Epidemiology and Global Health*, **2019**, *7* (3), 317–321.

22. Javaid, M.; Haleem, A. Using Additive Manufacturing Applications for Design and Development of Food and Agricultural Equipments. *International Journal of Materials and Product Technology*, **2019**, *58* (2–3), 225–238.

23. Jiang, J.; Zhang, M.; Bhandari, B.; Cao, P. Current Processing and Packing Technology for Space Foods: A Review. *Critical Reviews in Food Science and Nutrition*, **2019**, https://doi.org/10.1080/10408398.2019.1700348; Accessed on December 31, 2020.

24. Lin, C. 3D Food Printing: A Taste of the Future. *Journal of Food Science Education*, **2015**, *14* (3), 86–87.

25. Lipton, J.I. Printable Food: The Technology and Its Application in Human Health. *Current Opinion in Biotechnology*, **2017**, *44*, 198–201.

26. Liu, Y.; Liu, D.; Wei, G.; Ma, Y.; Bhandari, B.; Zhou, P. 3D Printed Milk Protein Food Simulant: Improving the Printing Performance of Milk Protein Concentration by Incorporating Whey Protein Isolate. *Innovative Food Science & Emerging Technologies*, **2018**, *49*, 116–126.

27. Liu, L.; Meng, Y.; Dai, X.; Chen, K.; Zhu, Y. 3D Printing Complex Egg White Protein Objects: Properties and Optimization. *Food and Bioprocess Technology*, **2019**, *12* (2), 267–279.

28. Melchels, F.P.; Feijen, J.; Grijpma, D.W. A Review on Stereolithography and its Applications in Biomedical Engineering. *Biomaterials*, **2010**, *31* (24), 6121–6130.

29. Muehlenfeld, C.; Roberts, S.A. 3D/4D Printing in Additive Manufacturing: Process Engineering and Novel Excipients. In: *3D and 4D Printing in Biomedical Applications: Process Engineering and Additive Manufacturing*; Maniruzzaman, M. (Ed.); Weinheim: Wiley-VCH; **2019;** pages 1–23.

30. Nachal, N.; Moses, J.A.; Karthik, P.; Anandharamakrishnan, C. Applications of 3D Printing in Food Processing. *Food Engineering Reviews*, **2019**, *11* (3), 123–141.

31. Park, S.M.; Kim, H.W.; Park, H.J. Callus-Based 3D Printing for Food Exemplified with Carrot Tissues and its Potential for Innovative Food Production. *Journal of Food Engineering*, **2020**, *271*, 109781.

32. Periard, D.; Schaal, N.; Schaal, M.; Malone, E.; Lipson, H. Printing food. *Proceedings of the 18th Solid Freeform Fabrication Symposium* . Austin, TX. **2007**, *2007*, 564–574

33. Pinna, C.; Ramundo, L.; Sisca, F.G.; Angioletti, C.; Taisch, M.; Terzi, S. Additive Manufacturing Applications within Food Industry: An Actual Overview and Future Opportunities. *21st Summer School Francesco Turco 2016*. AIDI (Italian Association of Industrial Operations Professors), **2016**, 18–24

34. Sachs, E.; Curodeau, A.; Gossard, D.; Jee, H.; Cima, M.; Caldarise, S. Surface Texture by 3D Printing. *1994 International Solid Freeform Fabrication Symposium*. **1994**, *1994*, 56–64.

35. Sher, D.; Tutó, X. Review of 3D Food Printing. *Temes de disseny*, **2015**, *31*, 104–117.

36. Su, A.; Al'Aref, S.J. History of 3D Printing. In: *3D Printing Applications in Cardiovascular Medicine*; Al'Aref, S.J.; Mosadegh, B.; Dunham, S.; Min, J.K. (Eds.); London: Academic Press, **2018**, 1–10.

37. Sun, J.; Peng, Z.; Yan, L.; Fuh, J.Y.; Hong, G.S. 3D Food Printing—An Innovative Way of Mass Customization in Food Fabrication. *International Journal of Bioprinting*, **2015,** *1* (1), 27–38.

38. Terfansky, M.L.; Thangavelu, M. 3D Printing of Food for Space Missions. *AIAA SPACE 2013 Conference and Exposition.* **2013,** *2013*, 5346–5357.

39. Tibbits, S. The Emergence of 4D Printing. **2013;** https://www.ted.com/talks/skylar_tibbits_the_emergence_of_4d_printing?language=en. Accessed on January 25, 2020.

40. Van Wijk, A.J.M.; Van Wijk, I. *3D Printing with Biomaterials: Towards a Sustainable and Circular Economy.* Amsterdam: IOS press, **2015,** 87.

41. Vancauwenberghe, V.; Mbong, V.B.M. 3D Printing of Plant Tissue for Innovative Food Manufacturing: Encapsulation of Alive Plant Cells into Pectin Based Bio-Ink. *Journal of Food Engineering,* **2019,** *263*, 454–464.

42. Wang, W., Yao, L., Zhang, T., Cheng, C.Y., Levine, D.; Ishii, H. Transformative Appetite: Shape-Changing Food Transforms from 2D to 3D by Water Interaction through Cooking. *Proceedings of the 2017 CHI Conference on Human Factors in Computing Systems.* **2017,** *2017*, 6123–6132

43. Wu, X.; Douglas, G.; Late, K.G. 3D Printed Prototype to Enhance the Space Food System. **2019;** https://ntrs.nasa.gov/archive/nasa/casi.ntrs.nasa.gov/20190033356.pdf.; Accessed on January 5, 2020.

44. Zelinski, P. Additive Manufacturing and 3D Printing Are Two Different Things. **2017;** https://www.additivemanufacturing.media/blog/post/additive-manufacturing-and-3d-printing-are-two-different-things.; Accessed on December 29, 2019.

45. Zhang, L.; Lou, Y.; Schutyser, M.A. 3D Printing of Cereal-Based Food Structures Containing Probiotics. *Food Structure,* **2018,** *18*, 14–22.

46. Zhang, Z.; Demir, K.G.; Gu, G.X. Developments in 4D-Printing: A Review on Current Smart Materials, Technologies, and Applications. *International Journal of Smart and Nano Materials,* **2019,** *10* (3), 205–224.

CHAPTER 5

OZONIZATION OF FOOD PRODUCTS

SUBHASHINI SUNDARAMOORTHY*,
NIVEADHITHA SUNDRAMOORTHY, RAKESH RAMALINGAM, and
MADHUMITHA MARAN

ABSTRACT

Ozonation is an advanced technology for the food processing industry. Ozone is being used for water decontamination and surface sanitizing processes. Inactivation of microorganism among the fresh produces ranging from meat, seafood, egg, fruits, vegetables, poultry, and dry produce like pulses, spices, and cereals has drawn the interest of various researchers to use ozone technology. Ozone plays a key role in the fumigation of food warehouses. This chapter broadly elaborates on the concentration of ozone, time of exposure, and further explains the factors that influence the food parameters, such as food quality and food safety.

5.1 INTRODUCTION

Ozone is now a preferred replacement for the traditional processing techniques as it has several advantages, such as destruction of microbes, leaving zero residues, antimicrobial, antifungal, antiviral, antiparasitic, sanitizing agent, fumigant, disinfectant, and much more. Ozone not only safeguard the product quality being a disinfectant in the aqueous stage, but it also ensures the sanitation of food packaging materials.

The formation of ozone happens when the oxygen reacts with lightning and ultraviolet (UV) irradiation, which is in an allotropic form of oxygen [64]. Ozone possesses both delta negative and positive electric charges and it is a triatomic molecule of oxygen (O_3). Ozone is formed naturally by some

---

*Corresponding author: E-mail: s.subhashini10@gmail.com

chemical reactions where an additional singlet oxygen atom gets added to oxygen. The smell that arises naturally during rain is due to ozone formation. Ozone molecule has poor stability and shorter half-life; therefore, it gets destructed into oxygen (O_2), the original form. The endothermic reaction below explains the formation of ozone [57]: $3O_2 \leftrightarrow 2O_3$ (ΔH_0 at 1 atm = +284.5 kJ/mol).

This chapter focuses on the ozonation of food products.

5.2   GENERATION OF OZONE

Ozone can be generated by letting free radical oxygen to react with another of the same molecule to produce a triatomic ozone molecule (Figure 5.1). The application of UV radiation and corona discharge makes the free radical oxygen commencement and larger energy is needed to split the oxygen bonding [32]. Further to this, ozone formation is possible through thermal, chemo nuclear, electrolytic and chemical methods [53].

5.2.1   ELECTRICAL (CORONA) DISCHARGE METHOD

When dried or concentrated oxygen/air is let through between electrodes spaced by a dielectric material possessing high voltage it results in ozone formation where the dry air provides 1%–3% of ozone and pure oxygen provides 3%–6% of ozone [69]. The application of higher voltage to electrodes, formation of corona discharge happens which makes the oxygen undergo splitting atoms diacritic that further combines with various other molecules to generate ozone [70].

5.2.2   ELECTROCHEMICAL (COLD PLASMA) METHOD

This method is more easier and cost-effective to produce ozone where an electrolytic solution where the dry air provides 1%–3% of ozone, and pure oxygen provides 3%–6% of ozone [69].

5.2.3   RADIOCHEMICAL OZONE GENERATION

Even to the present stage, the use of ozone production by high energy irradiation method is not used widely but this process of producing ozone will come

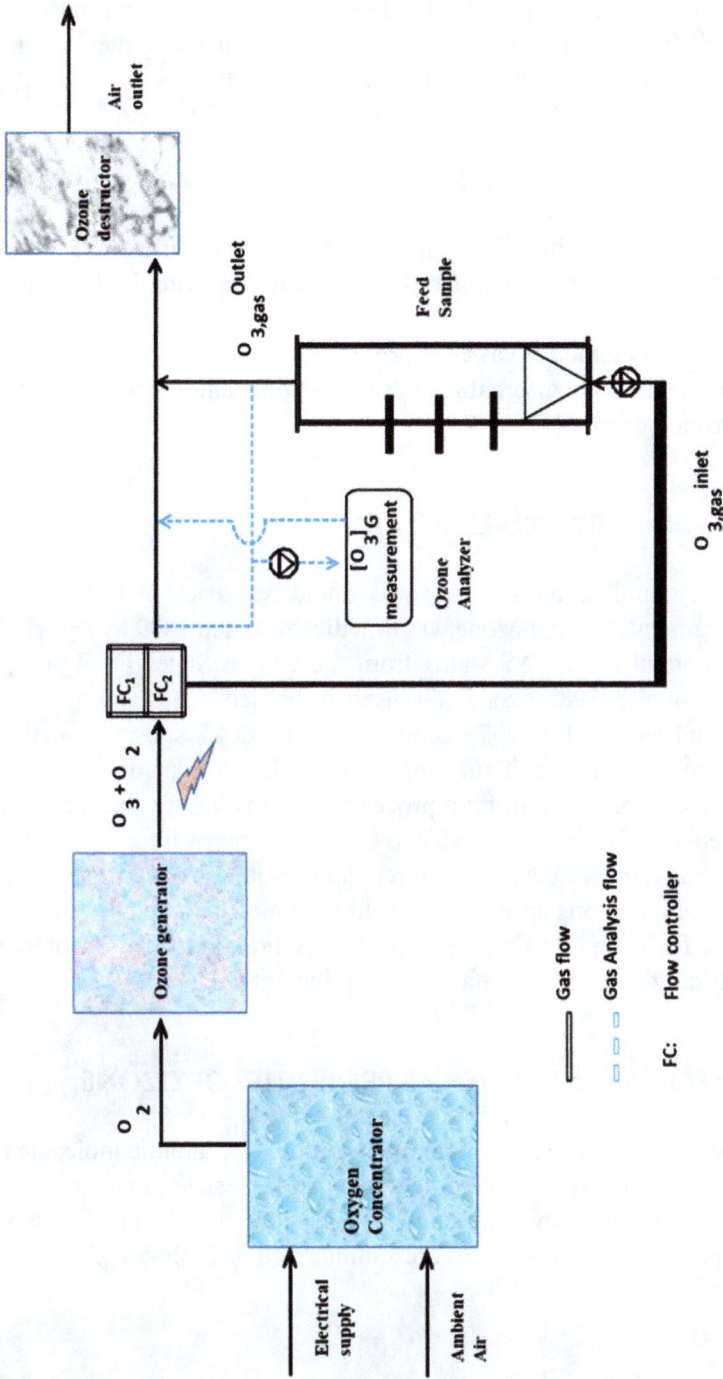

FIGURE 5.1 Ozone generation equipment.

into play in future applications [70]. The perplexity prevailing in the use of cheminuclear ozone generation though it has commending thermodynamic yield makes it an unfavorable method to choose [82].

5.2.4   ULTRAVIOLET METHOD

Application of UV light to oxygen makes the splitting of its molecules and further makes it combine with similar molecules resulting in the formation of ozone [69].

Though this method is easier to produce ozone but the higher cost of the UV bulbs makes it not affordable to use commonly and it also does not yield higher ozone quantity [92].

5.3   REGULATORY AND LEGISLATIVE ISSUES

The primary application of ozone was found way back in 1907 for water supply treatment, thought ozone was first discovered in 1839 by Kogelschatz [55]. Post-acquiring GRAS status from the United States Food and Drug Administration in 1982 ozone was used in bottled water and after expert scrutiny and testing, by 1997 ozone became a GRAS substance making it fit for disinfecting and sanitizing applications [32]. Among developed countries, the usage of ozone in food processing seems improved a lot after the announcement of ozone as GRAS and US FDAs approval for ozone to be an antimicrobial additive enabled its direct application on food products. Prior to this, the regulations around the world were not favorable or completely conducive for its application but were highly limited to waste water treatment and sanitizing the manufacturing equipment [31].

5.4   CHEMICAL AND PHYSICAL PROPERTIES OF OZONE

Ozone possesses a molecular mass of 48 g/mol is a triatomic molecule (O_3). Ozone decays quickly at ambient temperature and carries a pungent odor. It is measured at a very low ppm range of 0.01–0.05 ppm levels. Some of the other important ozone properties are summarized in Table 5.1 [32].

TABLE 5.1 Ozone Properties

Properties	Ozone
Chemical formula	O_3
Smell	Pungent odor
Color	Bluish color
Density of gaseous ozone	2.14 g/L at 0 °C and one atmosphere
Density of liquid ozone	1.354 g/mL at −111.9 °C
Boiling point	−111.9 ± 0.3 °C
Melting point	−192.5 ± 0.4 °C
Critical pressure	54.6 atm
Wavelength (visible spectrum)	560–620 nm
Wavelength (ultraviolet spectrum)	240–280 nm

5.5 FACTORS AFFECTING OZONE PROCESSING EFFICIENCY

5.5.1 EXTRINSIC PARAMETERS

5.5.1.1 FLOW RATE

The size of the bubble formed due to the rate of gas flow has a vital role over the rate of solubilization and disinfection efficiency of ozone. The smaller bubble sizes facilitate larger interfacial areas that make considerable improvement in ozone mass transfer and disinfection efficiency [1]. Lesser the bubble diameter makes the contact area vaster by 32 times [74]. Higher ozone concentration environs at the impinging of bubble's gas–liquid nature, dominantly diminishes the attraction of bacteria by ozone because of the surface-active property [38]. Conducting a study with a rectangular water tank to estimate the mass transfer process of ozone dissolution over bubble plumes indicated quicker divergence at the attainment of initial bubble's optimal size, whereas, on the other hand, it was evidently sluggish. It is further evinced the influence of gas and liquid phase's contact surface, contact time, and ratio of mass transfer plays a vital role in shrinking the diameter of the bubble, which is more preferred to improvise the prominence of ozone usage [30].

5.5.1.2   CONCENTRATION

The possessed concentration of ozone in the medium establishes an important role in defining Ozone strength. Through various researches it is obtained that increased ozone concentration to a higher value forcefully severs down the growth of microbes [80, 98, 113].

5.5.1.3   TEMPERATURE

Maintenance of temperature is considered a much-needed parameter for ozone that it becomes highly soluble and strongly potent of inactivating microbes when the temperature is kept lower which also influences gaseous to liquid ozone conversion [5, 23, 43, 88, 89, 105].

5.5.2   INTRINSIC PARAMETERS

5.5.2.1 pH

Having a lower pH, the possibility of microbial inactivation gets vastly prominent as the ozone decomposition rate gets influenced by any alterations in pH [23, 81, 113]. Another scope of study with the use of free phosphate carbonate buffer revealed that the higher pH attenuates ozone's competence [105], whereas ozone with varied acidic pH levels enriched ozone's efficiency [25]

5.5.2.2   ORGANIC MATTER

The prevalence of any organic matter of any nature in the treated medium exposed to ozone experiences insignificance of ozone's disinfection property over the microbial activity likewise the contained higher level of organics in wastewater treatment has oppresses the effect of ozone [35]. Various other researches were carried out with higher residual presence showcased the abating of ozone's disinfection property [42, 46, 80, 108].

5.6   FOOD PROCESSING AND PRESERVATION UTILIZING

5.6.1   OZONE TECHNOLOGY

Ozone is considered as a powerful and safe disinfectant that can be applicable in food processing industries as it is competent in controlling the biological growth of undesirable organisms in the products and equipment. Be it any purpose like disinfecting microorganisms with no use of any chemical by-products, or to the water used or to the storage atmosphere of the processed food, ozone suitability is found so predominant because of its superior properties. Food industry has scope for ozone application so copiously like surface hygiene, waste water recycling, equipment sanitation, treatment for reducing biological oxygen demand (BOD), and chemical oxygen demand (COD) levels of plant waste (Figure 5.2). Product shelf-life is found on the increasing graph when water, fruits and vegetables, meat, poultry, seafood, and spices are put to ozone treatment. Decomposition of ozone to oxygen is found remarkably when it is applied to food and also leaves no hazardous residues.

5.6.2 OZONE FOR WATER TREATMENT IN FOOD INDUSTRY

In the food processing industry, for various unit operations like washing, chilling, blanching, heating, cooling, pasteurizing, soaking, rinsing, steam production, etc., water is found essentially a primary ingredient and it is crucial for all the operations mentioned [7, 85]. Significant economic consideration is found prevailing in the food industry as extensive water recycling is required by the industry. On the other hand, the end product's quality is given higher importance by keeping an equivalent concentration on the product's safety post to the water treatment [54].

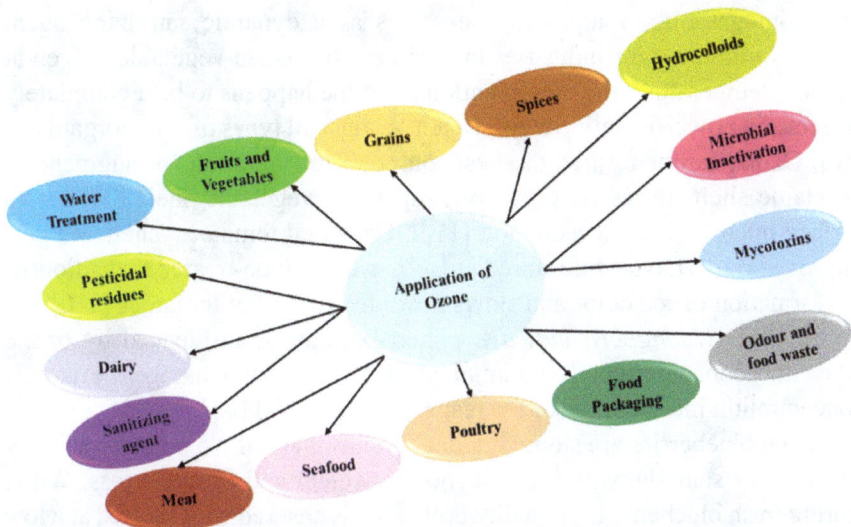

FIGURE 5.2   Ozone's application in food processing and preservation.

For accomplishing acceptable drinking standards, various treatment procedures like bioreactors, nanofiltration, and reverse osmosis that is membrane based are developed for the reduction of microbial load, BOD, and COD levels [22, 72] yet the double-staged process is involved with lower efficiency. A greater level of interest is found demonstrated by ozone treatment that depicts a 3000 times higher antimicrobial activity above that of chlorine treatment and gets dissolved in water efficiently [63].

At a lower pH, direct reaction of ozone with water is found but production of reactive oxygen species results due to the nonselective and ozone's indirect reaction with water at a higher pH [40]. These ideal properties of ozone have brought it to be the most preferred option for contaminants in a basement in wastewater [41, 52, 62, 86, 100] and drinking water [17, 63, 100, 106].

For washing carcasses with a higher microbial population, conventional usage of water amalgamated with the sanitizing agent is done. For eliminating or killing *Escherichia coli* and *Enterococcus faecalis*, and various other food-borne pathogens such as *Listeria monocytogenes, Bacillus cereus, Staphylococcus aureus, Yersinia enterocolitica*, and *Salmonella typhimurium,* the use of ozonized water is highly preferred as it is found a good combination of water sanitizer [51, 53, 87]. Successful application of ozone dissolution can eradicate 1,4-dioxane present in drinking water [17].

5.6.3   PROCESSING OF FRUITS AND VEGETABLES USING OZONE

Ozone in gaseous or aqueous states acts as a dynamic sanitizing agent which is found apt for industries that process fruits and vegetables. Even at ambient temperature, the identification of ozone happens to be a completely perfect and powerful antimicrobial agent against all types of microorganisms [10]. Ozone can be termed the best source of refrigeration for augmenting vegetable shelf-life as its property to increase vegetable shelf life due to surface microbial count reduction [110]. Ozonated tomatoes' shelf life gets improved by 12 days while stored at 15 °C when ozone-treated that adjourns the formation of red color and slows down the rotting of tomatoes [110].

Mangoes (*Mangifera indica* L.) shelf-life can be influenced by ozone when the concentration is maintained at 10 µL/L and has ozone at this concentration intensely make the respiration rate fall [103].

When blueberries are stored under O_3 atmosphere of 2.5 ppm at 12 °C by day 10 they start showing the reduction in weight and firmness loss. While storing fresh blueberries, its quality protection is ensured while stored at a low concentration of ozone gas along with proper refrigeration temperature [11].

It is been witnessed by the worldwide industry of fruit juice production that *Alicyclobacillus acidoterrestris* to be a big apprehension. An assessment revealed the contained *A. acidoterrestris* spores in apple juice has been reduced using ozone and thus making bubbling of ozonation in fruit juices to be a prospective method. [101].

Ozone, a powerful disinfectant and oxidizing agent, was applied to papaya fruit at different concentrations and periods which revealed that total mesophilic bacteria on papaya fruit were reduced by 83.3% to 99.7% after exposure to ozone concentration of 0.05–5.8 ppm for 24 h. Significant increase in soluble solids concentration along with the reduction in weight loss and retaining firmness is found while storing, whereas no effect in peel color and titratable acidity were observed [56].

Carrot quality was put to test under application of water dissolved ozone and gaseous ozone which evinced both stages having no effect on the color, weight loss percentage, and firmness of the vegetable but ozonated water showed impact on carrots pH. On the other hand, ozone as gas helped carrots' shelf life to be extended by ensuring a decrease of soluble solids during storage [14].

Contigiani et al. [12] stated that an extension of postharvest shelf-life of strawberry is ensured when it was exposed to ozonized water washing with the concentration of 3.5 mg/L for a period of 5–15 min which diminished the fungal decay and water loss along with cold storage with no alteration in mechanical parameters.

5.6.4   GRAIN PROCESSING USING OZONE

Many studies carried about ozone revealed ozone as a natural agent offering superior advantages for grain processing thus redressing the problem of pesticide usages. Ozone has a very marginal or zero effect on the quality of the grain as it effectively acts as a perfect fumigant that helps in destroying insects, inactivates microorganisms along with the elimination of mycotoxins.

Higher level concentration is laid over ensuring human health and monetary loss, mycotoxin deoxynivalenol (DON) found in wheat and its products need to be detoxified using ozone. A study about wheat put under ozonation at various conditions, like altered concentration of ozone, moisture content (MC), type of raw material, and exposure time to assess the effectiveness of eradication of DON. This study revealed that significant improvement of DON reduction was evident with improved ozone concentration and exposure time, showcasing the easiness of whole wheat flour degrading than that of wheat kernels. Major decrease of DON is witnessed at higher MC (20.10%). This study brought to attention that degradation of DON can be swiftly achieved with higher effectiveness when it is ozonated [107].

Christ et al. [9] put dry maize grains both in stored and naturally contaminated conditions to ozone treatment for the evaluation of antifungal properties of ozone gas which resulted in the finding that ozone destroyed the inhibition of spores by 88%, whereas the same spores did not germinate completely at a stage of 60 μmol/mol for 180 min.

Shah et al. [95] research concluded proving a range varying between 0.1 and 0.4 ppm of ozone dose being successful in reducing the count of *B. cereus* among rice both cooked and uncooked, though the ozone dose imparts physicochemical characteristics of rice. Proper care has to be taken to ensure the applied ozone concentration is limited and the physicochemical characteristics of rice being kept with reduced detrimental effects through the exposure time.

Obadi et al. [73] carefully examined and viewed a decrease in fatty acid value along with an evident increase in peroxide and p-inisidine value when whole-grain flour was ozone treated. It resulted in oil extraction from the flour which had low lipase activity, higher antioxidant activity, and higher phenolic content.

The study of ozonation as a biocide agent to treat soybean seeds was carried out by Rodrigues et al. [91] for the evaluation of its efficiency as a controller agent of plant pathogens and to study the changes in soybeans seeds physiological quality. The research revealed an achievement in the reduction of the fungi along with an uncompromising physiological quality of soybean through sanitary treatment with ozone gas.

There is a high demand prevailing in the food industry and grain market regarding the quality of wheat grains where wheat grains and grain mixtures have the tendency of attracting high toxicity. The fine structure of grain mixtures is viewed to be more toxic and to keep the scientific and practical interest research was conducted to find ways of reducing the overall toxicity. The results revealed that the grains and grain mixtures experienced a quanti-fiable reduction in their toxicity when exposed to ozone treatment [3].

5.6.5 SPICE PROCESSING USING OZONE

Spices sprayed with aflatoxin B1 witnesses a greater decrease percentage in their presence among red peppers when it is treated with ozone which also helps in reducing the microbial growth [48].

Effective degradation of Salmonella is evidenced when spices like chilli and pepper are put to a combined treatment of using ozone technology followed with pulsed UV, where the initial ozonation at various concentrations helps in reducing the growth of Salmonella, and then a comprehensive reduction takes place due to UV treatment [99].

The extension in retention of red cayenne peppers physical parameters like weight, color, and texture can be achieved along with a prolonged storage time of 30 days of cold storage through spraying of ozonated water [93].

5.6.6 HYDROCOLLOID PROCESSING USING OZONE

Hydorcolloids play a crucial role in thickening and retaining the texture of food products and their nature depends on their structural properties and has an influence on gel formation and viscosity. Ozonated potato starch exhibits greater viscosity and improves gel strength and also shows good water retention as the contained electronegative groups react with their glycosidic bonds and lead to the lessening of the oxidized starch compact structure which further shows positives about the use of potato starch in industrial applications [8].

3D printing main constituent starch when treated with ozone, exhibits the superior quality of printing nature with lesser viscosity with all temperatures, and makes its gelatinizing property perfect as ozonation enables the starch to gain hydrogels with improvised pasting qualities [61].

Treatment using ozone over flour and starch of waxy rice produced a result with increasing the viscosity of pasting of waxy rice flour with 2 h of treatment having its gelatinizing temperature and enthalpy remaining unaltered, whereas on other hand, waxy rice starch attracted a boost in pasting viscosity with just 0.5 h of treatment but a gradual decline of the same happened as when further treated also with its gelatinizing temperature and enthalpy falling [19].

5.6.7 MEAT PROCESSING USING OZONE

The growing trend of meat consumption among the public has been found to be a promising market for the meat industry where the processing of the meat is given higher weightage enabling the processed meat to reach the desired consumers in perfect conditions with no compromise in the quality and freshness.

The contaminations are highly eradicated by the use of ozone. A study about turkey meat exposed to the treatment of ozone has shown major changes in acid reactive substances, carbonyl contents, color, pH, thiobarbituric acid, and microbial inactivation were highly demonstrated. A significant increase in the treated turkey meat's water-holding capacity was noted along with cooking yield [4].

Lyu et al. [59] presented through his research over beef treated by utilizing carbon monoxide (CO) and ozone to evaluate quality of the meat, when it was vacuum packed following a refrigerated storage for 46 days at 0 °C. The results affirmed that pretreatment combination of O_3 and CO is a most favored choice for meat quality preservation as there is zero impact on the color development nature of CO and further CO plays a vital role in the maintenance of beef meat quality.

Iacumin et al. [44] investigation carried out to control the mold growth on meat products through ozonated air produced a result which states that ozone in gaseous form vetoed *Aspergillus ochraceus* growth and the presence of ochratoxin on the surface of sausages without prompting the physiochemical parameters, ripening, and sensory characteristics or peroxide value (PV) of sausages.

5.6.8   POULTRY PROCESSING USING OZONE

Poultry processing attracts a varied range of applications related to ozone as it is being identified as a good disinfecting agent for the hatching of eggs, contaminated eggs, poultry carcass, and chilling water.

The shelf life of chicken meat was evaluated where combination treatment of letting chicken meat to be washed with ozonated water for 10 min and refrigerated at 2–7 °C with and without ozonation separately. The results showed significant difference being noted about the color score intensity for brightness, PV, and average values of pH, whereas chicken meat washed using combination treatment having an ozone concentration of 3 ppm delayed the decomposition process of chicken meat up to a period of 10 days and thus making sure the extension of shelf-life of chicken meat is achieved using this process [109].

A study conducted over breast fillets of chicken and duck with gaseous ozone adequately diminished the growth of bacteria both aerobic and anaerobic and coliforms in both the breast fillets. Comparing to chicken breast fillets the duck breast fillets faced quicker lipid oxidation with a higher note of discoloration. This study evidenced the microbial growth in a bounded manner in both breast fillets but at the same time, there is zero effect on oxidation deterioration of chicken breast fillet [67].

A comparative analysis exposing lyophilized chicken meat fillets that are stored at a temperature of 21 ± 1 °C was done to measure the shelf-life extension of the product under normal conditions with that of exposure to gaseous ozone. Ozonation helped the extension of shelf-life up to 8 months for freeze-dried chicken with the lesser microbial count, whereas the untreated freeze-dried chicken experienced a less shelf-life of 4 months [6].

Extension of the shelf-life was tested for fresh chicken legs, packed, and stored using bags of polyamide/polyethylene at 4 ± 1 °C, for a period of 16 days by combining the ozone and vacuum packaging and the test resulted in the finding of significant variation of physicochemical parameters based on the ozonation dose and storage times. Simultaneously the packaging technology had a greater influence on the population of spoilage flora and sensory properties. A major advantage is achieved in an extension of shelf-life of chicken legs for a period of 6 days when treated with the combination of gaseous ozone and vacuum packing under refrigeration [27].

Ozone being perceived as a perfect replacement for chlorine sanitizers in the modern food industry, where chicken meat treated with ozonated water has shown complete ability to disinfect the total aerobic mesophilic bacteria but on the other side, a marginal decrease of 1% in protein content is noticed with the water content of chicken and its pH experiencing no significant effect [49].

5.6.9 SEAFOOD PROCESSING USING OZONE

The most preferred and a potential product of the food industry is seafood as its protein-rich content and omega-3 fatty acids attract the consumers. As seafoods are highly perishable in nature, its preservation is taken the highest care through technologies like freezing and chilling. In the modern industrial trend, the usage of ozone is also found suitable for ensuring the maintenance of product quality.

Assessment of ozonated water's efficiency for removal of microbes from freshwater fish depicted that watered ozone between 1 and 1.5 ppm resulted highest decrease of 77.2% and 79.49%, respectively, which implied the efficiency of ozonated water in reducing microbes and ensure the quality of fish [13].

Research over pacific white shrimp (*Litopenaeus vannamei*) subjected to ozone treatment shortly after harvest and iced storage has revealed that the quality characteristics of the shrimp have extended considerably when it minimally treated with ozone [75].

Similarly, the expansion of Pacific white shrimp (*L. vannamei*) of shelf life can be achieved with proper maintenance of safety and quality when it is ozone treated combining with modified atmosphere packaging [29].

Oysters exposed to ozone treatment with super-chilled storage can ensure human consumption safety up to 9 days whereas after 14 days its microbial safety gets exonerated [77].

Ozonated water- and ozone-flotation treatments over bighead carp to get the geosmin taken off from the fish muscle conceded that physicochemical properties of fish protein intensified when washed with ozonized water and it also helped eliminating muddy flavor in the fish [111].

5.6.10   *PROCESSING OF DAIRY PRODUCTS USING OZONE*

Ozonated water is playing a vital place in diary processing where the highest consideration prevails to meet the requirements of production quality and microbiological safety of the milk. Ozone is a perfect oxidizing agent to act against viruses, yeasts, protozoa, and other microorganisms [26, 51, 53, 65, 79, 87]. The modern dairy industry is attracting advanced and innovative technologies to avoid the use of the chemical is the cleansing of the pipelines used for carrying milk from bulk tanks to other areas and thus reducing the cost and the effects due to the chemicals used. A patented system and apparatus by Heacox [36] where ozonated water with a desired level of 0.04–1.2 ppm can be found are useful to sanitize animals, utensils, and other places of dairy processing.

Ozone when configured conveniently with the concentration and time factors in mind, it can help the protecting the dairy products like yoghurt and white cheese. Yoghurt cups applied with ozone stream prior to curd development can refrain the contamination due to the airborne microorganisms and helps in mound count devaluation about 0.6 log cfu/g (25.1%) without major effect on the sensory properties but on the other hand, the application of brine solution bubbled with ozone over white cheese reduces the mold load with no major advantage above the traditional brine (NaCl 7%) and also its sensory properties get changed [2].

A study on the ozonation of milk both in skimmed and whole along with Pseudomonas contaminant had an outcome where the microorganism degradation was on a higher degree for the skim milk compared to whole milk. The time and milk composition decide the effectiveness of microbial reduction when ozone is treated with milk with *Pseudomonas* contamination [68].

5.6.11   FOOD PACKAGING USING OZONE

Ozone technology in addition to the use of NaOH can help the food processing environment free from bio-films, a sort of static microorganisms present on the surfaces of stainless steel and other packing materials used over the existing environment of food processing. This is due to the fact that traditional practices of cleaning in the production do not help to protect the hygiene and reduce the contamination and also these biofilms will also led to food contamination and food-related illness.

The use of ozone generators can double up the storage life of fruits and vegetables in storehouses along with barrier and corona types of discharge. Ozone-air mixture and ozone decomposition define the ozone concentration and a 100% sterilization of surface microflora contaminants is so effectively achieved due to the ozone effect [97].

Ozone sterilization guarantees best sensory, microbial, and chemical evaluation samples of vacuum-packed tiger and sardine fish packages made of polyamide/polyethylene by lengthening the shelf life of the fish and thus making the period of storage extended better than compared to that of the same package with no ozone treatment [50].

5.6.12   OZONE AS A SANITIZING AGENT IN FOOD INDUSTRY

Making use of ozonated water is highly viewed as a best upcoming technology for carrying out the disinfectant activities in wine production where piping and equipments face major contaminations due to the use of conventional disinfectants. The application of ozonated water is found highly effective for sanitizing the pipes in combination with peracetic acid, whereas the same does not remove *Saccharomyces cerevisiae* among the microorganisms in the bottling machine [21].

Batch wash ozone sanitation system (BWOSS) with its major dominance as water temperature, organic load, and free chlorine shows amplified efficacy helping the retail industry where the fresh produces (strawberry, romaine head lettuce, tomato, cilantro) are more exposed to major defiling of their nature. BWOSS is superiorly coherent in disinfecting the foodborne pathogens that arise both in the production and retail environment [28].

Application of ozone washing pork lard proved to be a perfect alternative and eco-friendly detergent and helps enzyme residue deactivation ensuring the quality of the product as the washing results in the formation of surface-active molecules which gave on to noteworthy washing performance contrasting to conventional water treatment [47].

The existence of Listeria in food processing facilities especially in the cheese industry can be lessened by incorporating ozonation as a best practice that ensures the Listeria insolation to a greater level and the cheese processing equipment evinced no inimical effect [20].

5.6.13   OZONE FOR ODOR AND FOOD WASTE TREATMENT

The use of ozone pretreated wastewater in fastening the composting of organic matter is found highly effective as the process requires the influence of organic waste, moisture, microorganism, and oxygen. The application of ozone pretreated wastewater resulted in decreasing the bio-oxidation time and also pacing composting to a greater extend compared to that of untreated waste water [60].

Ozone is highly sought in industrial wastewater treatment as ozonation exhibits greater efficacy due to its comparatively greater oxidation potential with that of customary hypochloric acid or chlorine that helps better culminating color removal and algae elimination [34].

Tofu waste water is highly impacted by ozone treatments in tofu industries, which induces the reduction of COD present in tofu water thus enabling the microalgae growth considerably increased [33].

5.6.14   OZONE ON PESTICIDE RESIDUES

There is a vast interest inculcated among the researchers to tackle the effect of the residues from that of the pesticides used in agriculture, animal, and food industry as they add the major concern to the human health through the growing cancers, diabetes, reproductive disorders, and other chronic diseases [66, 78].

Ozonized water is used for treating vegetables as it tends to reduce the pesticide residue levels when washed and ozone in gaseous form that helps the cereals to be treated. Deterioration of these residues is done by the O_3 reaction with food that creates reactions with unsaturated and aromatic hydrocarbons, which is a direct pathway or through the indirect pathway by indulging higher oxidation potential radicals that can destroy molecules of both nature with nonselective reactions [45, 76].

Putting strawberries into normal and watered ozone through cleaning and boiling using ultrasonic help is the diminishing of the residue levels present in the berries. The effectiveness of ozonated washing is found as the residues reduction raises to 75.1% from that of 36.1% that is far above than normal tap water washing where reduction is from 19.8% to 68.1%. Both boiling and ultrasonic cleaning helps to decrease the pesticide residues to a considerably higher level of 92.9% and 91.2%, respectively, and thus helps to reduce the health risk that would cause due to the raw strawberries [58].

The presence of difenoconazole and linuron in carrots can be easily eradicated by using an optimal treatment with gaseous and watered ozone. Ozone's temperature has no impact whereas an increased concentration and treatment time has resulted in 80% removal of pesticides and after storage, it was found that difenoconazole and linuron faced a greater degradation of 98% and 95% each [15].

Ozone gas treatment is found a probable and effective method to control the pesticide residues in wheat grains as it makes the reduction of organophosphate (fenitrothion) by 66.7% and pyrethroid (deltamethrin) pesticides by 80.6% to 85.7% [94].

Fungicides present in tomatoes can be found reduced between 70% and 90% by using ozone treatment with bubbling O_3 and watered ozone which is having greater effectiveness over the conventional immersion procedure into distilled water and detergents. Higher the ozone concentration, the greater the pesticide residue removal but while storage a lower loss of fruit mass occurs due to lower ozone concentration [90].

Postharvest tomatoes are treated with the process of washing using tap water, ozonated water, detergent solution followed by ultrasonic cleaner greatly decreases the level of Dichlorvos, a vastly applied insecticide where a greater 91.9% removal was evinced by ozone water treatment that is way above of tap water's 30.7%, 70.7% of detergent solution and 88.9% of ultrasonic power treatments. This treatment also ensures in lowering the dietary exposure with no impact on the quality of the product [37].

Dried chilies exposed to ozone fumigation ensure the reduction of organophosphate pesticides like ethion, malathion, profenofos, and chlorpyrifos with a shorter half-life of 99.9% when compared to nonozone degradation without facing changes in water activity and capsaicin content and color of the dried chilies [96].

5.6.15 OZONATION IN MYCOTOXINS

Mycotoxins have higher and deleterious effect on human health as they consume the foodstuffs with its major presence. The removal of DON from durum wheat, semolina, and pasta can be accomplished greatly when they are exposed to ozone treatment at optimal conditions without altering the rheological, microbial, and chemical parameters [84].

Trombete et al. [104] indicated that the prevalence of aflatoxins and DON can be avoided to a larger extent when ozone treatment is performed over contaminated grains, where the O_3 concentration drastically persuade the mycotoxins along with total fungal count but this has no impact on grain mass.

Fruit industry's key requirement to overcome the major fungal contaminant and ochratoxin A (OTA) is attracting the usage of ozone which helps in attaining 60% of OTA downgrading and with no pronounced lessening of phenolic compounds in dried vine fruits [102].

Malting barley seeds having the presence of Fusarium can cause major diseases when consumed as the level of contamination is high. Treating barley seeds during germination with gaseous ozone at various concentrations will help in suppressing the filamentous fungi [83].

5.6.16   OZONATION IN MICROBIAL INACTIVATION

Food products are more prone to greater contamination while transporting, handling, storage, and also during the use of wash water if they are uncontrolled. These microbial hazards can be controlled by the application of ozone as its higher and improvising oxidizing nature creates permanent impairment to the fatty acids and to other cellular macromolecules like DNA and proteins [24, 39, 71].

Ozone treatment in domestic well drinking water helps the demolition of microorganisms in acidic and basic forms efficiently as ozone disinfects microorganisms when the increase in contact time occurs and in addition to this, application of higher dose of ozone leads to greater inactivation of microbes [16].

Ozonation of fresh and frozen strawberries for a period of 30 and 5 min produced a result where the log reduction levels of MNV-1 (1.8 and 0.7), MS-2 (3.3 and 1.8), *Enterococcus faecium* (1.5 and 0.7), and Salmonella (2.1 and 1.6) being noted. It is revealed that Salmonella was highly sensitive and MNV-1 was showing hesitance toward gaseous treatment [112]. Ozonation does not affect any sensory parameters and it is considered to be a dependable technology for decontamination.

The effect of chlorine disinfectant in drinking water treatment results in a severe threat to human health, and it is highly needed to eradicate it. Ozone actively reduces the bacteria's DNA content (*Aeromonasjandaei* < *Voges-ellaperlucida* < *Pelomonas* < *B. cereus* < *Aeromonassobria*) and spores (*Bacillus alvei* < *Lysinibacillus fusiformis* < *B. cereus*) drastically leaving out a mere 1/4th of the target genes, though it was highly damaged by ozone disinfection [18].

5.7 SUMMARY

Ozone has attracted wider implementation in the modern food industry as it does not involve higher temperature and is viewed as an attractive energy-saving model. Modern equipment that can generate, apply, monitor, and control ozone in both phases such as aqueous and gas are available for food processing industries. It is important to have a complete feasibility study for ozone application for every product and comparing it with the existing food preservation techniques. Indian Institute of Food Processing Technology—Tanjore, Tamil Nadu Agricultural University—Coimbatore, the Ohio State University—USA, AINIA Centro Technológico—Valencia, Ecoled'ingéniers dePurpan—Toulouse, Purdue University—USA, among others around the world are devoting significant efforts to encourage research studies on ozonation in agriculture. Augmented research is needed for conducting thorough study of ozone usage in food preservation to reduce infection due to microbes and insect infestation. Proper standardized conditions are needed for each food product under ozonation. Ozonation in combination with other nonthermal processes can provide additional benefits for the inhibition of microbes and extension of product shelf-life. The future scope of the study should concentrate on people awareness, legal provisions, and cost-effectiveness.

KEYWORDS

- disinfectant
- food safety
- fumigant
- greener technology
- nonthermal
- ozone
- preservation

REFERENCES

1. Ahmad, M.; Farooq, S. Influence of Bubble Sizes on Ozone Solubility Utilization and Disinfection. *Water Science and Technology*, **1985**, *17* (6–7), 1081–1090.
2. Alexopoulos, A.; Plessas, S. Experimental Effect of Ozone Upon the Microbial Flora of Commercially Produced Dairy Fermented Products. *International Journal of Food Microbiology*, **2017**, *246*, 5–11.
3. Avdeeva, V.; Starodubtseva, G.; Bezgina, J.; Zorina, E.; Lysakov, A. Ozone Effects on Toxicity of Wheat Grain and Grain Mixtures. *Engineering for Rural Development*, **2019**, *2019*, 764–767.
4. Ayranci, U. G.; Ozunlu, O.; Ergezer, H.; Karaca, H. Effects of Ozone Treatment on Microbiological Quality and Physicochemical Properties of Turkey Breast Meat. *Ozone: Science and Engineering*, **2019**, *42*(1), 95–103.
5. Bablon, G.; Bellamy, W. D.; Bourbigot, M. Fundamental Aspects of Ozone. In: *Ozone in Water Treatment: Application and Engineering*; Bruno Langlais, David A. Reckhow and Deborah R. Brink, (Eds.). AWWA Research Foundation: Denver, CO, **1991**; 11–132.
6. Cantalejo, M. J.; Zouaghi, F.; Pérez-Arnedo, I. Combined Effects of Ozone and Freeze-Drying on the Shelf-life of Broiler Chicken Meat. *LWT—Food Science and Technology*, **2016**, *68*, 400–407.
7. Casani, S.; Rouhany, M.; Knochel, S. A Discussion Paper on Challenges and Limitations to Water Reuse and Hygiene in the Food Industry. *Water Research*, **2005**, *39* (6), 1134–1146.
8. Castanha, N.; e Santos, D. N.; Cunha, R. L.; Augusto, P. E. D. Properties and Possible Applications of Ozone-Modified Potato Starch. *Food Research International*, **2019**, *116*, 1192–1201.
9. Christ, D.; Kreibich, H. H.; Valmorbida, R. Antifungal Properties of Ozone Gas in Stored Naturally Contaminated Dry Maize (*Zea mays* L.) Grains. *Scholars Journal of Engineering and Technology*, **2017**, *5*(4), 146–52.
10. Clark, J. P. Ozone—Cure for Some Sanitation Problems: Processing. *Food Technology*, **2004**, *58* (4), 75–77.
11. Concha-Meyer, A.; Eifert, J. D.; Williams, R. C. Shelf-life Determination of Fresh Blueberries (*Vaccinium corymbosum*) Stored Under Controlled Atmosphere and Ozone. *International Journal of Food Science*, **2015**, *2015*, 1–10.
12. Contigiani, E. V.; Jaramillo-Sánchez, G. Postharvest Quality of Strawberry Fruit (*Fragaria x Ananassa Duch* cv. Albion) as Affected by Ozone Washing: Fungal Spoilage, Mechanical Properties, and Structure. *Food and Bioprocess Technology*, **2018**, *11* (9), 1639–1650.
13. de Mendonça Silva, A. M.; Gonçalves, A. A. Effect of Aqueous Ozone on Microbial and Physicochemical Quality of Nile Tilapia Processing. *Journal of Food Processing and Preservation*, **2017**, *41* (6), 1–7.
14. de Souza, L. P.; Faroni, L. R. D. A.; Heleno, F. F. Effects of Ozone Treatment on Postharvest Carrot Quality. *Lebensmittel-Wissenschaft & Technologie*, **2018**, *90*, 53–60.
15. de Souza, L. P.; Faroni, L. R. D. A. Ozone Treatment for Pesticide Removal from Carrots: Optimization by Response Surface Methodology. *Food Chemistry*, **2018**, *243*, 435–441.
16. Demir, F.; Atguden, A. Experimental Investigation on the Microbial Inactivation of Domestic Well Drinking Water Using Ozone under Different Treatment Conditions. *Ozone: Science & Engineering*, **2016**, *38* (1), 25–35.

17. Dietrich, M.; Andaluri, G.; Smith, R. C.; Suri, R. Combined Ozone and Ultrasound for the Removal of 1,4-Dioxane from Drinking Water. *Ozone: Science & Engineering,* **2017,** *39* (4), 244–254.

18. Ding, W.; Jin, W.; Cao, S.; Zhou, X.; Wang, C.; Jiang, Q. Ozone Disinfection of Chlorine-Resistant Bacteria in Drinking Water. *Water Research,* **2019,** *160,* 339–349.

19. Ding, W.; Wang, Y.; Zhang, W.; Shi, Y.; Wang, D. Effect of Ozone Treatment on Physicochemical Properties of Waxy Rice Flour and Waxy Rice Starch. *International Journal of Food Science & Technology,* **2015,** *50* (3), 744–749.

20. Eglezos, S.; Dykes, G. A. Reduction of Environmental Listeria Using Gaseous Ozone in a Cheese Processing Facility. *Journal of Food Protection,* **2018,** *81* (5), 795–798.

21. Englezos, V.; Rantsiou, K.; Cravero, F. Minimizing the Environmental Impact of Cleaning in Winemaking Industry by Using Ozone for Cleaning-In-Place (CIP) of Wine Bottling Machine. *Journal of Cleaner Production,* **2019,** *233,* 582–589.

22. Fahnrich, A.; Mavrov, V.; Chmiel, H. Membrane Processes for Water Reuse in the Food Industry. *Desalination,* **1998,** *119* (1–3), 213–216.

23. Farooq, S.; Engelbrecht, R. S.; Chian, E. S. Influence of Temperature and UV Light on Disinfection with Ozone. *Water Research,* **1977,** *11* (8), 737–741.

24. Fetner, R. H.; Ingols, R. S. Bactericidal Activity of Ozone and Chlorine against *Escherichia coli* at 1 °C. *Ozone Chemistry and Technology, Advances in Chemistry,* **1959,** *21,* 370–374.

25. Foegeding, P. M. Ozone Inactivation of Bacillus and Clostridium Spore Populations and the Importance of the Spore Coat to Resistance. *Food Microbiology,* **1985,** *2* (2), 123–134.

26. Fontes, B.; Heimbecker, A. M. C. Effect of Low-Dose Gaseous Ozone on Pathogenic Bacteria. *BMC Infectious Diseases,* **2012,** *12* (1), 358.

27. Gertzou, I. N.; Karabagias, I. K. Effect of Combination of Ozonation and Vacuum Packaging on Shelf-life Extension of Fresh Chicken Legs During Storage under Refrigeration. *Journal of Food Engineering,* **2017,** *213,* 18–26.

28. Gibson, K. E.; Almeida, G.; Jones, S. L.; Wright, K.; Lee, J. A. Inactivation of Bacteria on Fresh Produce by Batch Wash Ozone Sanitation. *Food Control,* **2019,** *106,* 106747.

29. Gonçalves, A. A.; Santos, T. C. L. Improving Quality and Shelf-life of Whole Chilled Pacific White Shrimp (*Litopenaeus vannamei*) by Ozone Technology Combined with Modified Atmosphere Packaging. *Lebensmittel-Wissenschaft & Technologie,* **2019,** *99,* 568–575.

30. Gong, X.; Takagi, S.; Huang, H.; Matsumoto, Y. A Numerical Study of Mass Transfer of Ozone Dissolution in Bubble Plumes with Euler–Lagrange Method. *Chemical Engineering Science,* **2007,** *62* (4), 1081–1093.

31. Graham, D. M. Use of Ozone for Food Processing. *Food technology (Chicago),* **1997,** *51* (6), 72–75.

32. Guzel-Seydim, Z. B.; Greene, A. K.; Seydim, A. C. Use of Ozone in the Food Industry. *LWT-Food Science and Technology,* **2004,** *37*(4), 453–460.

33. Hadiyanto, H. Ozone Application for Tofu Waste Water Treatment and Its Utilisation for Growth Medium of Microalgae Spirulina sp. In *E3S Web of Conferences, EDP Sciences,* **2018,** *31,* 1–4.

34. Hajiali, A.; Pirumyan, G. Study on Oxidation Mechanism of Ozonated Wastewater Treatment: Color Removal and Algae Elimination. *Chemistry and Biology,* **2018,** *52* (3), 167–173.

35. Harakeh, M. S.; Butler, M. Factors Increasing the Ozone Inactivation of Enteric Viruses in Effluent. *Ozone: Science & Engineering*, **1984**, *6* (4), 235–243.

36. Heacox, D. Use and Generation of Ozone as a Disinfectant of Dairy Animal Tissues, Dairy Equipment, and Infrastructure. **2013**, Patent US 8609120B2; 11.

37. Heshmati, A.; Nazemi, F. Dichlorvos (DDVP) Residue Removal from Tomato by Washing with Tap and Ozone Water, a Commercial Detergent Solution and Ultrasonic Cleaner. *Food Science and Technology*, **2018**, *38* (3), 441–446.

38. Hill, A.G.; Spencer, H.T. Mass Transfer in a Gas Sparged Ozone Reactor. Presented at the First International Symposium on Ozone for Water and Wastewater Treatment, Washington, DC, **1974**, 367–380.

39. Hoffman, R. K. Ozone. In: *Inhibition and Destruction of the Microbial Cell*; Hugo W.B. (Ed.); London: Academic Press; **1971**; 251–253.

40. Hoigne, J.; Bader, H. The Role of Hydroxyl Radical Reactions in Ozonation Processes in Aqueous Solutions. *Water Research*, **1976**, *10* (5), 377–386.

41. Huber, M. M.; Canonica, S.; Park, G. Y.; Von Gunten, U. Oxidation of Pharmaceuticals during Ozonation and Advanced Oxidation Processes. *Environmental Science & Technology*, **2003**, *37* (5), 1016–1024.

42. Hunt, N. K.; Mariñas, B. J. Inactivation of *Escherichia coli* with Ozone: Chemical and Inactivation Kinetics. *Water Research*, **1999**, *33* (11), 2633–2641.

43. Hunt, N. K.; Mariñas, B. J. Kinetics of *Escherichia coli* Inactivation with Ozone. *Water Research*, **1997**, *31* (6), 1355–1362.

44. Iacumin, L.; Manzano, M.; Comi, G. Prevention of *Aspergillus ochraceus* growth on and Ochratoxin A Contamination of Sausages using Ozonated Air. *Food Microbiology*, **2012**, *29* (2), 229–232.

45. Ikehata, K.; Jodeiri Naghashkar, N.; Gamal El-Din, M. Degradation of Aqueous Pharmaceuticals by Ozonation and Advanced Oxidation Processes: A Review. *Ozone: Science and Engineering*, **2006**, *28* (6), 353–414.

46. Janex, M. L.; Savoye, P.; Roustan, M.; Do-Quang, Z.; Laine, J. M.; Lazarova, V. Wastewater Disinfection by Ozone: Influence of Water Quality and Kinetics Modeling. *Ozone: Science & Engineering*, **2000**, *22* (2), 113–121.

47. Jurado-Alameda, E.; García-Román, M. Assessment of the Use of Ozone for Cleaning Fatty Soils in the Food Industry. *Journal of Food Engineering*, **2012**, *110* (1), 44–52.

48. Kamber, U.; Gülbaz, G.; Aksu, P.; Doğan, A. Detoxification of Aflatoxin B1 in Red Pepper (*Capsicum annuum* L.) by Ozone Treatment and its Effect on Microbiological and Sensory Quality. *Journal of Food Processing and Preservation*, **2017**, *41* (5), e13102.

49. Karamah, E. F.; Wajdi, N. Application of ozonated water to maintain the quality of chicken meat: effect of exposure time, temperature, and ozone concentration. In *E3S Web of Conferences EDP Sciences*. **2018**, *67* (04044), 1–7.

50. Kdous, M. F.; Nagy, K. S.; Sorour, M. A. The Effect of Modern Packaging Systems on the Quality and Shelf-life of Salted Fish. *Research Journal of Food and Nutrition*, **2018**, *2* (3), 39–47.

51. Khadre, M. A.; Yousef, A. E.; Kim, J. G. Microbiological Aspects of Ozone Applications in Food: A Review. *Journal of Food Science*, **2001**, *66* (9), 1242–1252.

52. Kianmehr, P.; Kfoury, F. Prediction of Methane Generation of Ozone-Treated Sludge from a Wastewater Treatment Plant. *Ozone: Science & Engineering*, **2016**, *38* (6), 465–471.

53. Kim, J. G.; Yousef, A. E.; Dave, S. Application of Ozone for Enhancing the Microbiological Safety and Quality of Foods: A Review. *Journal of Food Protection*, **1999**, *62* (9), 1071–1087.

54. Kirby, R. M.; Bartram, J. Water in Food Production and Processing: Quantity and Quality Concerns. *Food Control*, **2003**, *14* (5), 283–299.

55. Kogelschatz, U. Advanced Ozone Generation. In: *Process Technologies for Water Treatment*; Stucki, S. (Ed.); Springer: Boston, MA.; **1988**; 87–118.

56. Kying, O. M. Effect of Ozone Exposure on Microbial Flora and Quality Attributes of Papaya (*Carica papaya* L) fruit. *Journal of Agronomy and Agricultural Aspects*, **2016**, *2014*, 104–113.

57. Loeb, B. L. Ozone: Science & Engineering: Thirty-three years and growing. *Ozone: Science & Engineering,* **2011**, *33* (4), 329–342.

58. Lozowicka, B.; Jankowska, M.; Hrynko, I.; Kaczynski, P. Removal of 16 Pesticide Residues from Strawberries by Washing with Tap and Ozone Water, Ultrasonic Cleaning and Boiling. *Environmental Monitoring and Assessment*, **2016**, *188* (1), 51–59.

59. Lyu, F.; Shen, K.; Ding, Y.; Ma, X. Effect of Pretreatment with Carbon monoxide and Ozone on the Quality of Vacuum Packaged Beef Meats. *Meat Science*, **2016**, *117*, 137–146.

60. Malik, S. N.; Ghosh, P. C.; Vaidya, A. N.; Mudliar, S. N. Ozone Pre-Treatment of Molasses-Based Biomethanated Distillery Wastewater for Enhanced Bio-Composting. *Journal of Environmental Management*, **2019**, *246*, 42–50.

61. Maniglia, B. C.; Lima, D. C.; Junior, M. D. M. Hydrogels Based on Ozonated Cassava Starch: Effect of Ozone Processing and Gelatinization Conditions on Enhancing 3D-Printing Applications. *International Journal of Biological Macromolecules*, **2019**, *138*, 1087–1097.

62. Mella, B.; Barcellos, B. S. D. C.; da Silva Costa, D. E.; Gutterres, M. Treatment of Leather Dyeing Wastewater with Associated Process of Coagulation-Flocculation/Adsorption/Ozonation. *Ozone: Science & Engineering*, **2018**, *40* (2), 133–140.

63. Miguel, N.; Lanao, M.; Valero, P.; Mosteo, R.; Ormad, M. P. Enterococcus Sp. Inactivation by Ozonation in Natural Water: Influence of H_2O_2 and TiO_2 and Inactivation Kinetics Modeling. *Ozone: Science & Engineering*, **2016**, *38* (6), 443–451.

64. Mohammadi, H.; Mazloomi, S. M.; Eskandari, M. H. The Effect of Ozone on Aflatoxin M1, Oxidative Stability, Carotenoid Content and the Microbial Count of Milk. *Ozone: Science & Engineering*, **2017**, *39* (6), 447–453.

65. Moore, G.; Griffith, C.; Peters, A. Bactericidal Properties of Ozone and its Potential Application as a Terminal Disinfectant. *Journal of Food Protection*, **2000**, *63* (8), 1100–1106.

66. Mostafalou, S.; Abdollahi, M. Pesticides and Human Chronic Diseases: Evidences, Mechanisms, and Perspectives. *Toxicology and Applied Pharmacology*, **2013**, *268* (2), 157–177.

67. Muhlisin, M.; Utama, D. T.; Lee, J. H.; Choi, J. H.; Lee, S. K. Effects of Gaseous Ozone Exposure on Bacterial Counts and Oxidative Properties in Chicken and Duck Breast Meat. *Korean Journal for Food Science of Animal Resources*, **2016**, *36* (3), 405.

68. Munhõs, M. C.; Navarro, R. S.; Nunez, S. C. Reduction of Pseudomonas Inoculated into Whole Milk and Skin Milk by Ozonation. *XXVI Brazilian Congress on Biomedical Engineering,* **2019**, *2019*, 837–840.

69. Muthukumarappan, K.; Halaweish, F.; Naidu, A. S. Ozone. In: *Natural Food Antimicrobial Systems*; Naidu, A. S (Ed.); CRC Press: Boca Raton, FL, **2000**, 796–813.

70. Muthukumarappan, K.; O'Donnell, C. P.; Cullen, P. J. Ozone Treatment of Food Materials. In: *Food Processing Operations Modeling: Design and Analysis*; Soojin Jun and Joseph M. Irudayaraj (Eds.); CRC Press; Boca Raton, FL, **2008**, 263–280.

71. Naitoh, S. Inhibition of Food Spoilage Fungi by Application of Ozone. *Journal of Food Microbiology*, **1994**, *11*, 11–17.

72. Noronha, M.; Britz, T.; Mavrov, V.; Janke, H. D.; Chmiel, H. Treatment of Spent Process Water from a Fruit Juice Company for Purposes of Reuse: Hybrid Process Concept and On-Site Test Operation of a Pilot Plant. *Desalination*, **2002**, *143* (2), 183–196.

73. Obadi, M.; Zhu, K. X.; Peng, W.; Noman, A.; Mohammed, K.; Zhou, H. M. Characterization of Oil Extracted from Whole Grain Flour Treated with Ozone Gas. *Journal of Cereal Science*, **2018**, *79*, 527–533.

74. Ogden, M. Ozonation Today. *Industrial Water Engineering*, **1970**, *7* (6), 36–42,

75. Okpala, C. O. R. Quality Evaluation and Shelf-life of Minimal Ozone-Treated Pacific White Shrimp (*Litopenaeus vannamei*) Stored on Ice. *Journal of Consumer Protection and Food Safety*, **2015**, *10* (1), 49–57.

76. Ormad, M. P.; Miguel, N.; Claver, A.; Matesanz, J. M.; Ovelleiro, J. L. Pesticides Removal in the Process of Drinking Water Production. *Chemosphere*, **2008**, *71* (1), 97–106.

77. Pardío Sedas, V. T.; López Hernández, K. M. Improved Microbial Safety of Direct Ozone-Depurated Shellstock Eastern Oysters (*Crassostrea virginica*) by Superchilled Storage. *Frontiers in Microbiology*, **2018**, *9*, 2802.

78. Parrón, T.; Requena, M.; Hernández, A. F. Environmental Exposure to Pesticides and Cancer Risk in Multiple Human Organ Systems. *Toxicology Letters*, **2014**, *230* (2), 157–165.

79. Patil, S.; Bourke, P. Ozone Processing of Fluid Foods. In: *Novel Thermal and Non-Thermal Technologies for Fluid Foods*; Cullen, P. J., Tiwari, B. K. and Valdramidis, V. P., (Eds.); Elsevier: London, **2012**, 225–261.

80. Patil, S.; Bourke, P.; Frias, J. M.; Tiwari, B. K.; Cullen, P. J. Inactivation of *Escherichia coli* in Orange Juice using Ozone. *Innovative Food Science & Emerging Technologies*, **2009**, *10* (4), 551–557.

81. Patil, S.; Valdramidis, V. P.; Cullen, P. J.; Frias, J.; Bourke, P. Inactivation of *Escherichia Coli* by Ozone Treatment of Apple Juice at Different pH Levels. *Food Microbiology*, **2010**, *27* (6), 835–840.

82. Pekarek, S. Ozone Generation Enhanced by TiO_2 Photocatalyst. *The European Physical Journal D*, **2008**, *50* (2), 171–175.

83. Piacentini, K. C.; Savi, G. D.; Scussel, V. M. The Effect of Ozone Treatment on Species of Fusarium Growth in Malting Barley (*Hordeum vulgare* L.) grains. *Quality Assurance and Safety of Crops & Foods*, **2017**, *9* (4), 383–389.

84. Piemontese, L.; Messia, M. C.; Marconi, E. Effect of Gaseous Ozone Treatments on DON, Microbial Contaminants and Technological Parameters of Wheat and Semolina. *Food Additives & Contaminants: Part A*, **2018**, *35* (4), 761–772.

85. Poretti, M. Quality Control of Water as Raw Material in the Food Industry. *Food Control*, **1990**, *1*(2), 79–83.

86. Quero-Pastor, M.; Garrido-Perez, C. Toxicity and Degradation Study of Clofibric Acid by Treatment with Ozone in Water. *Ozone: Science & Engineering*, **2016**, *38* (6), 425–433.

87. Restaino, L.; Frampton, E. W.; Hemphill, J. B.; Palnikar, P. Efficacy of Ozonated Water against Various Food-related Microorganisms. *Applied and Environmental Microbiology*, **1995**, *61* (9), 3471–3475.

88. Rice, R. G. Applications of Ozone in Water and Wastewater Treatment. In: *Analytical Aspects of Ozone Treatment of Water and Waste Water*; Rice, R. G. and Browning, M. J. (Eds.); The Institute: Syracuse, NY, **1986**, 7–26.

89. Rice, R. G.; Robson, C. M.; Miller, G. W.; Hill, A. G. Uses of Ozone in Drinking Water Treatment. *Journal-American Water Works Association*, **1981**, *73* (1), 44–57.

90. Rodrigues, A. A. Z.; de Queiroz, M. E. L. Use of Ozone and Detergent for Removal of Pesticides and Improving Storage Quality of Tomato. *Food Research International*, **2019**, *125*, 108626.

91. Rodrigues, V. O.; Penido, A. C. Sanitary and Physiological Quality of Soybean Seeds Treated with Ozone. *Journal of Agricultural Science*, **2019**, *11* (4), 183–196.

92. Sacco, A. Introduction to Ozone Generation Techniques-Corona, UV and Electrochemistry, **2009**.   https://ezinearticles.com/?Introduction-to-Ozone-Generation-Techniques---Corona,-UV-and-Electrochemistry&id=1686734; Accessed on January 28, 2020.

93. Sasmita, E.; Restiwijaya, M.; Yulianto, E. Effect of Ozone Technology Applications on Physical Characteristics of Red Cayenne Pepper (*Capsicum frutescens* L.) preservation. *Journal of Physics: Conference Series*, **2019**, 1217 (1), 1–6.

94. Savi, G. D.; Piacentini, K. C.; Scussel, V. M. Reduction in Residues of Deltamethrin and Fenitrothion on Stored Wheat Grains by Ozone Gas. *Journal of Stored Products Research*, **2015**, *61*, 65–69.

95. Shah, N. N. A. K.; Rahman, R. A.; Hashim, D. M. Changes in Physicochemical Characteristics of Ozone-Treated Raw White Rice. *Journal of Food Science and Technology*, **2015**, *52* (3), 1525–1533.

96. Sintuya, P.; Narkprasom, K.; Jaturonglumlert, S. Effect of Gaseous Ozone Fumigation on Organophosphate Pesticide Degradation of Dried Chilies. *Ozone: Science & Engineering*, **2018**, *40* (6), 473–481.

97. Smirnov, A.; Ukhanova, V.; Ershova, I. G.; Koshoeva, B. Optimization of Processing Modes of Disinfection of Vegetable Storehouses with the Use of Ozone. In: *Handbook of Research on Smart Computing for Renewable Energy and Agro-Engineering*; IGI Global: Hershey, PA, **2020**; 27–52.

98. Steenstrup, L. D.; Floros, J. D. Inactivation of *E. coli* 0157: H7 in Apple Cider by Ozone at Various Temperatures and Concentrations. *Journal of Food Processing and Preservation*, **2004**, *28* (2), 103–116.

99. Sudershan Rao, V.; Chakravarthy, D. P.; Naveen Kumar, R. Development of Ingenious Technique for *Salmonella* Decontamination of Spices Using Hybrid Technology of Ozone-Pulsed UV. *Indian Journal of Nutrition and Dietetics*, **2017**, *54* (4), 377–387.

100. Ternes, T. A.; Stüber, J.; Herrmann, N. Ozonation: A Tool for Removal of Pharmaceuticals, Contrast Media and Musk Fragrances from Wastewater? *Water Research*, **2003**, *37* (8), 1976–1982.

101. Torlak, E. Efficacy of Ozone Against *Alicyclobacillus acidoterrestris* Spores in Apple Juice. *International Journal of Food Microbiology*, **2014**, *172*, 1–4.

102. Torlak, E. Use of Gaseous Ozone for Reduction of Ochratoxin A and Fungal Populations on Sultanas. *Australian Journal of Grape and Wine Research*, **2019**, *25* (1), 25–29.

103. Tran, T. T. L.; Aimla-or, S.; Srilaong, V. Fumigation with Ozone to Extend the Storage Life of Mango Fruit cv Nam Dok Mai No. 4. *Agricultural Science Journal*, **2013**, *44*, 663–672.

104. Trombete, F. M.; Porto, Y. D.; Freitas-Silva, O. Efficacy of Ozone Treatment on Mycotoxins and Fungal Reduction in Artificially Contaminated Soft Wheat Grains. *Journal of Food Processing and Preservation*, **2017**, *41* (3), 1–10.

105. Vaughn, J. M.; Chen, Y. S.; Lindburg, K.; Morales, D. Inactivation of Human and Simian Rotaviruses by Ozone. *Applied and Environmental Microbiology*, **1987**, *53* (9), 2218–2221.

106. Vieno, N. M.; Härkki, H.; Tuhkanen, T.; Kronberg, L. Occurrence of Pharmaceuticals in River Water and Their Elimination in a Pilot-Scale Drinking Water Treatment Plant. *Environmental Science & Technology*, **2007**, *41* (14), 5077–5084.

107. Wang, L.; Luo, Y.; Luo, X.; Wang, R.; Li, Y.; Li, Y.; Chen, Z. Effect of Deoxynivalenol Detoxification by Ozone Treatment in Wheat Grains. *Food Control*, **2016**, *66*, 137–144.

108. Williams, R. C.; Sumner, S. S.; Golden, D. A. Inactivation of *Escherichia coli* O157: H7 and Salmonella in Apple Cider and Orange Juice Treated with Combinations of Ozone, dimethyl dicarbonate, and hydrogen peroxide. *Journal of Food Science*, **2005**, *70* (4), M197-M201.

109. Yuliani, M. M.; Mahfudz, L.; Nurwantoro, N.; Nur, M. The Effect of Plasma Generated Ozone for Cold Storage the Broiler Chicken Meat. *Nusantara Bioscience*, **2019**, *11* (1), 12–17.

110. Zambre, S. S.; Venkatesh, K. V. Tomato Redness for Assessing Ozone Treatment to Extend the Shelf-life. *Journal of Food Engineering*, **2010**, *96* (3), 463–468.

111. Zhang, T.; Xue, Y.; Li, Z.; Wang, Y.; Yang, W.; Xue, C. Effects of Ozone on the Removal of Geosmin and the Physicochemical Properties of Fish Meat from Bighead Carp (*Hypophthalmichthys nobilis*). *Innovative Food Science & Emerging Technologies*, **2016**, *34*, 16–23.

112. Zhou, Z.; Zuber, S.; Cantergiani, F.; Sampers, I. Inactivation of Foodborne Pathogens and their Surrogates on Fresh and Frozen Strawberries Using Gaseous Ozone. *Frontiers in Sustainable Food Systems*, **2018**, *2*, 51–59.

113. Zuma, F.; Lin, J.; Jonnalagadda, S. B. Ozone-Initiated Disinfection Kinetics of Escherichia Coli in Water. *Journal of Environmental Science and Health, Part A*, **2009**, *44* (1), 48–56.

PART II
Process Interventions for Food Processing and Preservation

CHAPTER 6

PRETREATMENTS AND DRYING OF FOOD PRODUCTS

RITESH B. WATHARKAR* and SADHANA SHARMA

ABSTRACT

Novel drying technologies have better performance than conventional drying methods. However, there is an urgent need for more research and development to create a technology that is industrially stable, cost-effective, and eco-friendly. There is tremendous scope for drying heat-sensitive liquid or semiliquid food for developing good quality dried products that are unattainable through conventional drying methods. Nowadays, consumers are much conscious regarding the quality and safety of the food. Moreover, more work is needed to develop the continuous pilot-scale design for drying freshly harvested perishable agricultural products. The knowledge of computational fluid dynamics and other advance engineering tools is crucial for developing and implementing these research innovations.

6.1 INTRODUCTION

Drying is a physical method mainly to preserve fruits, vegetables, spices, and other perishable food products. Drying is a process known since ancient times to removal of moisture from various foodstuffs. However, nowadays the concept of drying is not just limited to preservation but also to alternate applications, such as development of new dried food items or value-added products that can be incorporated with desirable ingredients. Drying reduces moisture content from 90% to 2%, necessary for enhanced shelf life. In addition to improving the shelf life, the dried food offers ease in transportation and handling. Moreover, such high-quality dried foods have

*Corresponding author. E-mail: watharkarritesh@gmail.com.

good reconstitution properties and seasonal food may be made available for consumption throughout the year.

Usually, the commercial driers work on the principle of a convective mode of heat and mass transfer. Such conventional driers include drum dryer, spray dryer, tray dryer, vacuum dryer, fluidized dryer, etc. These dryers though being cost-effective pose a disadvantage of damaging the physical attributes of the product. The final dried product quality and its nutritional traits are adversely affected with the conventional mode of drying [39]. Energy consumption and environmental safety are also serious issues in the drying of foodstuffs.

Convective type of industrial dryer uses either hot air or combustion gases as the heat transfer medium [57]. Therefore, researchers are constantly developing innovative drying techniques, equipments, various pretreatments for providing high-quality dried products [19]. For the increasing drying rate and produce, desirable quality of dried product is necessary. The treated fruits and vegetables with chemicals or dipping in hot water make them free from surface resistance [59].

Application of sound energy, microwaves, radiation, and others provides an innovative alternative to the conventional drying operations. These novel approaches promise remarkable success in the food industry.

This chapter discusses various advanced pretreatments and novel dryers. The chapter also includes some pretreatments, which have still not been studied independently for actual application in the food industry.

6.2   PRETREATMENTS AND THEIR AFFECTS

Pretreatment of agricultural goods effectively reduces the drying time and energy consumption. The drying rate of sample and quality of dried products are directly related to the type of pretreatment before drying [25]. In the case of fruit and vegetables, physical and chemical pretreatment can be used to shorten the drying rate, reduce energy consumption, and preserve the quality [59].

The aim of pretreatments of fruits and vegetables is the modification of tissues to facilitate handling and to enhance the drying rate, preservation of flavor, and color and inactivation of enzymes and microorganisms. These pretreatments are classified as chemical treatments (hypertonic solution; alkali, acid, and sulfite liquor; SO_2, O_3, and CO_2 gases) and physical treatments (mild and severe thermal treatment by different means and ultrasonic).

Chemical pretreatments lead to enhance the drying rate, but it causes loss of soluble nutrients of fruits and vegetables. Conventional hot water blanching has a great impact on the control of enzymatic reactions, microbial growth control, texture (softening), and drying rate.

These pretreatments also have several disadvantages, such as high time and temperature of processing, loss of texture, nutrients, and pigment leaching. The approaches of novel pretreatments can overcome the drawbacks of conventional pretreatments. Some novel pretreatments are discussed in this section.

6.2.1 NOVEL PRETREATMENTS

6.2.1.1 POULTICE-UP PROCESS

Poultice-up process (PUP) is used for drying fish paste. In this process, the product is first dried to a certain level of moisture content of drying (i.e., 20%–30%), followed by tempering for several hours under controlled temperature (2 ± 0.5 °C) conditions. The product can be tempered for some time thus facilitating moisture distribution uniformly throughout the food material. The second stage of drying helps to diffuse the moisture constantly from inside to the surface and then subsequently from surface to the atmosphere. In general, this method is beneficial in predrying of high moisture foods, such as meat, fruits, vegetables, and fish products.

The surface of high moisture foods shows a firm appearance during the initial stage of drying due to the removal of surface moisture, which is called case hardening. It causes stress cracks on the surface to promote loss of quality of dried products [33]. After case-hardening, the mass transfer of food material is reduced and the time of drying is increased [51], due to uneven distribution of moisture and porous structure of food material throughout the food material. Therefore, some nonuniform pores resist the moisture transfer during drying which leads to high residual moisture in food. The quality of the final product has deteriorated when the product is not properly dried.

PUP pretreatment results in uniform moisture distribution during relaxation time after the storage period under controlled temperature. This helps to reduce the issues of case hardening, reduction of drying time, and the process cost of drying. Konishi [32, 33] observed the effect of PUP on drying of fish paste sausage in a cylindrical shape and observed uniform distribution of moisture [32].

6.2.1.2   CONTROLLED PRESSURE DROP

The controlled pressure drop treatment is also called *Detente Instantanee Controlee* (DIC), which is used for treating postharvest goods before hot air drying. The DIC treatment can improve the drying rate and rehydration property of the final dried product [1]. The expansion in the new structure after controlled pressure drop leads to an increase in effective moisture diffusivity of a product during subsequent drying. This technology is an eco-friendly process and uses also less energy [4].

The DIC unit comprises mainly three units: treatment vessel, vacuum tank, and instant valve. The treatment vessel is a high-temperature–high-pressure steam vessel and the instant valve connects the vessel and vacuum tank. The food sample is placed in the treatment vessel and steam or hot air is allowed to circulate throughout the treatment vessel for 5–60 s at 100–180 °C. The pressure is instantly released through an instant valve up to 5 kPa of vacuum. The vacuum is generally held for 5–20 s. Once the vacuum is removed, the sample is placed in a hot air oven for drying at atmospheric pressure [1, 49].

The auto-vaporization, which takes place during DIC treatment, is responsible for puffing in the food material. Application of DIC treatment is highly effective in treating heat-sensitive food materials, such as fruits and vegetables. Reduction in the drying time of DIC-treated apple slice to 1 h was observed instead of 6 h during the second stage of hot air drying [41, 42]. In the case of DIC-treated paddy, more yield of white rice and reduced amount of broken material was observed [49]. Apart from the reduction in the cooking time of rice, it also resulted in an overall energy-efficient process. The processing time of DIC treatment is normally 30 s. The drying time of DIC treated products is only 3–4 h instead of 12–15 h during hot air drying. This treatment can be applied for drying of fresh food materials, such as meat, egg white, egg yolk, milk, fruits, and vegetables [41].

6.2.1.3   INFRARED

These days, infrared radiation (IR) technology has wide use in the food industry from drying, roasting, blanching, pasteurization, etc. This technology is based on the principle of using radiant electromagnetic energy, which may induce changes in the electronic, vibrational, and rotational states of atoms and molecules during striking on a food surface. As the food is subjected to IR radiation, the radiation is absorbed, reflected, or scattered throughout the thin layer of the food surface thus generating heat energy. The food absorbs this heat energy from the surface to the inner surface.

Compared to conventional treatment, IR drying has many beneficial aspects, such as short processing time, energy-efficient and better quality of final product in terms of color, texture, etc.

There are different types of IR emitters that are used in food applications. These are reflector type IR incandescent lamps, quartz tube type, ceramic IR emitter, direct flame IR radiator, and carbon twin IR emitter [18]. In general, the metal filament is fed inside the element and the element is situated in a sealed enclosure. A heat-resistant wire of metal alloy (nickel–chrome or iron–chromium) or tungsten filament is used to generate the radiant energy in electric IR heaters. It can also be done by passing an electric current through the wire (Nickel chrome) and the element; and the surrounding material gets heated to an extreme temperature to generate IR.

The application of IR as a pretreatment before drying offers enormous potential in the food industry. Recently, many researchers have used IR for blanching fruits and vegetables prior to drying. IR inactivates the activity of the enzyme, which deteriorates the quality of fruits and vegetables. Galindo [26] suggested that IR treatment was better than hot water blanching carrot slices because the former helps to preserve the texture attributes of carrot slices. Zhu et al. [58] reported successful inactivation of PPO and POD enzymes and improved the quality of thin apple slices with IR treatment [58]. There was a reduction in drying time of 23% and 17% for IR-treated apple slices in low humidity air and hot air type dryers, respectively [54]. There was the retention of 82%–90% of ascorbic acid and 72%–74% of phenolic content during the low humidity air drying.

6.2.1.4 ULTRASONICATION

The application of ultrasound is a nonthermal method that has been in use since the 1960s for the processing and preservation of foods [40]. Ultrasound is a sound wave having a frequency of above 20 kHz. Ultrasound waves are classified into two types, such as (1) high frequency-low energy (<20 kHz) and (2) low frequency-high energy wave (20–40 kHz). Mohapatra and Mishra [38] suggested the use of high-frequency–low energy waves for quality measurement and low-frequency–high energy waves (power ultrasound) for pretreatments [38].

The ultrasonic device comprises mainly of three components, that is, generator, transducer, and application system. Transducers are of three types, namely: (1) fluid-driven transducers, (2) magnetostrictive transducers, and (3) piezoelectric transducers. A transducer helps to convert mechanical energy, which is produced by the generator, into sound energy [45]. Among the three types of transducers, the piezoelectric type is more efficient because it can convert 85%–90% of mechanical energy into sound energy. During the pretreatment, the food material is placed in water or air later to which the ultrasound is applied.

Ultrasound creates microscopic channels or pores in the food material that assist in faster diffusion of moisture from the core toward the food surface. The treatment almost doubles the mass transfer rate. Fernandes and Rodrigues [25] investigated drying of different ultrasound-treated fruits, such as banana, papaya, melon, strawberry, sapota, apples, star fruit, and pineapple [25]. Researchers suggested that the ultrasound pretreated fruits were generally more efficient in terms of reduced drying time by 20% and drying cost by 30%, respectively.

The power ultrasound (20–40 kHz) can be used in processes, such as lactose crystallization, drying, freezing and thawing, dehydration, meat tenderization, etc. In addition, pretreatment of raw foods with power ultrasound can improve processing efficiency. Moreover, ultrasound can rupture the microbial cell-wall inhibiting the growth of spoilage microbes and release of undesirable enzymes, which in turn assist in improving the shelf life of the product. Thus, drying with ultrasound offers significant advantages including reduced loss of flavor, maximum homogeneity in the product, and increased saving in energy [23].

6.3   NOVEL DRYING TECHNOLOGIES

6.3.1   REFRACTIVE WINDOW DRYING

Refractive window (RW) drying is a less exploited drying technique, which is also known as indirect or film drying. This technology was introduced by MCD Technologies, Inc., in Tacoma, Washington. In this method, 1–2 mm layer of pulp is preferred on the conveyer for drying. Due to the indirect heating, RW drying is not suitable for drying bulky product. In this technique, product temperature does not rise above 70 °C, which in turn is highly beneficial for reducing nutritional loss and improving physical appearance.

According to research studies, the drying time during refractance window drying is not more than 2 h. In RW drying, heat transfer occurs via three modes, that is, radiation (infrared), conduction, and convection, thus resulting in increased heat and mass transfer through thin-film products.

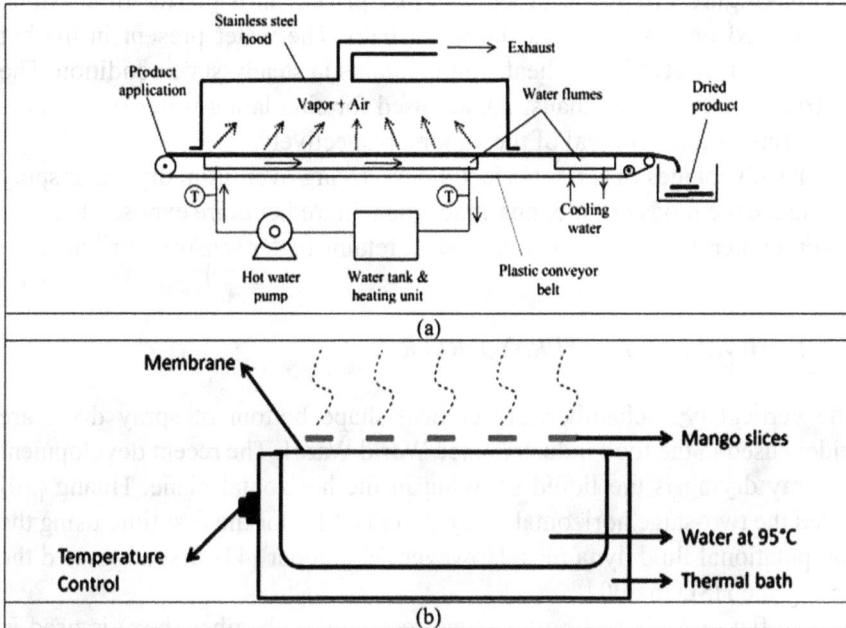

FIGURE 6.1   (a) Schematic diagram of an RW dryer; (b) lab-scale RW dryer.

Among several drying techniques (solar-based, tray, freeze, fluidized bed), refractance window drying method is more effective to offer low maintenance-cost and better quality of the final product. This technique was patented by Magoon in 1986 for the successful dehydration of heat-sensitive foods [35]. Apart from this, drying of herb-mixture and food extracts by applying RW dryer was patented by Jones [31]. The beta-carotene, lycopene, spirulina, teas, herbs, and barley grass blends are produced by using the RW technique [14]. Compared to freeze-drying, the RW drying technique requires less than half the energy necessary for drying of the same quantity of foodstuffs [47].

The product to be dried is spread over the conveyer belt in 2–3 mm layers. The hot water (90–95 °C) is circulated under the plastic conveyer belt. This helps to transfer thermal energy to the material to be dehydrated. The dried product gets cooled at the end of the process.

In the continuous RW dryer (Figure 6.1a), the product is dried over a moving plastic conveyer belt. The product temperature does not rise above 70 °C and is dried in few minutes in the form of flakes [44]. In the lab-scale system (Figure 6.1b), the product is just placed on a plastic film, which is tightened on both sides of the water bath. The water present in the hot water bath is heated by the heater and remains in steady-state condition. The centrifugal pump and exhaust fan are used for circulating water through the water flumes and removal of the vapors, respectively.

The advantages of refractance window drying over drum drying or spray drying are the foods or other pharmaceutical ingredients are exposed to much lower temperatures and thus the product retains better sensory attributes.

6.3.2   HORIZONTAL SPRAY DRYER

The vertical type chamber and conical shape bottom of spray dryer are widely used in the food industry since World War II. The recent development in spray drying is the liquid spraying in the horizontal plane. Huang [30] tested the two-stage horizontal spray dryer (HSD) for the first time using the computational fluid dynamics. However, Mujumdar [43] first suggested the idea of the HSD in 2004.

The flat or V-shaped bottom type rectangular chamber (box) is used in an HSD. The nozzle sprayed-liquid-suspension is horizontally distributed in the rectangular box (drying chamber) and the powder is collected at the bottom of the chamber. The screw conveyor or sweep conveyor situated at the bottom assists to remove the powder. In horizontal spray drying, low flow nozzles are mandatory, because the product is properly dried in a short residence time [6, 44].

Horizontal spray drying systems (Figure 6.2) are not so popular. However, few commercial applications have been using it. Only a few manufacturers are involved to supply such flat-bottom box dryers including CE Rogers, Marriott Walker, Henningsen Foods, Food Engineering Co. Apart from these, Henszey Co., Blaw-Knox, Bufflovak and Mora Industries are involved in the manufacture of V-bottom dryers.

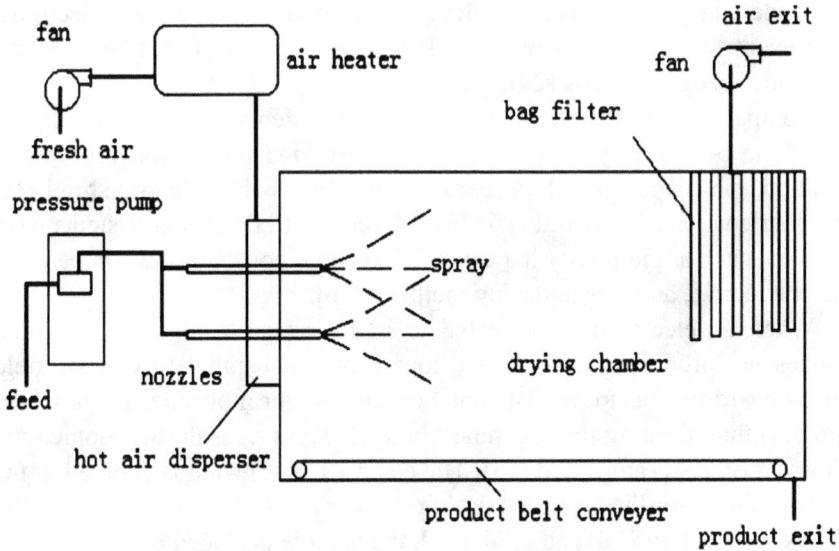

FIGURE 6.2 Schematic diagram of horizontal spray dryer.

Limited information is available on the application of horizontal spray-drying to heat-sensitive products. The study shows minimum degradation of amylase activity at 155 °C using horizontal spray drying [11]. There are some commercial applications for drying of skim milk, whey protein, cheese powder, egg albumin, whole egg powder, etc. Moreover, HSD is feasible for the drying of dairy products and vegetable products, such as soymilk, chocolate, and soy protein.

6.3.3 CORONA WIND DRYING OR ELECTROHYDRODYNAMIC DRYING

Corona wind drying is a nonthermal drying method also known as Electrohydrodynamic drying (EHD) or ionic drying. An electric field is created between two electrodes by injecting accelerated ionic species. The corona wind is directly allowed over the surface of the moist material, which results in the increased overall efficiency of the drying process.

According to Sakai et al. [53], corona wind is the movement of uncharged air particles moving in an electrostatic field; and an appreciable wind is produced due to collisions with ionized air particles. Gas ionization takes place as a result of the high electric field at the exterior of discharge (corona) electrode and the air ions are drifted to the opposite (grounded) electrode. This event forms a space-charge and an electric current flows between the electrodes because of this [29].

Compared to hot air (convective) drying systems, corona wind drying offers the best quality product in a low-cost food production process. In addition, physicochemical properties (such as: color, flavor, shrinkage, and nutrients) are improved [16, 36]. Moreover, EHD drying systems have benefits, such as simplicity in design, less energy consumption compared to the convective, and freeze-drying methods [16].

When an electrode is subjected to high voltage, a high electric field ionizes gas-producing ions. These ions move in parallel to electric field lines toward the electrode plate and collide with air molecules (Figures 6.3 and 6.4) thus forming the secondary bulk flow. As a result, the momentum transport of gas is amplified [13]. The electric (ionic) wind is denoted as the primary thrust for the escalated drying rate.

As reported by Dalvand et al. [16], the sample is placed on a steel plate in the drying chamber. Drying chamber is of transparent plastic, including two air vents on both sides of the chamber. A blower is placed to regulate the air velocity. The electrode is positioned vertically above the horizontal base, where the sample is placed. The electric field is applied to the sample by tuning the voltage and the electrode spacing. The needles or multiple pin-type electrodes are used for allowing the ions on the sample and are connected to the discharge electrode. The weight losses are measured by automatic electrical weighing balance. The temperature is adjusted by regulating the voltage to the electrode.

Alemrajabi et al. [3] reported that EHD offers low energy consumption and better color retention in carrots compared to conventional drying [3]. They also found that the moisture content of EHD dried carrots was reduced by 79.5% compared to 77% in oven drying (55 °C) and 22.5% in ambient air drying, respectively. Cao and Nishiyama [12] suggested an increase in the drying rate of wheat samples when the voltage was increased, and the discharge gap was decreased [12].

Esehaghbeygi et al. [24] reported that drying of tomato powder by the EHD method retained the red color and also the drying rate was increased compared to air drying. There is not enough literature on the effect of Corona drying on sugar-rich fruit puree or juices [24]. Hence, EHD can be used for further research in the drying of fruit puree or juices.

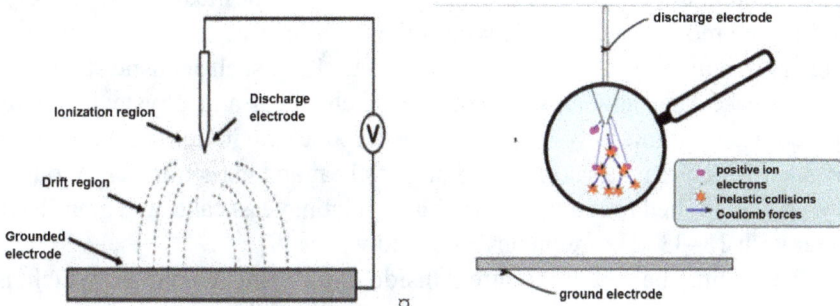

FIGURE 6.3 Mechanism of corona wind drying.

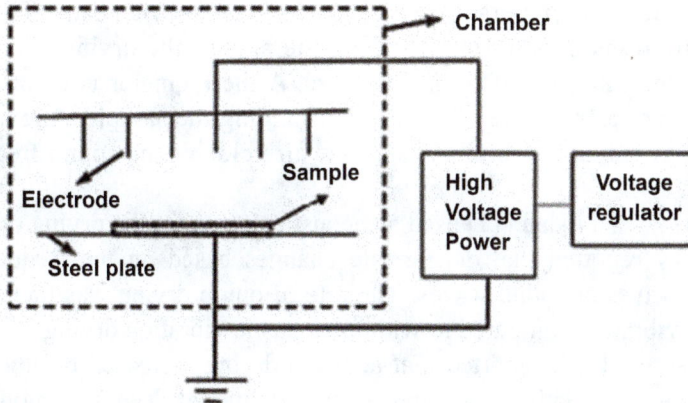

FIGURE 6.4 Corona wind drying.

6.3.4 ACOUSTIC DRYING

The concept of acoustic drying was first given by Boucher in 1961 and later he developed this novel technology in 1965 for the first time [9, 10]. The high-powered sound waves are used in acoustic drying. Acoustic waves are classified in three different categories based on different frequencies: (1) infrasound (<20 Hz), (2) audible sounds (20 Hz–18 kHz), and (3) ultrasounds (>18 kHz).

Boucher [9, 10] first suggested the hypothesis on increasing drying rate by acoustic waves. He reported that rapid, successive compressions and rarefactions take place when desired frequency sound waves or acoustic energy travels through the air. Due to these elevated and reduced pressure sites at gas–liquid interfaces, the rate of moisture evaporation is increased.

The sound waves generate acoustic pressure over the medium, which helps to promote motion and mixing within the fluid, that is, called acoustic stirring or streaming [7]. According to Ensminger [22], such acoustic streaming can increase heat and mass transfer in a chemical and physical system. Sometimes high-intensity acoustic waves produce cavitation throughout the product, which may assist in increasing the heat and mass transfer. Acoustic drying was applied first time for drying of gelatin, yeast cake, and granulated sugar with 18–33 kHz frequency of sound waves [9].

The drying basket is situated inside the drying chamber, which is hanged with an electronic balance. The sonic field is generated, when compressed air is passed through the siren (Figure 6.5). In addition, the pressurized air also serves as a moisture-carrier in the dryer. The thermostat maintains air temperature by controlling an electric heater. The probe of sound pressure meter moves along the axis of the drying chamber to regulate the intensity of sound pressure. A thermometer is connected to measure the temperature of the air surrounding the sample. Anemometer and tachometer are used to measure the air velocity and sound frequency, respectively.

Da-Mota and Palau [17] used an acoustic design for the drying of onions [17]. They reported that drying rate changes based on the frequency of acoustic waves or sound waves. The rate of onion drying was increased in acoustic vibration compared to without acoustic vibration drying.

Aversa et al. [7] reported that acoustic drying increased drying rate of cylindrical carrot cubes compared to the traditional drying method. Sonic waves can also be applied as hybrid drying in combination with solar drying, or freeze-drying or spray drying. According to García-Perez et al. [27], the acoustic energy is more effective on porous material in terms of moisture diffusivity. In addition, acoustic drying can also be used for drying of grains, vegetables and fruits, etc.

6.3.5 ATMOSPHERIC FREEZE-DRYING

Atmospheric freeze-drying (AFD) was defined by Harold Meryman in 1959. AFD integrates both benefits and limitations of convective drying (economical) and freeze-drying (high product quality). According to Meryman, the mass transfer flux from the ice-front to the drying space depends on the pressure gradient in the dried food, and not on the absolute pressure of the system. As a consequence, if a water vapor pressure gradient is maintained in the dried product, it can carry out the drying process at atmospheric pressure. In AFD, moisture or solvent is removed as vapors in a vacuum chamber from the frozen sample of moist food or solution by sublimation.

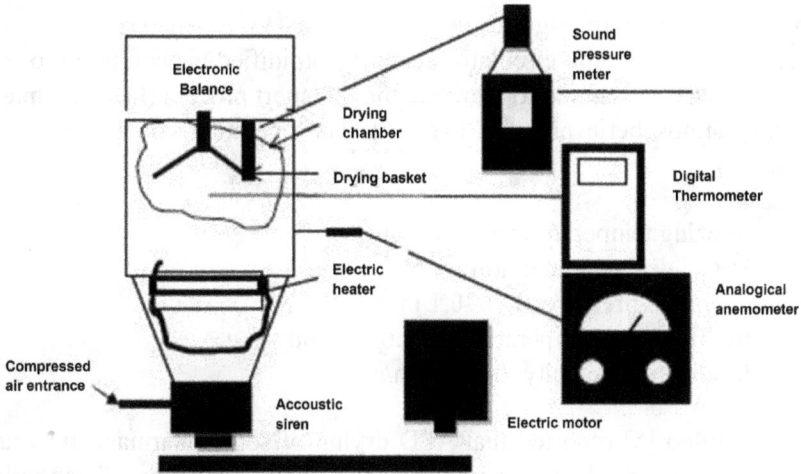

FIGURE 6.5   Acoustic drying.

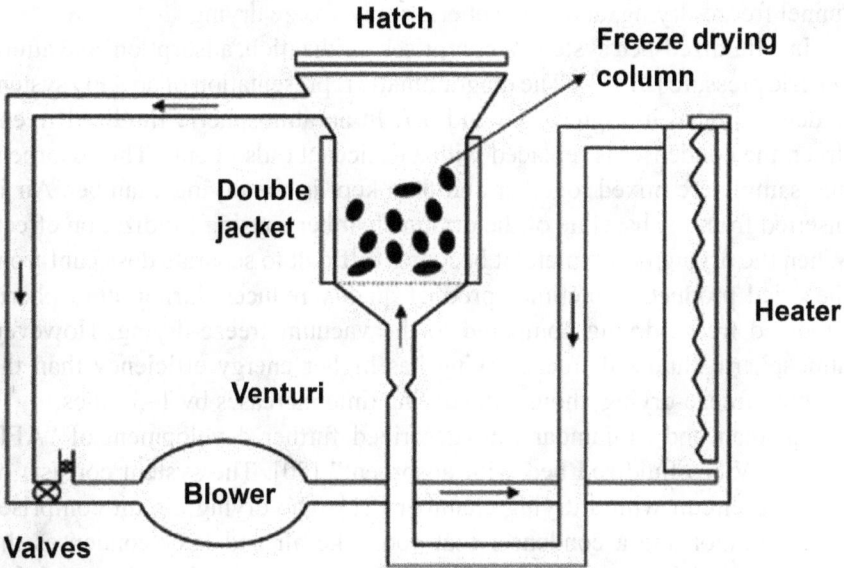

FIGURE 6.6   Atmospheric fluidized freeze dryer.

The first-time application of AFD in food was reported by Lavin and Matels in 1962 [51]. However, AFD was commercially developed at the beginning of 1990 [51]. This dryer was developed to replace the freeze dryer system as the latter was costly due to continuous vacuum requirements throughout the long drying process.

The food is dried by circulating cold dehumidified air below −6 to −10 °C over the frozen sample to facilitate the transport process (heat and mass) at nearly atmospheric pressure [51, 55]. The operating conditions are as follows:

1.   freezing temperature: −20 to −40 °C,
2.   freeze-drying temperature: −5 to −8 °C,
3.   operating pressure: 4.56–0.1 mmHg,
4.   fluidized bed temperature: 0–7.6 °C, and
5.   fluidization velocity: 0.1–0.5 m/s.

Alves-Filho [5] reported that AFD drying of some pharmaceutical and biotechnological products can improve the quality in terms of negligible shrinkage, better color, avoid protein denaturation, and minimum lipid oxidation [5].

The AFD operation may be performed by fluidized-bed freeze-drying, tunnel freeze-drying, and atmospheric spray-freeze drying.

In a fluidized-bed system, it comprises fluidization, adsorption, and atmospheric pressure [21, 51]. The diagrammatic representation of an AFD system is demonstrated in Figures 6.6 and 6.7. In an atmospheric fluidized freeze dryer, the condenser is replaced with a desiccant (adsorbent). The adsorbent and sample are mixed together and then kept in the drying chamber. Air is inserted from the baseline of the drying chamber causing fluidization effect. When the drying is complete, it becomes difficult to separate desiccant from the dried product. Sometimes product quality reduces during atmospheric fluidized freeze-drying compared to the vacuum freeze-drying. However, atmospheric fluidized freeze-drying has higher energy efficiency than the vacuum freeze-drying, though the drying time increases by 1–3 times.

Rahman and Mujumdar [50] described further development of "AFD assisted Vibro-fluidized bed with adsorbent" [50]. The system consists of a drying circuit with a drying chamber [21]. The drying circuit comprises an evaporator and a condenser that cools the air and later condenses the moisture, which can be easily drained out (Figure 6.7). They reported that the proposed system was capable of reducing processing time at a low cost and can provide better product quality than the AFD and FD.

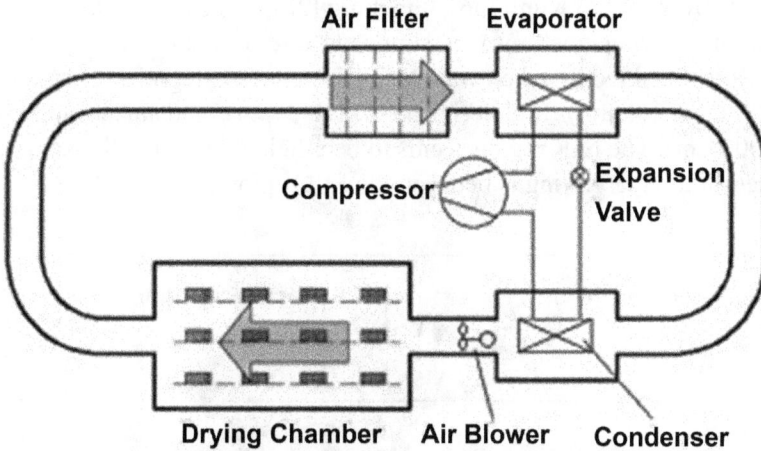

FIGURE 6.7   Atmospheric freeze dryer.

Both condenser and heating coils help to heat the cold air to keep inlet drying temperature at -10 to 40 °C. The containing inlet air humidity and air velocity is about 30% and 1.5 m/s, respectively. The cubes of the product are kept on perforated shelves, which are mounted in the drying chamber. Sensors are placed at different locations of the AFD to allow the monitoring and continuous recording of the inlet–outlet conditions for each relevant component. All drying circuits are made of stainless steel for easy cleaning. Wolff and Gibert [55] observed that AFD can save 34% of energy, compared to vacuum freeze-drying.

Claussen [15] reported that atmosphere freeze drying provides better quality of food products than those under vacuum freeze-drying [15]. Duan [21] observed a reduction in drying time by 50% during AFD of apple cubes, while also retaining the product quality as in freeze-drying [21]. According to Matteo [37], the heat and mass transfer coefficients for drying of food material were higher in AFD than those during vacuum freeze-drying.

6.3.6   INFRARED DRYING (IR)

The use of infrared drying has been reported for the drying of agricultural products by Yagi and Kunii [56], Ginzburg [28], and Zarein et al. [57]. IR is based on electromagnetic waves similar to microwave or ultraviolet radiation. IR is classified based on wavelength into three categories: near-infrared (NIR) (0.78–1.4 μm), middle-infrared (1.4–3 μm), and far-infrared (FIR) (3–1000 μm). The NIR region seems to be efficient in drying thicker bodies, whereas thin layer drying is better in the FIR region [52].

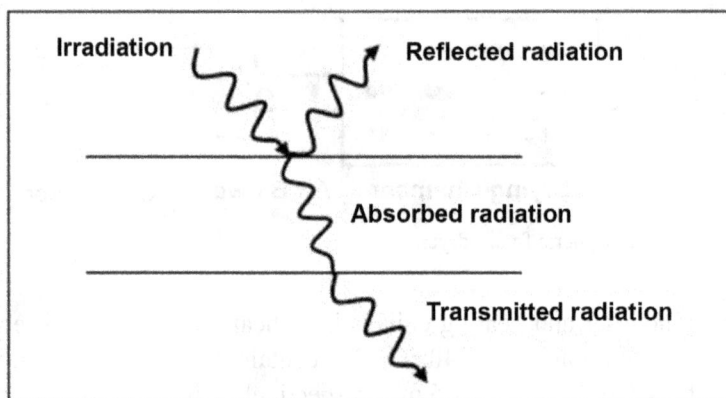

FIGURE 6.8   Characteristics of IR when applied on food.

The infrared drying is highly efficient drying method due to higher heat transfer rate and short drying time [46]. Other main advantages offered by infrared drying are energy effective, temperature consistency of the product, characteristic product, and eco-friendly [34]. The drying efficiency of IR depends on the density and wavelength of the IR radiation. Moreover, the distance between the food sample and the emitters affects the drying rate. Shorter the distance, higher is the drying rate.

It is well known that hot air-dried vegetables have a low rehydration rate and therefore are not suitable for preparing instant vegetable mixes [34]. On the other hand, freeze-drying techniques are best to produce instant vegetables; however, this technique has high cost of operation among all supplementary methods. Consequently, IR has become popular for drying fruits and vegetables. Recent applications of IR include drying of nuts, sweet potato, potato, kiwifruit, etc.

Figure 6.8 demonstrates the basic operation of radiation for drying of food materials. Introduction of IR on the food is followed by absorption, reflection, and transmission of radiation throughout the material. The IR drying process is suitable for thin-layered food material due to the wide surface area that is exposed to the radiation. The absorption of radiation results in molecular vibrations within the food material, which ultimately generates heat [53]. The process creates maximum absorption to generate adequate amount of heat resulting in an increased mass transfer during drying. The IR heating results in fairly uniform drying of fruits to give better quality characteristics compared to alternate drying methods [46, 48].

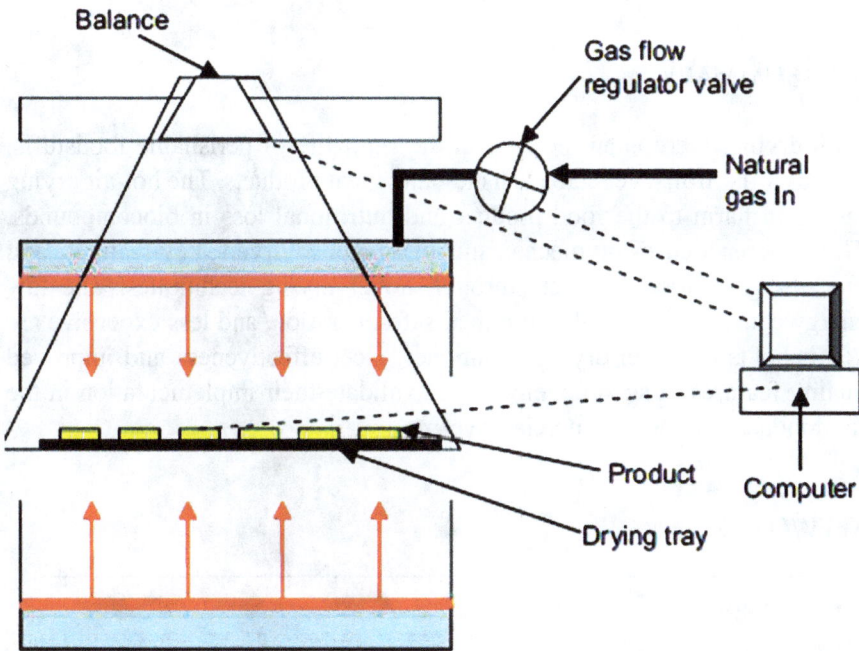

FIGURE 6.9 Schematic diagram of infrared drying method.

An infrared dryer (Figure 6.9) mainly comprises an IR emitter, a wave guard, a gas flow regulator valve, trays, and a digital balance. A catalytic infrared dryer/dehydrator contains two catalytic IR emitters that are powered with natural gas and these are placed at both sides of the drying tray. The tray and the emitter are in complete parallel positions. The wave guard prevents loss of radiation being produced by the emitter to prevent heat loss and to allow uniform heating. Undisrupted supply of natural gas makes the IR heating a continuous process. Digital weighing balance is used for measuring the weight loss in the sample.

Recently, many researchers have compared infrared drying with other conventional methods for the quality and drying performance of various fruits and vegetables. Adak [2] reported increased drying rate and retention of nutrients during increased IR drying of strawberries [2]. According to Baysal [8], infrared dried carrots resulted in the best rehydration properties compared to microwave and hot air-drying of carrots. Doymaz [20] also suggested that the moisture diffusion in carrots increases with increase IR power [20]. Also, increase in the intensity of IR was able to increase the color loss during drying of carrots.

6.4   SUMMARY

With drying, there is an increase in the shelf life of perishable foodstuffs, such as milk, fruits, vegetables, fishes, and meat products. The hot-air drying results in harm to the food product and nutritional loss in biocompounds. This chapter focuses on mechanisms of advanced dryers, key features, and processing of food products through novel drying techniques. The low energy consumption, quality product, safe-operation, and less expensive are key benefits of novel drying techniques. Cost-effectiveness and improved quality features of such novelty dryers validate their implementation in the food industry at the commercial level.

KEYWORDS

- drying
- freezing
- heat and mass transfer
- hybrid drying
- infrared
- spray drying
- ultrasonic

REFERENCES

1. Abdulla, G.; Belghit, A.; Allaf, K. Impact of the Instant Controlled Pressure Drop Treatment on Hot Air Drying of Cork Granules. *Drying Technology*, **2010**, *28* (2), 180–185.

2. Adak, N.; Heybeli, N.; Ertekin, C. Infrared Drying of Strawberry. *Food Chemistry*, **2017**, *219*, 109–116.

3. Alemrajabi, A. A.; Rezaee, F.; Mirhosseini, M.; Esehaghbeygi, A. Comparative Evaluation of the Effects of Electro Hydrodynamic, Oven, and Ambient Air on Carrot Cylindrical Slices During Drying Process. *Drying Technology*, **2012**, *30* (1), 88–96.

4. Al-Haddad, M.; Mounir, S. Drying Process Combining Hot Air, DIC Technology & Microwaves. In: *5th Proceeding of the Asia-Pacific Drying Conference*, Hong Kong, China, **2007**, 1064–1068.

5. Alves-Filho, O. Innovative and Conventional Dryers for Production of Powders for Pharmaceutical and Biotechnological Industrial Applications. *Paper Presented at International Meeting on Pharmaceutics Biopharmaceutics and Pharmaceutical Technology*, Nuremberg, **2004**, 6.

6. Aundhia, C. J.; Raval, J. A.; Patel, M. M.; Shah, N. V. Spray Drying in the Pharmaceutical Industry: Review. *Journal of Pharmaceutical Research*, **2011**, *2*, 63–65.

7. Aversa, M.; Van der Voort, A. J.; de Heij, W. Experimental Analysis of Acoustic Drying of Carrots: Evaluation of Heat Transfer Coefficients under Different Drying Conditions. *Drying Technology*, **2011**, *29* (2), 239–244.

8. Baysal, T.; Icier, F.; Ersus, S. Effects of Microwave and Infrared Drying on the Quality of Carrot and Garlic. *European Food Research and Technology*, **2003**, *218* (1), 68–73.

9. Boucher, R. M. G. Drying by Airborne Ultrasonic. *Ultrasonic News*, **1961**, *3* (8), 14–16.

10. Boucher, R.M.G. US Patent 3175299; March 3e0, **1965**, 11.

11. Cakaloz, T.; Akbaba, H.; Yesügey, E. T.; Periz, A. Drying Model for α-amylase in a Horizontal Spray Dryer. *Journal of Food Engineering*, **1997**, *31* (4), 499–510.

12. Cao, W.; Nishiyama, Y. Electro Hydrodynamic Drying Characteristics of Wheat Using High Voltage Electrostatic Field. *Journal of Food Engineering*, **2004**, *62* (3), 209–213.

13. Chen, Y.; Barthakur, N. N.; Arnold, N. P. Electrohydrodynamic (EHD) Drying of Potato Slabs. *Journal of Food Engineering*, **1994**, *23*, 107–119.

14. Clarke, T. P. (2004). Refractance Window™—"Down Under". In: *Proceedings of the 14th International Drying Symposium*; São Paulo, **2004**, 813–820.

15. Claussen, I. C.; Ustad, T. S.; Strommen, I.; Walde, P. M. Atmospheric Freeze Drying: Review. *Drying Technology*, **2007**, *25* (6), 947–957.

16. Dalvand, M. J.; Mohtasebi, S. S.; Rafiee, S. Effect of Needle Number on Drying Rate of Kiwi Fruit in EHD Drying Process. *Agricultural Sciences*, **2013**, *4* (1), 27400. doi:10.4236/as.2013.41001.

17. Da-Mota, V. M.; Palau, E. Acoustic Drying of Onion. *Drying Technology*, **1999**, *17* (4–5), 855–867.

18. Das, I.; Das, S. K. Emitters and Infrared heating system design. In: *Infrared Heating for Food and Agricultural Processing*; Z. Pan and G. Atungulu (Eds.); CRC Press: Boca Raton, FL, **2010**, 58–87.

19. Dev, S. R.; Raghavan, V. G. Advancements in Drying Techniques for Food, Fiber and Fuel. *Drying Technology*, **2012**, *30* (11–12), 1147–1159.

20. Doymaz, I.; Kipcak, A. S.; Piskin, S. Characteristics of Thin Layer Infrared Drying of Green Beans. *Czech of Journal Food Science*, **2015**, *33*, 83–90.
21. Duan, X.; Ding, L.; Ren, G. Y.; Liu, L. L.; Kong, Q. Z. The Drying Strategy of Atmospheric Freeze-Drying Apple Cubes Based on Glass Transition. *Food and Bioproducts Processing*, **2013**, *91* (4), 534–538.
22. Ensminger, D. Acoustic and Electroacoustic Methods of Dewatering and Drying. *Drying Technology*, **1988**, *6* (3), 473–499.
23. Earnshaw, R. G.; Appleyard, J.; Hurst, R. M. Understanding Physical Inactivation Processes: Combined Preservation Opportunities Using Heat, Ultrasound and Pressure. *International Journal of Food Microbiology*, **1995**, *28*, 197–219.
24. Esehaghbeygi, A.; Basiry, M. Electrohydrodynamic (EHD) Drying of Tomato Slices (*Lycopersicon esculentum*). *Journal of Food Engineering*, **2011**, *104* (4), 628–631.
25. Fernandes, F. A. N.; Rodrigues, S. Ultrasound Application as Pre-Treatment for Drying of Fruits. *Proceedings of the International Congress on Engineering and Food*, **2011**, *3*, 1987–1989.
26. Galindo, F. G.; Toledo, R.T. Tissue Damage in Heated Carrot Slices: Comparing Mild Hot Water Blanching and Infrared Heating. *Journal of* Food Engineering, **2005**, *67*, 381-385.
27. García-Pérez, J. V.; Cárcel, J. A.; Riera, E. Influence of the Applied Acoustic Energy on the Drying of Carrots and Lemon Peel. *Drying Technology*, **2009**, *27* (2), 281–287.
28. Ginzburg, A. S. *Application of Infrared Radiation in Food Processing*. Leonard Hill: London, **1969**, 241.
29. Gourdine Systems Inc. Electro Gas Dynamic Method and Apparatus for Detecting the Properties of Particulate Matter Entrained in Gases. Gourdine Systems Inc., Washington, D.C., USA, **1969**; 42.
30. Huang, L. X.; Passos, M. L. Three-dimensional Simulation of Spray Dryer Fitted with a Rotary Atomizer. *Drying Technology*, **2005**, *23*, 1859–1873.
31. Jones, K. Dehydration of Food Combinations. US Patent US2006/0112584-A1, **2006**, 19.
32. Konishi, Y.; Horiuchi, J. I.; Kobayashi, M. Dynamic Evaluation of the Dehydration Response Curves Of Foods Characterized By Poultice-Up Process Using Fish-Paste Sausage, Part I: Determination of the Mechanism for Moisture Transfer. *Drying Technology*, **2001**, *19* (7), 1253–1270.
33. Konishi, Y.; Kobayashi, M. Characteristic Innovation of Food Drying Process Revealed by Physicochemical Analysis of Dehydration Dynamics. *Journal of Food Engineering*, **2003**, *59* (2), 277–283.
34. Krishnamurthy, K.; Khurana, H. K.; Soojin, J.; Irudayaraj, J. Infrared Heating in Food Processing: An Overview. *Comprehensive Reviews in Food Science and Food Safety*, **2008**, *7* (1), 2–13.
35. Magoon, R. E. Method and Apparatus for Drying Fruit Pulp and the Like. US Patent 4,631,837, **1986**, 18.
36. Martynenko, A.; Zheng, W. Electrohydrodynamic Drying of Apple Slices: Energy and Quality Aspects. *Journal of Food Engineering*, **2016**, *168*, 215–222.
37. Matteo, D. P.; Donsi, G., Ferrari, G. The Role of Heat and Mass Transfer Phenomena in Atmospheric Freeze-Drying of Foods in a Fluidized Bed. *Journal of Food Engineering*, **2003**, *59*, 267–275.
38. Mohapatra, D.; Mishra, S. Current Trends in Drying and Dehydration of Foods. *Food Engineering*, **2011**, *2011*, 311–351.

39. Moses, J. A.; Karthickumar, P. Effect of Microwave Treatment on Drying Characteristics and Quality Parameters of Thin-layer Drying of Coconut. *Asian Journal of Food and Agro-Industry*, **2013**, *6* (2), 72–85.

40. Moson, T. J. Large-scale Sonochemical Processing, Aspiration and Actuality. *Ultrasonic Sonochemistry*, **2000**, *7*, 145–149.

41. Mounir, S.; Allaf, T.; Mujumdar, A. S. Drying: Coupling Instant Controlled Pressure Drop DIC to Standard Convection Drying Processes to Intensify Transfer Phenomena and Improve Quality: An Overview. *Drying Technology*, **2012**, *30* (14), 1508–1531.

42. Mounir, S.; Besombes, C.; Al-Bitar, N. Study of Instant Controlled Pressure Drop DIC Treatment in Manufacturing Snack and Expanded Granule Powder of Apple and Onion. *Drying Technology*, **2011**, *29* (3), 331–341.

43. Mujumdar, A. S. Practical Guide to Industrial Drying. Colour Publications: Mumbai, **2004**, 215.

44. Mujumdar, A. S.; Huang, L. X. Global R&D Needs in Drying. *Drying Technology*, **2007**, *25* (4), 647–658.

45. Mulet, A.; Carcel, J.; Benedito, C. Ultrasonic Mass Transfer Enhancement in Food Processing. *Transport Phenomena in Food Processing*, **2003**, *18*, 265–278.

46. Nowak, D.; Lewicki, P. Infrared Drying of Apple Slices. *Innovative Food Science and Emerging Technologies*, **2004**, *5*, 353–360.

47. Ochoa-Martínez, C. I.; Quintero, P. T.; Ayala, A. A.; Ortiz, M. Drying Characteristics of Mango Slices Using Refractance Window Technique. *Journal of Food Engineering*, **2012**, *109* (1), 69–75.

48. Pan, Z.; Khir, R.; Godfrey, L. D. Feasibility of Simultaneous Rough Rice Drying and Disinfestations by Infrared Radiation Heating and Rice Milling Quality. *Journal of Food Engineering*, **2008**, *84* (3), 469–479.

49. Pilatowski, I.; Mounir, S. The Instant Controlled Pressure Drop Process: Post-Harvesting Treatment of Paddy Rice—Impacts on Drying Kinetics and End-product Attributes. *Food and Bioprocess Technology*, **2010**, *3* (6), 901–907.

50. Rahman, S. M. A.; Mujumdar, A. S. Novel Atmospheric Freeze-Drying System Using Vibro-Fluidized Bed with Adsorbent. *Drying Technology*, **2008**, *26* (4), 393–403.

51. Ratti, C. Freeze and Vacuum Drying of Food. *Drying Technologies in Food Processing*, **2008**, *2008*, 225–246.

52. Riadh, M. H.; Ahmad, S. A. B. Infrared Heating in Food Drying: An Overview. *Drying Technology*, **2015**, *33* (3), 322–335.

53. Sakai, N.; Fujii, A. Heat Transfer Analysis in Food Heated by Far-Infrared Radiation. *Nippon Shoku hin Ko Gyo Gakkaishi* (Japanese), **1993**, *40* (7), 469–477.

54. Shewale, S. R.; Hebbar, H. Effect of Infrared Pretreatment on Low-Humidity Air Drying of Apple Slices. *Journal of Drying Technology,* **2016**, *35* (4), 213–216.

55. Wolff, E.; Gibert, H. Atmospheric Freeze-Drying, Part 2: Modeling Drying Kinetics using Adsorption Isotherms. *Drying Technology*, **1990**, 8(2), 405–428.

56. Yagi, S.; Kunii. D. Infrared Drying Characteristics of Granular or Powdery Materials. *Ka gakuki Kai* (Japanese), **1951**, *15* (3) 108–123.

57. Zarein, M.; Samadi, S. H.; Ghobadian, B. Microwave Dryer Effects on Energy Efficiency during Drying of Apple Slices. *Journal of the Saudi Society of Agricultural Sciences*, **2015**, *14* (1), 41–47.

58. Zhu, Y.; Pan, Z.; McHugh, T. H. Processing and Quality Characteristics of Apple Slices Processed under Simultaneous Infrared Dry-Blanching and Dehydration with Intermittent Heating. *Journal of Food Engineering*, **2010**, *97* (1), 8–16.

59. Tunde-Akintunde, T. Y. Effect of Pretreatment on Drying Time and Quality of Chilli Pepper. *Journal of Food Processing and Preservation*, **2010**, *34* (4), 595–608.

CHAPTER 7

PRINCIPLES OF FOAMING AND FOAM MAT DRYING TECHNOLOGY: FRUITS AND VEGETABLES

RITESH B. WATHARKAR

ABSTRACT

In this chapter, foam mat drying (FMD) of fruits and vegetables focuses on the history of foam in the food processing sector, principle and method, formation of foam during whipping, basic concept and structure of foam, foam characteristics (stability, density, expansion), factors affecting the foaming parameter, foaming agents (egg albumin, milk protein, soya protein), economic importance, and commercial applications of FMD.

7.1 INTRODUCTION

According to Singh et al. [41], the total production of fruits and vegetables was around 67.67 and 1135.69 million tons (MT) worldwide, respectively. However, this production was increased to 97.35 and 187.5 MT in 2017–2018 [50]. Due to the perishable nature of fruits and vegetables, there is spoilage of 20%–40% during transportation from the field to the table [41]. Therefore to prevent such undesirable losses, it is necessary to reduce moisture content and water activity as the high moisture leads to microbiological spoilage. To confront these challenges, drying technique is used to remove moisture from the food system to an optimum level, which can reduce unnecessary microbial growth. Drying process in food industries can also maximize shelf-life of perishable foods, which get degraded due to poor handling, packaging, storage, transportation, and hygienic issues [14]. The principle of drying is mainly based on continuous heat and mass transfer operations.

*Corresponding author: E-mail: watharkarritesh@gmail.com

In this chapter, foam mat drying (FMD) has been highlighted and discussed in detail as a suitable technology for the preservation of food items.

7.2   FMD TECHNOLOGY

The FMD is considered a simple technology to dry liquid or semiliquid food materials. Heat-sensitive, high sugar, and viscous foods cannot be dried easily in convectional dryers as such foods become sticky or rubbery on circulating mild heat [26]. Drying of fruit puree or juices has same issues, which can be resolved by using FMD method in processing industry. FMD is also commercially used in milk, fruit concentrate, teas, and coffee industries [5].

In the FMD processing, whipping of liquid or viscous material takes place with the addition of a foaming agent. Furthermore, foam can be spread on the tray and placed under warm air. Formation of porous in final product helps to improve the food quality. Honeycomb structure of foam is responsible for easy removal of water and can result in a light-weight product. The short drying time and low production cost is an additional advantage of FMD [14]. The drying time depends on the amount of air incorporated in liquid or semiliquid food. Foam stability (FST) also determines the effectiveness in the drying of foamed material. The dried product is of better quality in terms of color, bulk density, and rehydration and reconstitution properties.

The increase in air–liquid interface increases the heat and mass transfer. Freeze drying is good for highly viscous and sugar-rich fruit juices. Spray drying is not suitable for highly viscous and sticky liquid products [43]. Therefore, FMD can yield reconstituted cost-effective fruit juice powder, which has a free-flowing behavior and can be used for the preparation of value-added products, such as weaning food, ice cream, snacks, bakery, and confectionery product.

7.3   HISTORY OF FOAM IN FOOD PROCESSING

The first aerated product (Swiss cheese) was developed in 1250. After that, whipped cream came in the early 16th century followed by aerated products, such as ice-cream in the early 17th century and Marshmallow in 1850 in France [46]. The researcher paid attention in the 19th century to execute the foaming concept in different categories of food processing. FMD was first patented by Campbell Food Company in 1917 for drying of milk [31, 46]

In 1950, Eddy mentioned that spray drying was not suitable for the production of free-flowing powder from fruits and vegetable juices. At that time, they suggested the use of methylcellulose for producing foam of grapefruit and orange juice and dry under spray drying method. This suggestion was not appropriate because it required some mechanical operation for the formation of foam. However, the FMD process was found to be suitable for such conditions and was successfully incorporated by the US Western Regional Research Laboratory [26]. Foam is heated by undehumidified or dehumidified air at atmospheric pressure and was cheaper than vacuum-, freeze-, and spray-drying methods.

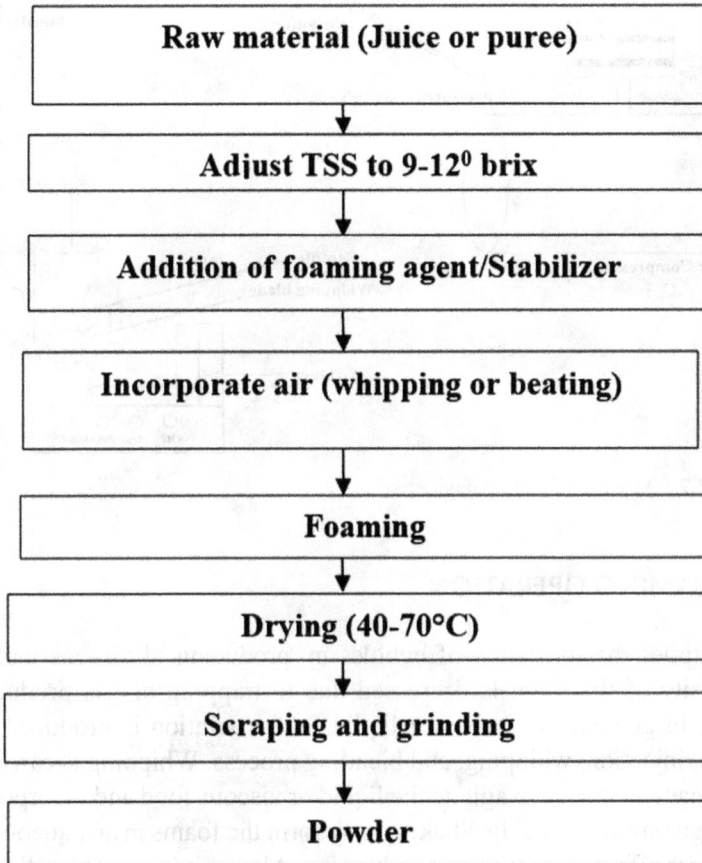

Raw material (Juice or puree)

↓

Adjust TSS to 9-12⁰ brix

↓

Addition of foaming agent/Stabilizer

↓

Incorporate air (whipping or beating)

↓

Foaming

↓

Drying (40-70°C)

↓

Scraping and grinding

↓

Powder

FIGURE 7.1 Steps in FMD method.

7.4   METHOD OF FMD

Figure 7.1 indicates steps that are employed in FMD. The fruit pulp is extracted from fruits like mango, banana, beal, etc., using a pulping machine. A specific level of consistency is necessary to maintain for effective foaming. The consistency (9°–12° brix) can be adjusted using water-based on "Pearson square method" [19]. Then, the addition of 1%–5% foaming agent is necessary for the formation of air–water interface in the pulp.

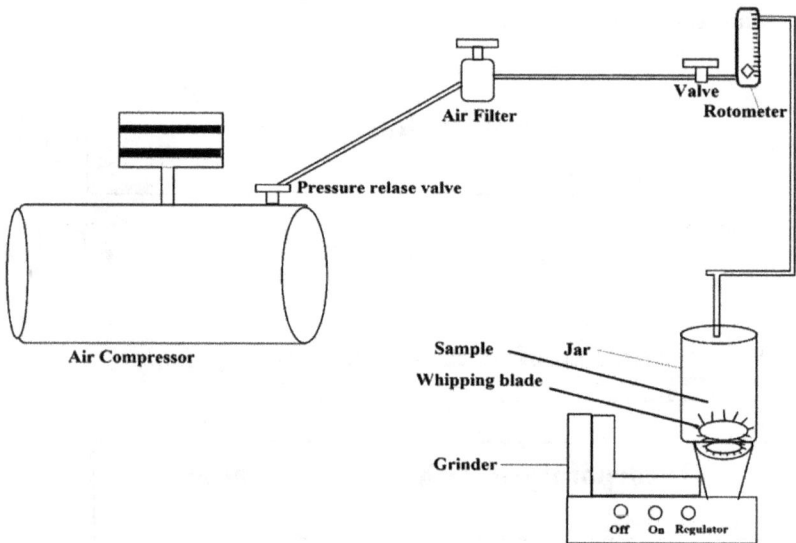

FIGURE 7.2   Modified foaming device.

7.5   FOAMING OPERATION

Foam implies the formation of bubbles by producing of air–gas interface. The density of the foam is decreased due to trapping of gas produced in aeration. In general in viscous or liquid foods, aeration is produced using agitation, injection, whipping, and blending process. Whipping is carried out in a mechanical device to agitate the liquid or viscous food and incorporation of air or gas simultaneously. Shaking may form the foams in an aqueous state but it is less effective compared to whipping. Also, mixing and blending does not produce gas–liquid interface effectively like whipping. Therefore, the foaming device consists of whipping and the aeration operation can be an effective choice for producing foam in viscous or liquid foods (Figure 7.1).

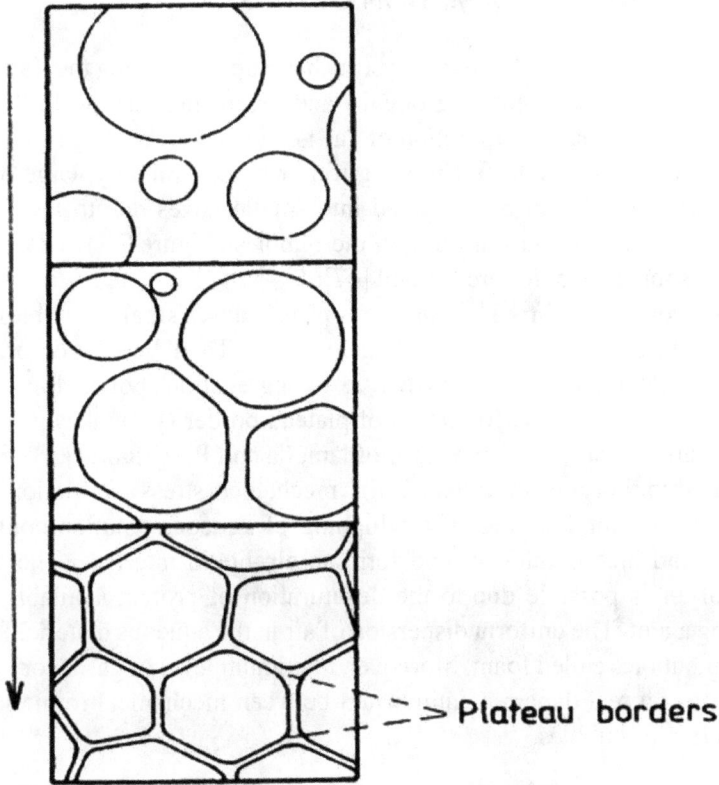

FIGURE 7.3 Formation of foam.

A modified device for FMD was developed by authors of this chapter in the Department of Food Engineering and Technology with slight modification in process reported by Bag et al. [5]. It was a combination of three operations, such as mixing, aeration, and whipping (Figure 7.2). The device consists of four main parts, such as grinder, compressor, rotameter, and air filter. A 1.6 mm thick circular stainless-steel blade in the jar was used for whipping. The compressed air was filtered using an air filter and was passed through rotameter to maintain the flow. The air was passed through the filter to remove unwanted materials and to produce clean air. The airflow rate of 10–50 lph was maintained using rotameter for whipping. The stroboscope (Strobotac, USA) was used to measure the rotational speed of the whipping blade, that is, 3000 rpm.

7.6 PRINCIPLE OF FORMATION OF FOAM

The foam can be formed in two ways, such as supersaturating the liquid with gas (pop drinks & fermenting dough) and mechanical action [47]. In the mechanical method, incorporation of gas is allowed by beating or whipping (bubble generated due to fast-moving rod or blade). Initially, large bubbles are formed, which can be disrupted into smaller sizes due to the velocity gradient formed by sharing force in the bubbles (Figure 7.3). The changes occur as soon as bubbles are formed [47].

Diffusion of gas through continuous phase causes smaller bubbles to get dissolved but increases the size of larger bubbles. Then, bubbles deform each other, and segregation is formed. Due to segregation, polyhedral bubbles are produced, where the formation of plateau border (PB) takes place. The bubbles are ruptured after breakage of lamella and PB (Figure 7.3).

The whipping or beating creates high mechanical stress and friction, which is responsible for foaming. The whipping of aqueous solution containing foaming-aid incorporates air and form an air–liquid interface. The entrapment of air is possible due to the denaturation of protein available in the foaming agent. The uniform dispersion of air in the aqueous material forms a layer of bubbles called foam. Moreover, maximum level of gas incorporation may reflect a real dynamic equilibrium between mechanical formation and destruction of bubbles.

7.7 BASIC CONCEPTS OF FOAMING

7.7.1 THEORY OF FORMATION OF FOAM

Foam is the formation of a gas–liquid dispersion or gas–solid interface during any mechanical action. Foams burst within few seconds and so are very unstable. The basic geometry and symbolic laws of foam equilibrium were first established in 1873 by Belgian physicist Joseph Plateau [45]. During foaming, surfactant makes connections between gas and liquid forming a thick or thin films.

The typical polyhedral arrangement forms PBs among three adjacent bubbles (Figure 7.4A). The noticeable highlighted edges of bubbles are called as PBs. According to plateau's rule to make 3D films within dry limit and equilibrium position, the three films need to be joined at 120° mutual angles, whereas four plateaus should form tetrahedral vertices [45]. The

3D structure of dry foam of low consistency material denotes an almost spherical shape (Figure 7.4B). This kind of wet foam achieves about 64% gas fractions, which are closed in bundles of spherical bubbles [13].

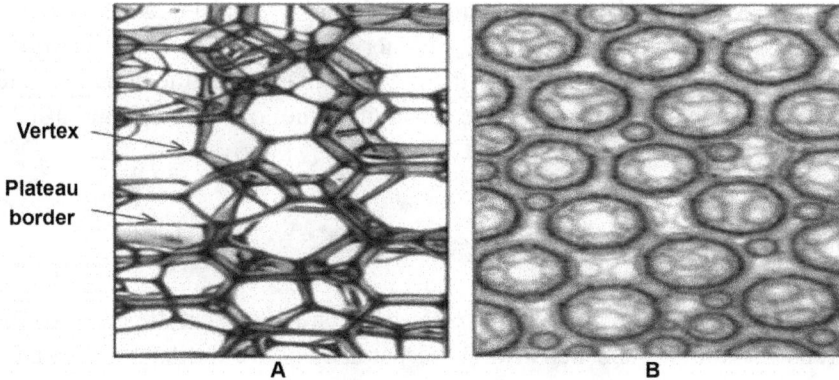

FIGURE 7.4 Structural arrangement of bubbles in foam.

The separation of liquid from the cavity of foam is called drainage [39]. The flows always move through the capillary toward the gravitational force (Figure 7.5a). Due to contact of flowing liquid with capillary, the nodes and plateau network borders can be clearly seen as a wet foam layer. It has been observed that the bubbles formed in the first layer of the interface are spherical but polyhedral at the top, where foam is very dry (Figure 7.5a). Thick film of bubbles can retard the movement of liquid and this kind of foam is considered as stable. The liquid flow, which occurs through interstitial spaces of bubbles, is called foam drainage. The driving force of flow toward gravity is resisted by viscous damping. The volume of bubble can be divided into three particular sections [20]:

1. Films: the one bounded by flat bubble faces and present between two adjacent bubbles.
2. Plateau border: three films of liquid connect together at one point (Figure 7.5c).
3. Node: four PBs connect and form one region, also known as a junction (Figure 7.5c).

The observation of single bubble flow is quite difficult because nodes eventually split and flow the liquid through different channels.

When liquid increases in the node curvature, then interfaces decrease and edges of the polygonal bubbles become more rounded [20]. The channels and nodes are formed due to liquid near the thin film (<100 nm) of foam [20]. The size of nodes is generally ≤1 mm wide. The shapes of bubble are changed with the change of quantity of liquid present in each node. The dryness in the liquid film leads to thinner and brittle film from the robust nature. The brittleness can break the foam easily [20]. In the aqueous state, surfactant plays an important role to stabilize the foam. Surfactant absorbs water at the surface and helps to avoid rapture [7].

7.8   FOAM CHARACTERISTICS

The several important foam characteristics are size, stability, expansion, drainage volume, and density. These parameters are the backbone of the foam in the processing sector. Foam quality refers to the percentage of space occupied by the gas in the whole volume of liquid at a specific pressure and temperature [15].

FST has a direct relationship with the drainage volume. The poor FST allows breakage of the film easily due to external stress and liquid can be drain easily. Schramm et al. [44] reported that the decrease in stability of foam is due to diffusion of gas, which increases with the increment in bubble size.

Foam expansion (FE) determines the incorporation of maximum air and decrease in foam density (FD). The finer-sized foam is necessary for getting spreaded during processing because this kind of foam is able to resist the flow and will minimize the drainage.

The methods for the measurement of FST, FD, and FE are presented in this section.

7.8.1   FOAM STABILITY

Stable foam is very important for drying of fruit pulp. The drying of fruit pulp takes place if the foam is stable for a longer time. The stability of foam is reduced for a thicker foam film. Drainage of liquid is occurred when thin film of bubbles is broken. The temperature, whipping time, whipping speed, and foaming agents are responsible for the stability of foam. FST can be measured using the following equation [36]:

$$\text{Foam stability}\left(\text{FST}\right) = V_0 \times \frac{T_1}{V_1} \tag{7.1}$$

where T_1 is the volume change of foam that occurred within time interval; and V_0 is the volume of the foam at zero time.

7.8.2 FOAM DENSITY

The density and expansion of foam depend on the volume of air incorporated in the fruit pulp. The decrease in density and increase in FE takes place with the increase in the air incorporation. FD can be calculated using the following formula [21]:

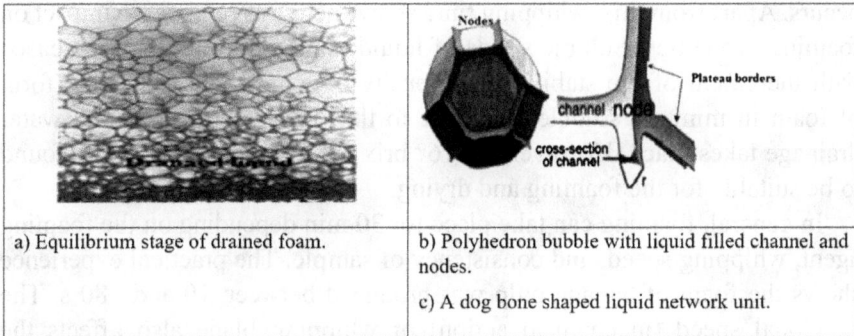

a) Equilibrium stage of drained foam.	b) Polyhedron bubble with liquid filled channel and nodes.
	c) A dog bone shaped liquid network unit.

FIGURE 7.5 Foam drainage concept.

$$\text{Foam density}\left(\text{FD}\right) = \frac{W_1}{W_2} \tag{7.2}$$

where W_1 is the weight of foam (g); and W_2 is the volume of foam (cm^3).

7.8.3 FOAM EXPANSION

FE can be calculated by the following formula [4]:

$$\text{Foam expansion}\left(\text{FE}\right) = \frac{V_1 - V_0}{V_0} \tag{7.3}$$

where V_0 is initial volume (cm^3); and V_1 is the final volume (cm^3).

7.8.4   FACTORS AFFECTING FOAMING PARAMETERS

The stability of foam also depends on the consistency of liquid, foaming agent, and stabilizer. When consistency is too low, quantity of additives (stabilizer) is necessary to increase the absorption of moisture and to have thick film for stability increment. The concentration of foaming agent also affects the foaming properties. For example, several chemical properties of egg white (such as molecular weight, glycosylation, phosphorylation, sulfhydryl /disulfide content) help the protein interaction during the formation of foam [24], thus making the egg white as strong foaming agent. However, additives (such as salts and sugars) may influence the functionality of egg white protein leading to changes in handling properties [35].

The quantity of protein in the foaming agent is also responsible for the expansion of foam. Generally, increment in the quality of protein with the FE occurs. Apart from this, whipping time shows negative or positive impact on foaming properties. Soluble solids of liquid (e.g., fruit juice) are increased with increment of the stability and density of foam. The less stable form of foam in minimal soluble solid due to thin film is collapsed and water drainage takes place. However, 9°–16° brix TSS of banana pulp was found to be suitable for the foaming and drying.

In general, foaming can take place for 30 min depending on the foaming agent, whipping speed, and consistency of sample. The practical experience shows the foam of banana pulp was produced between 10 and 180 s. The rotational speed (mechanical action) of whipping blade also affects the foaming characteristics. The authors of this chapter observed that highest speed of whipping was able to increase FE and decrease the stability. The stable foams do not collapse during spreading in the drying trays.

7.9   GENERAL INFORMATION ON FOAMING AGENTS

There are several foaming agents that are available in the form of surfactants and blowing agents. The proteins and lipids are the main surfactants used as foaming agents [8].

7.9.1   PROTEIN CONTAINING FOAMING AGENT

The protein can produce foam easily after the denaturation has been facilitated by a mechanical action and it helps occupy more air. There are more flexible and open structures with both hydrophobic and hydrophilic groups of protein required for the formation of better quality foam [28]. The film is obtained by polypeptide interaction with partial denaturation of the protein. The sources of food-grade protein are discussed in this section.

7.9.1.1 EGG ALBUMIN

The EA is a protein, which can help to form a stable foam. Due to the amphiphilic property of EA, continuous and dispersed phases are mixed properly. EA contains about 40 proteins; and among these three proteins like ovalbumin, globulins, and ovomucoid are involved in foaming. Ovalbumin is a globular protein that contributes about 54% of the total proteins in egg white and has gelling and emulsifying properties [12]. During foaming, ovalbumin forms a single layer and lysozyme generates a thick film of air–water interface [14]. The egg yolk is responsible for reducing the foaming capacity of EA [17]. The triglycerides contained in the yolk are harmful to foaming of egg white if it is still present during separation. One drop of yolk brings 135–40 mL reduction in volume provided by egg white [22, 42].

7.9.1.2 MILK PROTEIN

Whole milk and skim milk powders are good for foaming in the food industry. Both of these can form a better foam, when whipping is carried out. Milk contains mainly casein and whey proteins, which have the ability to interface with air and generate the foam. Casein generates more foam than whey protein. In the case of milk, β-casein, α-lactalbumin, and β-lactoglobulin are also important proteins for the incorporation of air [3]. β-lactoglobulin is a globular protein mainly involved in foaming because of the ability to entrap more air bubbles [16]. The random coiled β-casein absorbs air and spreads in the interface more rapidly than the globular protein, such as β-lactoglobulin [33].

7.9.1.3 SOYA PROTEIN

There are two types of soya proteins (i.e., soya protein concentrate and soya protein isolate (SPI)) that can be used as foaming agents. Liu [23] reported that SPIs have good functional properties, such as gelation, emulsification, viscosity, water binding, and foaming properties. The purified form of SPI contains 90% protein and low amount of fats. There are two major proteins, that is, β conglycinin and glycinin that are responsible for foaming of SPI. Rickert et al. [32] suggested that glycinin compared to β conglycinin has a better ability to unfold and stabilize the air–water film interface due to its larger size and disulfide linkage between acidic and basic subunits.

7.9.1.4 ADDITIONAL FOAMING AGENTS

There are many nonprotein-based foaming agents used for the production of foam. Methylcellulose is a chemical compound, which can be used as a stabilizer and foaming agent [19]. The surface activity and amphiphilic nature of methylcellulose show the air–water interface positively [29]. Glycerol monostearate is a foaming aid that can be used to create foam during processing. There are some plant proteins (such as solubilized soya protein, gums) and emulsifiers (propylene glycerol monostearate, carboxyl methylcellulose, and trichloro phosphate) that are typically added as foaming agents to juice, pulp, and concentrate. Moreover, modified soya protein and monoglyceride also have better foaming ability.

7.10 FOAM STABILIZERS

The stability of foam is very important for an effective drying process. The destabilized foam cannot dry properly because texture of the final product could be lost.

Soya protein albumin also acts as stabilizer that can be denatured immediately and proper balance can be retained between hydrophobicity and hydrophilicity. Egg white also has ability to stabilize the foam. The egg white containing globular protein helps to make stronger cohesive interactions and flexibility leads to form stable foam [38]. Labelle [25] used methocel as a stabilizer during foaming of orange juice. Methocel at 1.5% concentration revealed good impact on the stability of tomato and apple juice foam [9].

Recently, albumin has been extracted from guar meal, which is found to stabilize the foam up to 3 h [38]. The acid-precipitated protein called guar foaming albumin (GFA) assists to form stable foam. GFA is safe for people suffering from food allergy.

Methylcellulose comprises of same features of anionic polysaccharides, which are responsible for FST. The methylcellulose with other foaming agents has the good stabilizing ability in the star fruit juice [21]. Raharitsifa et al. [34] found that 0.2% methylcellulose with 2%–3% egg white was sufficient for obtaining stable apple juice foam. Anionic polysaccharides can provide viscosity to the aqueous solutions resulting in the formation of stable foam. Xanthan gum is also anionic polysaccharide that forms cohesive flexible films that are necessary for stable foam. Therefore, Xanthan gum is considered as a stabilizer, thickener, and foam enhancer.

Propylene glycol alginate (PGA) is also one of the polysaccharides that can be used as a foam stabilizer. Mathu-Kumaran et al. [27] investigated that 1% concentration of PGA provided minimal drainage (stable foam) of egg white foam. Also, sodium alginate, sodium carboxyl methylcellulose, soluble starch, gum Arabic, pectin, and dextrin are hydrophilic colloids accountable as good foaming stabilizers.

7.11 CLASSIFICATION OF FMD

The drying of foam can be carried out in vacuum, spray, freeze, and air-drying methods. The present scenario of research is not only directed for this type of convective drying but also application in conventional freeze-drying and microwave-assisted FMD [31]. The FMD in vacuum condition was first proposed by Bronstein in 1996. FMD of banana puree cannot take place successfully under vacuum conditions. Due to vacuum, foam was sucked and dispersed in the drying chamber. The bursting of foam also occurred, which is not feasible for drying.

Microwave-assisted FMD (MFMD) is a hybrid drying method that has been studied recently by many researchers. During MFMD, energy is penetrated through the foam, which helps to evaporate the moisture [48]. The volumetric heating and simultaneously enlarged of product surface area can help to remove water vapors easily [48]. The 70% of drying time was reduced in drying of blackcurrant pulp in MFMD.

The foam-assisted freeze-drying is also an existing technology being used in the drying of fruit and vegetable juices. Practically, foam-assisted freeze-drying can be used to reduce drying time [30]. Raharitsifa [30] also observed the reduction of time in the case of apple juice foam dried with this method. The spray drying processes have been developed around the 1870s, whereas the first spray foam drying was employed by Hanrahan and Webb for the drying of cottage cheese whey [10] followed by whole milk [11] and skim milk [2]. There are several advantages of foam spray drying according to Zbicinski et al. [49]: (1) increased dryer output and thermal efficiency; (2) reduced degradation and retention of volatiles in the product. Hanrahan et al. [10] observed that drying of foamed milk resulted in faster drying rate and lighter density of the final product. The foam of fruit puree can also be dried in drum-type dryers. As compared to nonfoamed apple puree, 10% saving of energy was possible during drying of foamed puree in drum type dryer.

7.12 ECONOMIC IMPORTANCE OF FMD

FMD is economically beneficial due to the reduced time of drying. The energy consumption is reduced by reducing drying load. The drying load was reduced in FMD because density was decreased from 0.85 g/cm³ to 0.30–0.60 g/cm³. The increased liquid–air interface of material is important for the reduced drying time. Rajkumar et al. [36] investigated that 22% dryer throughput was obtained in the case of FMD of mango puree. The foam mat dried final product like flakes, powder can be obtained at minimal cost to the consumers. The cost of foamed mango can be lowered by 11% to 10%, after drying on belt conveyer and drum dryer [18]. Also, energy consumption of FMD was lower compared to nonfoamed drying, which caused saving in the cost of process and final dried product. It is reasonable to expect financial benefits for generally regarded FMD by analogy to foam-spray drying [18]. Apart from this, simplicity of FMD shows that the process is economically affordable compared to spray-, freeze-, and vacuum drying techniques. Nonfoamed material cannot be scrapped easily and is subjected to loss of dried material. In the case of FMD, scraping and grinding can take place easily and loss of powder can be avoided.

7.13 COMMERCIAL APPLICATION OF FMD

The application of FMD is generally assessed on a small-scale basis. Small industrialists mostly use FMD for the drying of fruit juice, fruit concentrate, and food slurries. Foam can be prepared continuously in the mixer by the addition of foaming agent and sufficient stabilizer.

The existing technology of FMD on the perforated tray is presently being used for the drying of lemon juice and mango puree, etc., at commercial basis. There are several fruits and vegetable products (such as flakes, powder, and grits) that can be prepared by using high-speed continuous belt drier. FMD is a better alternative to freeze- and spray-drying in terms of saving time and minimal process cost. Foam mat dried product is porous in nature due to instant rehydration ability [14]. Apart from this, low bulk density, retention of nutrition quality, and better color of powder enhance the consumer acceptability.

A continuous type belt vacuum dryer was designed for the drying of milk foam [1, 40]. The milk foam was poured on the stainless-steel belt. Heater was mounted below and upper side of the belt to heat foam and dry it. A belt moved around the cooling drum for the scraping of dried material. Rajkumar et al. [36] developed the process of continuous type foamed mango drying process. They investigated that foamed mango pulp required only 35 min of drying compared with 75 min in conventional drying. There is a scope to develop more effective continuous types of FMD process for high viscous and sugar-rich fruit puree.

7.14 SUMMARY

FMD process is cost-effective and time-saving. It results in a lightweight, nutrient-rich, and good reconstitution property of a product. Compared to freeze-drying and spray drying, FMD is cheap and easy to operate. The convenient foaming agents and stabilizers can improve the foaming characteristics and drying behavior. This chapter also focuses on foam structure, foaming agents, and stabilizers. The combination of FMD with other dryers may be a motivation for more adaptations in the food industry.

ACKNOWLEDGMENT

The author acknowledges the MoFPI, Government of India for sponsoring this project; and Tezpur University to provide resources. This chapter is partially based on PhD Thesis by author of this chapter, "Ritesh B. Watharkar, *Foam-Mat Drying Characteristics of Wild (Musa Balbisiana) and Hybrid (Musa Nana Lour) Banana Pulp in Relation to Process Development.*" For the Department of Food Engineering & Technology, Tezpur University, Sonitpur, Napaam 784028, Assam, India; August 2019; pages 215; Thesis Advisor—Dr. Brijesh Srivastava, Associate Professor.

KEYWORDS

- bubbles
- foam mat drying
- foaming agent
- foaming device
- proteins
- stabilizer

REFERENCES

1. Aceto, N. C.; Schoppet, E. F.; Sinnamon, H. I. Continuous Vacuum Foam Drying of Whole Milk under Simulated Commercial Scale Conditions. *Journal of Dairy Science*, **1972**, *55* (6), 875–879.
2. Bell, R. W.; Hanrahan, F. P.; Webb, B. H. (1963). Foam Spray Drying Methods of Making Readily Dispersible Nonfat Dry Milk. *Journal of Dairy Science*, **1963**, *46* (12), 1352–1356.
3. Brooker, B. E. Observations on the Air-Serum Interface of Milk Foams. *Food Structure*, **1985**, *4* (2), 12–16.
4. Balasubramanian, S.; Paridhi, G.; Bosco, J. D.; Kadam, D. M. Optimization of Process Conditions for the Development of Tomato Foam by Box-Behnken Design. *Food and Nutrition Sciences*, **2012**, *3* (7), 925–930.
5. Bag, S. K.; Srivastav, P. P.; Mishra, H. N. Optimization of Process Parameters for Foaming of Bael (*Aeglemarmelos L.*) Fruit Pulp. *Food and Bioprocess Technology*, **2011**, *4* (8), 1450–1458.
6. Cox, S.; Weaire, D.; Brakke, K. Liquid foams—precursors for solid foams. In: *Cellular Ceramics: Structure, Manufacturing, Properties and Applications*; Michael Scheffler and Paolo Colombo (Eds.), John Wiley & Sons, Inc., New York, NY, **2005**, 18–29.
7. Durand, M.; Langevin, D. Physicochemical Approach to the Theory of Foam Drainage. *The European Physical Journal E*, **2002**, *7* (1), 35–44.
8. Eisner, M. D.; Jeelani, S. A. K.; Bernhard, L.; Windhab, E. Stability of Foams Containing Proteins, Fat Particles and Nonionic Surfactants. *Chemical Engineering Science*, **2007**, *62* (7), 1974–1987.
9. Glicksman, M. Utilization of Synthetic Gums in the Food Industry. *Advances in Food Research*, **1963**, *12*, 283–366.
10. Hanrahan, F. P.; Webb, B. H. USDA Develops Foam-Spray Drying. *Food Engineering*, **1961**, *33* (8), 37–40.
11. Hanrahan, F. P.; Tamsma, A.; Fox, K. K. Production and Properties of Spray-Dried Whole Milk Foam. *Journal of Dairy Science*, **1962**, *45*(1), 27–31.
12. Hagolle, N.; Relkin, P.; Popineau, Y.; Bertrand, D. Study of the Stability of Egg White Protein-Based Foams: Effect of Heating Protein Solution. *Journal of the Science of Food and Agriculture*, **2000**, *80* (8), 1245–1252.

13. Hohler, R.; Cohen-Addad, S. Rheology of Liquid Foam. *Journal of Physics: Condensed Matter*, **2005**, *17* (41), R1041-R1045.

14. Hardy, Z.; Jideani, V. A. Foam-mat Drying Technology: Review. *Critical Reviews in Food Science and Nutrition*, **2017**, *57* (12), 2560–2572.

15. Join, S. P. E. Foams Properties. 2015. <https://petrowiki.org/Foam_properties>; Accessed on November 9, 2019.

16. Kailasapathy, K. Chemical Composition, Physical and Functional Properties of Milk and Milk Ingredients. In: *Dairy Processing & Quality Assurance*; R.C. Chandan (Ed.). John Wiley & Sons Inc.: Ames, IA, **2008**, 75–103.

17. Kim, K.; Setser, C. S. Foaming Properties of Fresh and of Commercially Dried Eggs in the Presence of Stabilizers and Surfactants. *Poultry Science*, **1982**, *61* (11), 2194–2199.

18. Kudra, T.; Ratti, C. Foam-mat Drying: Energy and Cost Analysis. *Canadian Biosystems Engineering*, **2006**, *48*, 3–10.

19. Kandasamy, P.; Varadharaju, N.; Kalemullah, S.; Maladhi, D. Optimization of Process Parameters for Foam-Mat Drying of Papaya Pulp. *Journal of Food Science and Technology*, **2014**, *51* (10), 2526–2534.

20. Koehler, S. A.; Hilgenfeldt, S.; Stone, H. A. Generalized View of Foam Drainage: Experiment and Theory. *Langmuir*, **2000**, *16* (15), 6327–6341.

21. Karim, A. A.; Wai, C. C. Characteristics of Foam Prepared from Starfruit (*Averrhoa carambola* L.) Puree by using Methyl Cellulose. *Food Hydrocolloids*, **1999**, *13* (3), 203–210.

22. Lomakina, K.; Mikova, K. (2006). Study of the Factors Affecting the Foaming Properties of Egg White: Review. *Czech Journal of Food Sciences*, **2006**, *24* (3), 110–118.

23. Liu, K. *Soybeans as Functional Foods and Ingredients*. 1st Edition; AOCS (The American Oil Chemists' Society) Publishing: Urbana, IL, **2004**, 331.

24. Li-Chan, E. Biochemical Basis for the Properties of Egg White. *CRC Crit. Rev. Poultry Biol.*, **1989**, *2*, 21–58.

25. Labelle, R. L. Characterization of Foams for FMD. *Food Technology*, **1966**, *20* (8), 1065–1070.

26. Morgan, A. I.; Ginnette, L. F.; Graham, R. P.; Williams, G. S. Recent Developments in Foam-Mat Drying. *Food Technology*, **1961**, *15* (1), 37–42.

27. Muthu-Kumaran, A.; Ratti, C.; Raghavan, V. G. (2008). Foam-mat Freeze Drying of Egg White—Mathematical Modeling, Part II: Freeze Drying and Modeling. *Drying Technology*, **2008**, *26* (4), 513–518.

28. Mleko, S.; Kristinsson, H. G.; Liang, Y.; Gustaw, W. Rheological Properties of Foams Generated from Egg Albumin after pH Treatment. *Food Science and Technology*, **2007**, *40*, 908- 909.

29. Nasatto, P. L.; Pignon, F.; Silveira, J. L. Methylcellulose: A Cellulose Derivative with Original Physical Properties and Extended Applications. *Polymers*, **2015**, *7* (5), 777–803.

30. Raharitsifa, N.; Ratti, C. (2010). Foam-mat Freeze-Drying of Apple Juice, Part 1: Experimental Data and ANN Simulations. *Journal of Food Process Engineering*, **2010**, *33* (s1), 268–283.

31. Ratti, C.; Kudra, T. Drying of Foamed Materials: Opportunities and Challenges. In: *Proceedings 11th Polish Drying Symposium*; **2005**, 1101–1108. https://doi.org/10.1080/07373930600778213; Accessed on November 9, 2019.

32. Rickert, D. A.; Johnson, L. A.; Murphy, P. A. Functional Properties of Improved Glycinin and β-nglycinin Fractions. *Journal of Food Science*, **2004**, *69* (4), FCT303 - FCT311.

33. Ridout, M. J.; Mackie, A. R.; Wilde, P. J. Rheology of Mixed β-casein/ β-Lactoglobulin Films at the Air–Water Interface. *Journal of Agricultural and Food Chemistry*, **2004**, *52* (12), 3930–3937.

34. Raharitsifa, N.; Genovese, D. B.; Ratti, C. Characterization of Apple Juice Foams for Foam-Mat Drying Prepared with Egg White Protein and Methylcellulose. *Journal of Food Science*, **2006**, *71* (3), E142-E151.

35. Raikos, V.; Campbell, L. Effects of Sucrose and Sodium Chloride on Foaming Properties of Egg White Proteins. *Food Research International*, **2007**, *40* (3), 347–355.

36. Rajkumar, P.; Kailappan, R.; Viswanathan, R.; Raghavan, G. S. V. FMD of Alfonso Mango Pulp. *Drying Technology*, **2007**, *25* (2), 357–365.

37. Shimoyama, A.; Kido, S.; Kinekawa; Y. I. Guar Foaming Albumin: Low Molecular Mass Protein with High Foaming Activity and Foam Stability Isolated from Guar Meal. *Journal of Agricultural and Food Chemistry*, **2008**, *56* (19), 9200–9205.

38. Shimoyama, A.; Doi, Y. Guar Foaming Albumin—Foam Stabilizer. In: *Food Additive*. Yehia El-Samragy (Ed.); **2012**, 115–118. https://www.intechopen.com/books/food-additive/guar-foaming-albumin-a-foam-stabilizer; Accessed on November 9, 2019.

39. Saint-Jalmes, A. Physical Chemistry in Foam Drainage and Coarsening. *Soft Matter*, **2006**, *2* (10), 836–849.

40. Schoppet, E. F.; Aceto, N. C.; Eskew, R. K.; Craig, J. C. Continuous Vacuum Drying of Whole Milk Foam, II: Modified Procedure. *Journal of Dairy Science*, **1965**, *48* (11), 1436–1440.

41. Singh, V.; Hedayetullah, M.; Zaman, P. Postharvest Technology of Fruits and Vegetables: An Overview. *Journal of Postharvest Technology*, **2014**, *2* (2), 124–135.

42. St. John, J. L.; Flor, I. H. Study of Whipping and Coagulation of Eggs of Varying Quality. *Poultry Science*, **1930**, *10* (2), 71–82.

43. Sangamithra, A.; Venkatachalam, S.; John, S. G. FMD of Food Materials: A Review. *Journal of Food Processing and Preservation*, 2015, *6* (39), 3165–3174.

44. Schramm, L. L.; Wassmuth, F. Foams: Basic Principles. *Advances in Chemistry Series*, **1994**, *242*, 3–10.

45. Weaire, D. L.; Hutzler, S. (Eds.). *The Physics of Foams*. Oxford University Press: London, **2001**, 264.

46. Webb, C.; Pandiella, S. S.; Niranjan, K. (Eds.). *Bubbles in Food*. Eagan Press: St Paul, MN, **1999**, 384.

47. Walstra, P. Principles of Foam Formation and Stability. Chapter 1; In: *Foams: Physics, Chemistry and Structure*. Springer Link: London, **1989**, 1–15.

48. Zheng, X. Z.; Liu, C. H. Optimization of Parameters for Microwave-Assisted FMD of Blackcurrant Pulp. *Drying Technology*, 2011, *29* (2), 230–238.

49. Zbicinski, I.; Rabaeva, J.; Lewandowski, A. Drying of Foamed Materials. *Modern Drying Technology*, **2014**, *2014*, 163–190.

50. https://www.businesstoday.in/top-story/horticulture-crops-production-at-31467-million-tonnes-in-2018–19-higher-than-past-5-yrs-average/story/315259.html; Accessed on November 9, 2019.

CHAPTER 8

MECHANIZATION OF MANUFACTURING PROCESSES FOR TRADITIONAL INDIAN DAIRY PRODUCTS

GAJANAN P. DESHMUKH*, REKHA R. MENON, and NAVEEN JOSE

ABSTRACT

Good volume of the fluid milk is converted to value-added traditional Indian dairy products, which help the dairy industry in product diversification and profit augmentation. Organized manufacture of these products entails mechanization and controlled production for uniform quality and safety of a product. Several efforts have been reported for the mechanical production and process upgradation of different classes of traditional dairy products, which have been reviewed in this chapter.

8.1 INTRODUCTION

Traditional Indian dairy products (TIDPs) include heat desiccated milk solids (*khoa*) and sweets prepared from *khoa* as a base material, acid coagulated milk solids (*chhana*), and sweets (*shrikhand, paneer, payasam, kheer,* etc.) prepared from *chhana* as a base material.

Recently, the TIDPs are becoming more popular throughout India and all over the world due to the presence of a large diaspora from India. Traditional dairy products have a large consumer base due to socioeconomic, cultural, nutritional, and religious activities. Each product in this category has its own unique color, texture, flavor, and appearance and is developed over an extended period of time with the culinary skills of *halwais* and homemakers [54].

*Corresponding author. E-mail: gajanannnn@gmail.com.

The traditional milk products in India are conventionally prepared on a domestic level or on a cottage industry scale in either iron/stainless steel (SS) *karahi* or in the double jacketed kettle supplied with steam. The small to medium scale cottage industries, engaged in the preparation of such TIPDs products, inherently suffer from numerous drawbacks, such as variation in the product quality, limited sanitation, and poor energy inefficiency, wherein the opportunities to optimize and control the heat treatment operation are very restricted [78]. All these inherent disadvantages of the traditional processes can be overcome with an innovative design of equipment with batch, semicontinuous, and continuous mode for manufacturing of these value-added milk products on the domestic and commercial level.

There is a paradigm shift in the structure of the dairy processing industry, which is witnessing a gradual movement to the organized dairy sector, and with this process upgradation of indigenous product technology, and design of product manufacturing equipment have also received the needed impetus. Small-scale manufacturing techniques for the production of TIPDs cannot be exploited for industrial-scale production without mechanization [58]. The mechanization of the process not only helps in efficient use of energy, water, and other engineering services but also facilitates hygienic production of quality TIDPs, with longer shelf-life [7].

Recently, several engineering interventions have been made by dairy engineers and scientists in the development of the innovative design of manufacturing equipment for these value-added TIDPs with hygienic processing and less labor-intensive production [64]. Another approach has been to adopt novel process equipment, which is already well-established in allied process industry, and re-engineering the technology for TIDPs.

This chapter presents a summary of various efforts to mechanize the unit operations involved in the processing of TIDPs.

8.2 DEVELOPMENT OF MANUFACTURING EQUIPMENT FOR TIDPS

Traditional Indian milk products have different nomenclature in various regions of India due to the variation in ingredients and manufacturing methods. In general, TIDPs are broadly categorized into five major classes of milk products, such as heat desiccated milk products, heat and acid coagulated milk products, Fermented/cultured dairy products, fat-rich dairy products, and milk-based puddings/desserts.

8.2.1 HEAT DESICCATED/CONCENTRATED MILK PRODUCTS

These products are prepared by heat desiccation of milk to intermediate moisture content. The total moisture level in the product gets reduced due to the evaporation of moisture from the product during the desiccation process. *Khoa* and its sweets, rabri, *basundi,* etc., are examples of heat desiccated milk products.

8.2.1.1 KHOA (HEAT DESICCATED DEHYDRATED MILK)

Khoa refers to heat desiccated, partially dehydrated whole milk product that is utilized as a raw material during the manufacturing of various traditional milk sweets, such as *kalakand, burfi, gulabjamun, peda,* etc. The traditional method, commonly employed by the unorganized sector (*halwais*) for the production of *khoa* on small scale, has been described by several authors [23, 24, 65]. The traditional process involves the concentration of milk in mild steel/iron shallow open pans on a direct fire with vigorous stirring and mixing using a steel or wooden ladle. Some manufacturers use a kerosene/diesel-fired stove (*chula*) for the small-scale production of *khoa*. This traditional method has numerous disadvantages, such as its unsuitability for commercial production, variation in batch-to-batch product quality, low heat-transfer coefficients (U-value) of equipment, unhygienic product conditions, limited shelf-life of the product, more strain on the operator, etc.

To overcome these drawbacks, process modifications through mechanized batch type *khoa* making process and new machines have been developed and reviewed by several researchers [3, 19, 52]. Few examples of these machines are batch type, continuous mode units, and multipurpose types. A timeline on the developments in mechanized production of *khoa* is tabulated in Table 8.1.

8.2.1.2 BURFI

Burfi is a popular indigenous *khoa*-based sweet, manufactured by admixing of *khoa* and sugar in different proportions along with fruit pulps, dry fruits, cereals, etc. Traditional process of *burfi* manufacturing involves vigorous blending of *khoa* with sugar and other ingredients in an open shallow pan till a homogeneous, fine, and smooth grain mass has formed, and then it is subjected to cooling and setting in a shallow tray [23].

TABLE 8.1　Developments in Mechanized Production of *Khoa*

Mechanized Method	Reference
1968: Designed and developed a *khoa* manufacturing machine (50 L of milk per hour) based on the concept of scraped surface heat exchanger (SSHE) for semicontinuous operation. It was composed of a steam jacked cylinder attached with rotary scraper, with provision for finish concentration in a semiopen steam jacketed cascade pans with reciprocating spring-loaded scraper	[9]
1970: Modified and upgraded the model developed by Banerjee et al. [9] and standardized the method for manufacturing of *khoa* using buffalo milk (BM)	[25]
1971: Developed a batch type, semicommercial model for manufacturing of khoa comprising of a steam jacketed mild steel kettle with built-in-stirrer	[63]
1975: Improved the design of khoa kettle by replacing mild steel with stainless steel (SS)	[65]
1980: Developed an improved khoa pan to overcome village level problems faced in the traditional method of manufacturing of khoa. The unit was composed of a steel pan (hemispherical shape) fused to a cylindrical shape water jacket. The water content inside the jacketed space was transformed into steam by employing the pan over the furnace	[70]
1986: Developed a khoa making pan with a facility of mechanized scraping of khoa mass in the pan during the heat desiccation process. The jacked column of pan was filled with water, in which steam is produced by placing it over the furnace. The mechanized scraper blade assembly comprising of central shaft with an anchor type arrangement was fitted with the different swinging Teflon blades	[69]
1987: Developed a semimechanized batch model for small and medium scale khoa production units that used the principle of SSHE. The unit consisted of a stationary jacketed drum attached with scraper assembly having steam inlet, steam trap, condensate outlet, pressure measure device, and vapor exhaust	[52]
1987: Developed a mechanical unit based on conical process vat (SS 316) having a cone angle of 60° for batch type production of khoa. For improving the thermal efficiency of unit, steam jacket is divided into 4-segments. The scraper assembly comprises of 3-equidistance arms supported on the main central shaft and each arm is carrying three different independent designs of spring loaded Teflon blades	[3]
1988, 1990, 1992: Extensively worked on mechanization of *khoa* based on SSHE model and developed three-stage unit consisting of a three jacketed cylinder placed in cascade agreement	[18, 19, 20]
1990: Developed a continuous model of *khoa* manufacturing machine working on the principle of thin-film scrapped surface heat exchanger	[32]
1990: Developed an Inclined Scraped Surface Heat Exchanger for continuous manufacturing of *khoa* at National Dairy Development Board, Anand.	[62]

TABLE 8.1 *(Continued)*

Mechanized Method	Reference
1992: Developed the continuous *khoa* making machine comprising of two SSHE connected in a cascade arrangement. The rotor in the first stage of SSHE is equipped with 4 variable clearance blades. While the second stage of SSHE is equipped with 2 variable clearance blades and 2 helical blades	[31]
1998: Developed and evaluated the performance of three stages continuous khoa manufacturing unit working on the principle of SSHE	[10]
2007: Developed new rotor blade assembly in SSHE for manufacturing of *khoa* by providing scraping blades with a leaf-type spring arrangement so that it can be operated with variable scraping angle	[43]
2009: Modified the third stage SSHE for continuous production of *khoa* developed by Bhadania (1998) with improved scraping blades	[44]
2012: Mechanized the process for production of danedar *khoa* in 3-stage SSHE	[30]
2013: Developed a contherm-convap SSHE system for manufacture of *khoa*. It was reported that *khoa* obtained with this process was more sticky with fine texture	[58]

The process parameters of the third-stage SSHE of *khoa* making machine during the continuous production of *burfi* has been optimized and the performance of the machine was tested using BM (15% total solids) and BM concentrated to 32% and 36% total solids [46]. The scraper speed used during the study was 200 rpm and 150 rpm in the first and second stages, respectively, whereas the third stage was operated at three variable scraper speeds of 100, 150, and 200 rpm with 2, 4, and 6 blades for best quality *burfi*. The optimum operating parameters were 150 rpm of scrapper speed and four blades in the third stage that resulted in the best quality of *burfi*.

Palit and Pal [56] developed the process for commercial manufacturing of *burfi*, which involved manufacturing of *khoa* using a thin film SSHE, followed by separate blending of *khoa* with sugar in a Stephan Processing Unit. The *burfi* produced by this unit was slightly chewy with a sticky body compared with the traditionally prepared product.

A unit for continuous manufacturing of *burfi* was designed using a rotor blade assembly comprising of a combination of flat blades and skewed blades [50]. During the optimization, the effect of skewed angles (20°, 30°, and 40°) and different rotary speeds (100, 140, and 180 rpm) on the product accumulation and quality were considered. However, the discharge of *burfi* mass was not satisfactory and the prepared *burfi* exhibited pastiness with a lack of characteristic flavor.

The conical process vat (CPV) of *khoa* making machine was also tested for the manufacturing of *burfi* [42]. The process involved the mixing and blending of *khoa* obtained from thin-film SSHE along with sugar and other ingredients in a CPV with continuous evaporation of moisture using scraper assembly.

Another approach in the mechanization of *burfi* manufacture has been through the exclusive processing in a thin film SSHE [53]. Single-pass operation was used for preheating with scraper speed of 110 rpm and rapid cooling with acidity level of milk 0.185% LA to produce high-quality *burfi*. Decrease in scraper speed was observed to increase the acidity of milk with a concurrent improvement in body and texture of *burfi*.

Inline production of *burfi* by integrating SSHE has been reported by integrating a CPV with a mechanized cooling system [82]. Continuous production of *burfi* was possible at a steam pressure of 1.5 kg/cm^2 (after sugar addition) and rotor speed of 8 rpm when the concentration of milk during the stage of sugar addition stage was set to 50%.

Continuous manufacturing of *burfi* using a three-stage SSHE with three types of sugar/sugar dosing methods (adding crystalline sugar into milk, adding caramelized sugar into milk, and adding crystalline sugar in an inlet of the third stage) were also compared [14]. The process combination of scraper speed during the first, second, and third stage set to 200, 175, and 15 rpm, respectively, along with the addition of crystalline sugar during the third stage was recommended. The mechanized preparation of *burfi* in a multipurpose SSHE at a steam pressure of 2.5 kg/cm^2 and scraper speed of 30 rpm for a batch size of 30 kg product gave a good quality product [15].

8.2.1.3   PEDA

Peda is a popular traditional *khoa*-based sweet that is prepared by blending *khoa* with a predetermined quantity of sugar. *Peda* is made on a small scale by *halwais* using *khoa* as raw material with the addition of sugar and flavoring agents. However, little efforts have been made for the development of continuous mechanized system for the production of *peda* from *khoa*. The industrial mechanized method for continuous production of *kesar Peda* was developed at National Dairy Development Board, Anand. The manufacturing process involved reprocessing of *khoa* by heating up to 60 °C with incorporation of sugar, flavor, and other ingredients in a planetary mixer. The *khoa*-sugar dough thus obtained was cooled and subsequently subjected to ball forming and shaping unit, and then was succeeded by packing and storing at cool conditions [53].

Amit and Kohli [4] developed a mechanized system for the production of *peda* using SSHE. It was observed that the best quality *peda* was obtained at a scraper speed of 125 rpm, 2 kg/cm^2 steam pressure, addition of sugar @30% on *khoa* basis, and feed rate of 25 kg/h. The Rajkot District Cooperative Milk Producers Union Ltd, Rajkot, Gujarat, has fabricated special equipment for the production of *peda* [53].

Singh [75] optimized the process parameters for the mechanized production of *khoa–peda* in an experimental setup that was designed on the principle of piston press technique. The *khoa peda* mass was compressed in the vertical cylindrical barrel using a piston and passed through a tapered die under pressure. A cutting mechanism was attached at the exit of the die.

8.2.1.4   RABRI

Rabri is a partially sweetened concentrated dairy product comprising layers of clotted cream. A commercially viable method standardized for the production of *rabri* consists of a concentration of BM in SSHE, with the addition of shredded *paneer* as an alternative to clotted cream to offer desirable texture to the end-product [34]. Large-scale production of *rabri* using a TSSHE has also been developed [56].

Saroj [68] studied continuous preparation of *rabri* in a three-stage thin-film SSHE at 127, 121, and 15 rpm in the first, second, and third stage of SSHE, respectively, with a flow rate of milk at 151 kg/h. In-line production of good quality *rabri* using a SSHE and CPV with initial concentration of milk to 29.9% TS and the final concentration of milk to 39% TS in CPV before the addition of sugar was also evaluated [17].

8.2.1.5   GULABJAMUN

The mechanical semicontinuous unit for the commercial manufacture of *gulabjamun* from *khoa* was practiced at Sugam Dairy, Baroda [8]. The *khoa* was continuously manufactured by convep–contherm process and then it was subjected to mechanical dough and ball forming unit, followed by frying of *gulajamun* balls and subsequently packaging in tin cans. Sub-baric thermal processor was developed for frying and soaking *gulabjamun* balls [46]. The unit consisted of soaking and frying chamber operated under a vacuum. In a research study, vacuum impregnation significantly reduced the soaking time in *gulabjamun* processing [71].

8.2.1.6   BASUNDI

A power-driven batch type SSHE for the concentration of milk for production of *basundi* has been attempted [80]. Tellabati [79] developed and evaluated three pilot models (such as conical type, *Karahi* type, and cylindrical type unit working on the principle of SSHE) for the manufacturing of *basundi*. The heat transfer behavior and energy consumption were estimated during optimization of process conditions for *basundi* making in the developed models.

The manufacture of *basundi* with standardized BM was also attempted using CPV. The product was found to be good in term of its body, color and appearance, texture, and overall acceptability [29, 66]

Patel [60] developed a continuous mechanized system for manufacturing of *basundi* using the principle of thin-film SSHE. The developed equipment was energy efficient, and under optimum processing conditions it gave processing cost 50% less than the conventional method. This machine resulted in the product of uniform quality and better sensory attributes under hygienic conditions and higher profit-margin compared to the traditional method.

Singh [74] reported the feasibility of *basundi* making in a three-stage SSHE. The best quality of *basundi* was prepared by keeping the scraper speed of both the SSHE at 2.461 m/s and flow rate of 165 kg/h and by using caramelized sugar syrup solution to give the product a characterized caramel color and flavor. Product of required consistency was obtained in two-stages, and therefore the third stage of SSHE was not used.

8.2.1.7   HALWA

An attempt to mechanize the manufacture of *halwa* was reported in a feasibility study for the preparation of bottle gourd *halwa* in a batch type SSHE, which was originally designed for the manufacture of *khoa*. The good quality product was obtained with better rheological properties compared to the traditional method [81].

8.2.1.8 KALAKAND

Deshmukh [27] reported the continuous production of *kalakand* in three-stage thin-film SSHE. It was found that *kalakand* with the best quality attributes was obtained using standardized milk with 0.19% LA acidity. The optimized parameters during the production of *kalakand* were scraper speed of 175, 125, and 15 rpm in the first, second, and third stage of SSHE, with milk flow rate of 195 kg/h.

8.2.1.9 KAJUKATLI

The process technology for mechanized production of *kajukatli* was developed with appropriate rolling, sheeting, and cutting system mechanisms. A mechanical unit comprised of moving bed, rollers, and feed plate for desired sheeting, cutting, and collection of the product after the processing. The unit had provision for desired speed control and characteristic diamond shape cutting mechanism [21].

8.2.2 HEAT AND ACID COAGULATED MILK PRODUCTS

These milk products are prepared by the addition of acidulants (citric acid, malic acid, lactic acid, vinegar, aged whey) to the heated milk. *Paneer* and *Channa* are examples of heat and acid coagulated milk products. *Channa* is a base material for the manufacturing of several types of sweets, such as *rossagolla, sandesh, pantao,* etc.

8.2.2.1 PANEER

The production of *paneer* at a pilot plant level was first standardized by Bhattacharya et al. [13], where BM was heated to 85 °C for 5 min and coagulated at 70 °C by adding 1% hot citric acid. A prototype model for the continuous curdling of milk was developed by Gupta et al. [35]. Process upgradation of *paneer* making using the ultrafiltration process was reported by Khan et al. [41] with an improvement in the yield by 25%. A batch mode microprocessor controlled automated *paneer* press working on pneumatic compression was designed and tested by Chitranayak et al. [16]. Suryawanshi and Ravindra [77] integrated a flooded type hot water jacket to the pressing unit of the *paneer* press to control the matting temperature of *paneer* while pressing, and they observed an improvement in the texture and acceptability of the product.

Continuous systems developed for the preparation of *paneer* included the use of a bucket centrifuge with a custom-designed cage-wall supported by compaction unit for pressing [12] and a double-flanged apron conveyor-cum-dewatering unit designed at NDRI that has a production capacity of 80 kg/h [41]

Sherawat [73] developed the coaxial cylindrical jacketed unit for multi-tasking operation, namely, heating, coagulation, and pressing of coagulum mass, etc., for the manufacturing of *Paneer*. The unit has the provision of thermostatically controlled electric heater (2 kW) to heat the heating medium in the jacketed space to 98–99 °C, which maintains the temperature of milk to 90 °C before the coagulation.

8.2.2.2   CHHANA AND CHHANA-BASED SWEETS

To overcome the difficulties associated with small-scale production of *channa*, mechanization has been attempted for continuous and quality production of *chhana* and its products. Aneja [5] reported among the earlier prototypes continuous unit with a capacity for 40 kg/h of *chhana*. It consisted of a tubular heat exchanger, acid injector, holder, and strainer unit. A lab-scale continuous *chhana* making (milk coagulating) machine was designed and fabricated by Singh [76] with a milk handling capacity of 60 lph.

Rasogolla is a popular succulent sweet prepared by the cooking of kneaded and rolled *channa* balls in a sugar syrup. The mechanization for its preparation was achieved by integrating the *chhana* kneader with *chhana* ball former in a continuous mode [40]. Extended shelf-life of *Rasagollas* has been attempted by applying pressure during the cooking process [12] or postprocessing thermal treatment, such as canning [37]. A pilot-scale unit for *sandesh* manufacturing based on the principle of SSHE was designed with a rotator type scraper assembly with four spring-loaded scraper blades, and the cylinder fitted with a half-round jacket, whereas the other half was left as space for removal of vapors [57].

Another popular *channa*-based sweet that is prepared by kneading and cooking *Channa* with a suitable sweetener (sugar or jaggery) is popularly called as *Sandesh*, and it was mechanized by using a custom-designed vented single screw extruder [48]. Afterward, this design was improved with a vented single-screw extruder integrated with the mechanized unit for cooking of *chhana* and sugar mixture for the continuous production of *sandesh* from cow milk [47].

Mechanization of the process for cooking of *chhana* balls in a SS cooker was evaluated by Jain [38]. Cooking time for the unit for satisfactory product quality was 15 min at an auger speed of 1.0 rpm. When the steam pressure in the cooker exceeded 1.2 kg/cm^2, the *channa* balls were observed to disintegrate.

8.2.3 CULTURED/FERMENTED DAIRY PRODUCTS

Cultured and fermented dairy products are an important category of TIDPs due to the presence of live organisms in the product, which will support gut microflora. Large scale mechanization in this product category has been commercially viable in the preparation and packaging of *shrikhand*, which is a creamy sweetened dairy product that is prepared using dewatered curd in western India,

Aneja [5] developed a semimechanized system for the production of *shrikhand* on a commercial scale. Skim milk (9% SNF) was heated to 85 °C for 116 s, cooled to 30 °C, before inoculation with a starter culture at 0.25%–0.50%. The mixture was incubated for 8 h for the setting of the curd. Several reports have standardized various technological parameters and processes related to the preparation of this product, such as homogenization, rate of sugar addition, moisture and fat content and in-pack thermization, and product characterization for chemical, and microbiological yardsticks [26, 61, 81].

Sugam dairy in Varodadara—Gujarat is a commercial dairy plant that is famous to prepare and pack different flavors of *shrikhand* on large scale by mechanical methods. The process-line employs a quarg separator, SSHE for thermization, and a continuous packaging machine [5]. The use of ultra-filtered milk as a raw material for curd setting has been attempted successfully to reduce the process time [72]. Mechanical continuous thermization of *shrikhand* using an SSHE has been described in detail by Dhotre [28]. The unit is made of a jacketed product tube (thermization and mixing zone) fitted internally with a specially designed scraper assembly (consisted of variable clearance spring-loaded scraper blades) along with a feeder assembly, variable frequency drive (VFD) for regulating scraper speed, and sensors (temperature and pressure) for monitoring the operating parameters during operation of the machine.

8.2.4   MILK-BASED PUDDING OR DESSERT

In order to overcome the disadvantages of the traditional method of *kheer* preparation (namely, time and labor intensive, and poor hygiene [39]), batch-to-batch variation in quality was able to optimize the process parameters for continuous manufacturing of *kheer*. Based on the sensory quality attributes, 0.27 MPa of operating pressure and 7.5 min cooking time was selected for the design of pressurized cooking chamber during the production of *kheer* with a continuous unit.

A batch-type multipurpose SSHE for manufacturing *kheer* with the SSHE comprised of a jacketed tube integrated with a feed hopper, Teflon edged spring-loaded scraper assembly, vapor hood with exhaust fan, VFD, and measurement devices [38].

8.3   SCRAPER DESIGN FOR MANUFACTURING OF DAIRY PRODUCTS

Since almost all TIDPs are conventionally prepared using process technology involving heating along with continuous stirring and scraping, the use of a mechanical scraper assembly built-in with a process vat for batch process or in an SSHE for continuous operation is a widely adopted process upgradation approach. The choice of the scraper design is dependent on the specific processing requirements (heating and cooling, mixing and bending, concentration, drying, and crystallization) for a particular product. Therefore, several scraper assemblies have been designed, developed, and tested for the manufacturing of numerous dairy products as summarized below:

1. Among the early reports in this direction, a reciprocating spring-loaded scraper design [10] was designed with two open-semijacketed cascade pans for the continuous manufacture of *Khoa*.
2. Sawhney et al. [70] designed a scraper assembly that comprised of a central shaft fitted with an anchor type mechanism with multiple swinging.
3. Teflon blades have been tested for continuous scraping of the product during the manufacturing of *khoa* in a mechanized *khoa* pan.

4. More [52] indicated a design of a scraper assembly unit incorporated within the steam jacketed drum of a *khoa*-making machine. The assembly was designed as a central shaft attached with Teflon-SS spring-loaded blades rotated using a geared electrical motor and a belt and pulley drive.

5. Agrawala et al. [3] developed a scraper assembly for a CPV for *khoa* manufacture. The scraper is comprised of three symmetrically placed equidistance arms (each having independent spring-loaded blades) supported through the main central shaft.

6. Gupta et al. [35] reported a multipurpose mixing and blending scraper assembly within a CPV suitable for the manufacturing of viscous dairy products. The scraper assembly was made of three different designs of standard agitator blades, namely, anchor type, turbine type, and helical type attached to the central main driving shaft for satisfactory performance.

7. A scraper assembly was attached to a pilot model SSHE [57] for the production of *sandesh*, consisting of a solid S.S. shaft on which four spring-loaded scraper blades were arranged alternately. The spring compression allowed uniform pressure on the blade to scrape out the deposited material from the effective heat transfer area. Teflon strips were fitted to the edge of the SS scraper blades to cushion the scraping surface of the SSHE.

8. Dhotre [28] designed the scraper assembly for mixing and blending of *chakka* and sugar during the processing of *shrikhand*. The scraper assembly was comprised of two butterfly wing-shaped structures joined to each other at four points through four SS bars. Helical ribbon-type blades were provided between the SS bars to facilitate the forward movement of the product. Six spring-loaded Teflon blades were also provided over the helical ribbon blades to wipe the product during the rotation to allow more effective scraping action.

9. Jain [38] developed a scraper assembly for a SSHE consisting of a central solid SS shaft with four custom-designed Teflon-SS scraper blades for the manufacturing of various indigenous dairy products. The spring-loaded Teflon-edged scraper blades were arranged in such a way that the whole surface was efficiently scraped during operation of the SSHE.

10. Mayur [51] designed a scraper assembly with vertical circular agitator with adjustable Teflon scraper blades for manufacturing of *halwa* in a batch type machine. The scraper assembly was comprised

of four Teflon scraper blades fixed on SS strips supported by arms and connected to the main shaft with an agitator stand.

8.4 ENERGY REQUIREMENT OF DAIRY EQUIPMENT DURING MANUFACTURING TIDPS

Primarily, two forms of energy have been used by equipment during the manufacture of dairy products, that is, thermal and electrical (mechanical) energy.

The requirement of thermal energy by a system mainly depends on its thermal efficiency, thermal properties of the material being handled, and heat losses during the processing. Steam is the major source of thermal energy in industrial units, and the consumption is determined as the amount of condensate and is expressed in terms of kg of steam required/per kg of water vapor being evaporated.

The mechanical energy to drive the scraper assembly depends on the rheological characteristics, composition of the product, and speed of rotation of scraper. The mechanical energy required by the system is provided using electricity and is estimated using an energy meter and is expressed in terms of kW/h [28].

In recent times, there have reports of solar energy being harnessed as an auxiliary source for cooling milk in storage units and for running refrigeration plants [33], the application of such an alternate source of energy has not yet been explored in processing equipment for TIDPs.

8.4.1 THERMAL ENERGY

Patel [59] studied the energy consumption in fabricated SSHE models for *khoa* manufacturing and concluded that the specific steam consumption was independent of the batch size. The average steam consumption for the entire process was 1.23 kg/kg of water evaporated and the relationship was linear between the steam consumption and operating steam pressure (from 1.0–3.0 kg/cm^2).

A continuous *ghee*-making equipment from butter comprised of a continuous butter melting unit followed by a horizontal SSHE for the final evaporation of moisture [2]. It was reported that the developed unit was thermally more efficient, consuming only 35 kg of steam/h in comparison with 68 kg steam/h for a jacketed *ghee* kettle.

Bhadania [10] developed a 3-stage SSHE for continuous manufacturing of *khoa* and reported that the total steam consumption during the production of the *khoa* was 65.97 kg/h; the specific steam consumption value varied from 1.446 to 1.618, 1.275 to 1.360, and 1.278 to 1.380 in the first, second, and third stage, respectively, under different operating parameters.

The steam consumption during the manufacturing of *sandesh* [57] in a mechanical unit varied from 1.212 to 6.753 kg of steam/h with a specific steam consumption between 1.692 and 2.707 kg of steam/kg of water evaporation under different operating parameters.

The specific steam consumption ranged from 1.30 to 1.40, 1.35 to 1.42, and 1.40 to 1.45 kg of steam per kg of water evaporated for the first, second, and third stage of SSHE, respectively, during the production of *basundi* [60].

In the second-stage, thin-film SSHE for production of *burfi*, a reduced steam consumption of 4.5 kg/kg of *burfi* was observed in comparison with 4.8 kg/kg of *burfi* for the traditional method [56]. A similar report on the reduction in steam consumption by the mechanical method was reported by Abhitosh [1] in a continuous *khoa* making machine, for which 4.15 kg steam per kg of *khoa* was observed compared with 5.05 kg steam per kg of *khoa* in the conventional method.

A study on the energy requirements in an SSHE for *khoa* manufacture under different operating conditions concluded that the third-stage SSHE was better suitable with an average steam consumption of 1.28–1.62 kg/kg of water [11].

Jain [38] reported that the specific steam consumption during the production of *kheer* in a batch type SSHE ranged from 1.568 to 1.702 kg steam/kg of water evaporated under different operating parameters. During the production of bottle gourd *halwa* in the same batch type SSHE, the specific steam consumption varied between 1.603 and 1.680 kg steam/kg of water evaporated.

8.4.2   ELECTRIC ENERGY

The requirement of electrical energy during processing in dairy equipment is mainly dependent on the product and process characteristics. The major parts of power consumption in SSHEs are mechanical components, such as the shear stress created by the liquid on the blade and fluid pumping, friction in the bearings, the scraping action of the blades, and the rotation of the mass of the fluid within the cylinder and blades [67]. The power required to rotate the shaft and blades is dependent on the design and configuration of the blades [36, 79].

More [52] evaluated the electrical energy consumption in a batch type *khoa* making machine and the consumption was 0.093 kWh electricity per kg of milk handled. In a study on power consumption in SSHE during evaporation of milk [2], the power was consumed mainly during the three principal operations: accelerating the product to the scraper speed, overcoming the surface tension, and viscous forces to form a film on the heat transfer surface and agitating the product film. They found the following correlation to relate power consumption with process parameters:

$$P_o = 1746.6\ We - 0.74\ Bf \qquad\qquad (8.1)$$

where Po = power number; We = Weber number; and Bf = Blade factor.

Patel et al. [57] reported the electrical energy consumption ranging from 320 to 495 W to drive the scraper assembly of a *sandesh* manufacturing unit. During the manufacture of *khoa* in continuous SSHE, the electrical energy required to drive the scraper assemblies ranged from 292.68 to 324.48 W, which was about 1% of the steam energy required in evaporation [11]. Dhotre [28] reported that the electrical energy requirement for thermization in a 20 kg batch of *shrikhand* varied from 64.4 to 127.8 W under different operating parameters.

Jain [38] computed the electrical power consumption in a batch type SSHE during the production of *kheer* and bottle gourd *halwa* and reported values from 230.1 to 260.5 W for *kheer* and 240.1 to 263.6 W for *halwa*, respectively. They observed that the electrical power consumption was lower than the thermal energy requirement of the machine.

8.5 SUMMARY

The large-scale industrial mechanized production of TIPDs has been identified as one of the key priority areas by the Government of India and policymakers. In view of this context, significant efforts have been made by dairy engineers, scientists, and industrial sector in developing novel processing equipment with energy-efficient designs for manufacturing of TIDPs.

KEYWORDS

- dairy industry
- *Khoa*
- mechanization
- SSHE
- traditional Indian dairy products

REFERENCES

1. Abhitosh, K. Performance Evaluation of Continuous *Khoa* Making Machine. M.Sc. Thesis, Bundelkhand University, Jhansi, **2005**, 115.
2. Abichandani, H.; Dodeja, A. K.; Sarma, S. C. Unique System for Continuous Manufacture of *Ghee* and *Khoa*. *Indian Dairyman*, **1997**, *48* (10), 11–12.
3. Agrawala, S. P.; Sawhney, I. K.; Kumar, B. Mechanized Conical Process Vat for Production of *Khoa*. Unpublished Paper Presentation at the National Seminar on Recent Advances in Dairy Processing; Karnal: NDRI, **1987**, 6.
4. Amit, R. K. Continuous Production of *Peda* by Using Scraped Surface Heat Exchanger. M. Tech. Thesis; National Dairy Research Institute (Deemed University), Karnal, **2001**, 214.
5. Aneja, R. P. *Dairy India*. 5th ed. New Delhi: Baba Barkhanath Printers, **1997**, 387–392.
6. Anonymous. *Milk Production in India*; National Dairy Development Board, **2018**, 22. https://www.nddb.coop/information/stats/milkprodindia; Accessed April 21, 2019.
7. Bandyopadhyay, P.; Khamrui, K. Technological Advancement on Traditional Indian Desiccated and Heat-Acid Coagulated Dairy Products. *Bulletin: International Dairy Federation*, **2007**, *415* (2), 4–8.
8. Banerjee, A. K. Processes for Commercial Production. In: *Dairy India*. 5th ed. R. P. Aneja (Ed.), New Delhi: Baba Barkhanath Printers, **1997**, 387–392.
9. Banerjee, A. K.; Verma, I. S.; Bagchi, B. Pilot Plant for Continuous Manufacture of *khoa*. *Indian Dairyman*, **1968**, *20* (1), 81–84.
10. Bhadania, A. G. Development and Performance Evaluation of Continuous *Khoa* Making Machine. M.Sc. Thesis, Anand Agricultural University (AAU), Anand, **1998**, 223.
11. Bhadania, A. G.; Shah, B. P. Energy Requirement of Scraped Surface Heat Exchanger (SSHE) During Manufacture of *Khoa*. *Indian Food Industry*, **2005**, *86*, 13–17.
12. Bhattacharya, D. C; Raj, D. Studies on the Production of *Rasogolla*: Traditional Method. *Indian Journal of Dairy Science*, **1980**, *33* (2), 237–243.
13. Bhattacharya, D. C.; Mathur, O. N.; Srinivasan, M. R.; Samlik, O. Studies on the Method of Production and Shelf-life of *Paneer* (Cooking Type of Acid Coagulated Cottage Cheese). *Journal of Food Science and Technology*, **1971**, *8*, 117–120.
14. Chauhan, I. A. Performance Evaluation of Three Stage Scraped Surface Heat Exchanger for Continuous Manufacture of *Burfi*. M. Tech. Thesis, National Dairy Research Institute (Deemed University), Karnal, **2009**, 236.

15. Chauhan, I. A.; Bhadania, A. G. Mechanization and Optimization of Parameters for the Preparation of *Burfi* in Multipurpose Scraped Surface Heat Exchanger. *International Journal*, **2017**, *5* (3), 385–387.

16. Chitranayak, M.; Rekha, M. R.; Magdaline, F. E. E. Physicochemical Characterization of *Paneer* Assessed by Varying Pressure-Time Combination. *Indian Journal of Dairy Science*, **2017,** *70* (3), 280–293.

17. Chopde, S; Bikram Kumar; Minz, P. S; Pravin, S. Feasibility Study for Mechanized Production of *Rabri*. *Asian Journal of Dairy and Food Research,* **2013**, *32* (1), 30–34

18. Christie, I. S.; Shah, C. M. Feasibility studies on Prototype of *Khoa* Making Machine. *Beverage and Food World*, **1988**, *3*, 15–16.

19. Christie, I. S; Shah, U. S. Development of *Khoa* Making Machine. *Indian Dairyman*, **1990**, *42*, 249–252.

20. Christie, I. S.; Shah, C. M. Development of a Three-Stage Continuous *Khoa* Making Machine. *Indian Dairyman*, **1992**, *44*, 1–4.

21. Darji, M. M. Development and Performance Evaluation of Continuous Rolling, Sheeting and Cutting System for Manufacture of *Kajukatli*. M. Tech. Thesis, SMC College of Dairy Science, Anand Agricultural University, Anand, **2016**, 169.

22. Das, S.; Das, H. Performance of an Impact Type Device for Continuous Production of *Paneer*. *Journal of Food Engineering*, **2009,** *95* (4), 579–587.

23. De, S. Indian Dairy Products. In: *Outlines of Dairy Technology*. 1st ed. New Delhi: Oxford University Press, **1980**, 380–389.

24. De, S.; Ray, S. C. Studies on the Indigenous Method of *Khoa*-Making. *Indian Journal of Dairy Science*, **1951**, 5, 147–165.

25. De, S.; Singh, B. P. Continuous Production of *Khoa*. *Indian Dairyman*, **1970**, *2* (12), 294–298.

26. Desai, H. K.; Gupta, S. K. Sensory Evaluation of *Shrikhand*. *Dairy Guide*, **1986**, *8* (12), 33–38.

27. Deshmukh, P. V. Mechanized Manufacture of *Kalakand* using Three Stage Thin Film SSHE. M. Tech. Thesis, National Dairy Research Institute (Deemed University), Karnal, **2012**, 125.

28. Dhotre, A. V. Development and Performance Evaluation of Scraped Surface Heat Exchanger for Continuous Thermization of *Shrikhand*. M. Tech. Thesis, Anand Agricultural University, Anand, **2006**, 245.

29. Dodeja, A. K.; Agrawala, S. P. Mechanization for Large Scale Production of Indigenous Milk Products—A Review. *Indian Journal of Animal Science*, **2005**, *75* (9), 1118–1125.

30. Dodeja, A. K.; Deep, A. Mechanized Manufacture of Danedar *Khoa* Using Three Stage SSHE. *Indian Journal of Dairy Science*, **2012**, *65* (4), 671–675.

31. Dodeja, A. K.; Abichandani, H.; Sharma, S. C.; Pal, D. Continuous *Khoa* Making System Design, Operation and Performance. *Indian Journal of Dairy Science*, **1992**, *45*, 671–674.

32. Dodeja, A. K.; Abichandani, H.; Sharma, S. C.; Pal, D.; Verma, R. D. Performance of Thin Film Scraped Surface Heat Exchanger for Continuous Manufacture of *Khoa*. *Indian Journal of Dairy Science*, **1990**, *43* (4), 625–627.

33. Foster, R.; Jensen, B.; Dugdill, B.; Knight, B. Solar Milk Cooling: Smallholder Dairy Farmer Experience in Kenya. *South Korea: Proceedings of the Solar World Congress*, **2015**, 180–188.

34. Gayen, D.; Pal, D. Studies on the Manufacture and Storage of *Rabri*. *Indian Journal Dairy Science*, **1991**, *44* (1), 84–88

35. Gupta, S. K.; Patel, A. A.; Sawhney, I. K. Development of Equipment's for Indigenous Dairy Products. *Indian Dairyman*, **1987**, *39* (9), 419–425.

36. Harrod, M. Scraped Surface Heat Exchangers—A Review. *Journal of Food Process Engineering*, **1986**, *9*, 1–62.

37. Jagtiani, J. K.; Iyengar, J. R.; Kapur, N. S. Studies on the Preparation and Preservation of Rasagollas. *Food Science*, **1960**, *9* (2), 46–47.

38. Jain, S. Development of Multipurpose Scraped Surface Heat Exchanger for Mechanization of Selected Indigenous Dairy Products. M. Tech. Thesis, SMC College of Anand Agricultural University, Anand, **2010**, 215.

39. Kadam, S.; Tushar, G.; Dattaa, A. K. Optimization of Process Parameters for Continuous *Kheer*-Making Machine. *LWT—Food Science and Technology*, **2013**, *51* (1), 94–103.

40. Karunanithy, C.; Varadharaju, N.; Kailappan, R. Studies on Development of Kneader and Ball Former for *Chhana* in *Rasogolla* Production, Part-II: Development of *Chhana* Ball Former and its Evaluation. *Journal of Food Engineering*, **2007**, *80* (3), 961–965.

41. Khan, S. U.; Pal, M. A. Paneer Production: A Review. *Journal of Food Science and Technology*, **2011**, *48* (6), 645–60.

42. Khojare, A.; Kumar, B. Process Engineering Studies to Upgrade Burfi Production. *Journal of Food Science and Technology*, **2003**, *40* (3), 277–279.

43. Kumar, A. R. Performance Evaluation of Third Stage Component of Scraped Surface Heat Exchanger for *Khoa* Production. M. Tech. Thesis, National Dairy Research Institute (Deemed University), Karnal, **2007**, 129.

44. Kumar, A. R.; Agrawala, S. P.; Dodeja, A. K.; Pal, D. Effect of Modifications in the Three Stage Scraped Surface Heat Exchanger for Continuous *Khoa* Production. *Indian Journal of Dairy Science*, **2009**, *62* (3), 175–181.

45. Kumar, G. Design and Development of Microcontroller-Based Sub-Baric Thermal Processor for Manufacture of Fried And Soaked Dairy Products. Ph.D. Thesis, National Dairy Research Institute (Deemed University), Karnal, **2016**, 265.

46. Kumar, K. B. Application of Scraped Surface Heat Exchanger for Manufacture of *Burfi*. M. Tech. Thesis, National Dairy Research Institute, Karnal, **1999**, 147.

47. Kumar, R. R.; Das, H. Performance Evaluation of Single Screw Vented Extruder for Production of Sandesh. *Journal of Food Science and Technology*, **2007**, *44* (1), 100–105.

48. Kumar, R. R.; Das, H. Optimization of Processing Parameters for the Mechanized Production of *Sandesh*. *Journal of Food Science and Technology*, **2003**, *40* (2) 187–193.

49. Kumar, R. R.; Das, H. Performance Evaluation of Single Screw Vented Extruder for Production of Sandesh. *Journal of Food Science and Technology*, **2007**, *44* (1), 100- 105.

50. Kunju, C. S.; Dodeja, A. K. Studies on the Manufacture of Burfi using Continuous *Burfi* Making System. *Indian Journal of Dairy Science*, **2004**, *57* (3), 167–170.

51. Mayur, V. R. Development and Performance Evaluation of Batch Type *Halwasan* Making Machine. M. Tech Thesis, Anand Agricultural University, Anand, **2011**, 127.

52. More, G. Development of Semi-Mechanized *Khoa* Making Equipment. *Indian Journal of Dairy Science*, **1987**, *40* (4), 246–248.

53. Pal, D. Technological Advances in the Manufacture of Heat Desiccated Traditional Indian Milk Products—An Overview. *Indian Dairyman*, **2000**, *52* (10), 27–35.

54. Pal, D.; Raju, P. N. Traditional Indian Dairy Products with Functional Attributes: Status and Scope. *Indian Food Industry*, **2010**, *29* (1), 13–21.

55. Pal, D.; Verma, B. B.; Dodeja, A. K.; Mann, B.; Garg, F. C. Upgradation of the Technology for Manufacture of Rabri. Annual Report, NDRI, Karnal, 2011, 99.

56. Palit, C.; Pal, D. Studies in Mechanized Production and Shelf-Life Extension of *Burfi.* *Indian Journal Dairy Science*, **2005**, *58* (1), 12–16.
57. Patel, J. S.; Bhadania, A. G.; Boghra, V. R. Chemical Composition and Sensory Evaluation of *Sandesh* Manufactured in Scraped Surface Heat Exchanger (SSHE). *Journal Dairying, Foods and Home Science*, **2005**, *25*, 28–33.
58. Patel, S.; Bhadania, A. G. Mechanized Production of Traditional Indian Dairy Products: Present Status, Opportunities and Challenges. In: *National Seminar on Indian Dairy Industry—Opportunities and Challenges*, Karnal, **2016**, 214–222.
59. Patel, S. M. Study on Heat Transfer Performance of SSHE during *Khoa* Making. M. Tech. Thesis, Anand Agricultural University, Anand, **1990**, 239.
60. Patel, S. M. Design and Development of Continuous *Basundi* Making Machine. Ph.D. Thesis, Anand Agricultural University, Anand, **2006**, 229.
61. Prajapati, J. P. Study on the Influence of Thermization of *Shrikhand* on its Quality and Shelf-life. MSc Thesis, Gujarat Agricultural University, Anand, **1989**, 176.
62. Punjrath, J. S.; Veeranjamlyala, B. Inclined Scraped Surface Heat Exchanger for Continuous *Khoa* Making. *Indian Journal Dairy Science*, **1990**, *43* (2), 225–230.
63. Rajorhia, G. S. Studies on the Yield and Chemical Quality of *Khoa*. *Indian Journal Animal Research*, **1971**, *5* (1), 25–28.
64. Rajorhia, G. S. Modernization of Traditional Indian Dairy Products and Improving Their Quality. *Indian Dairyman*, **2000**, *52* (10), 9–15.
65. Rajorhia, G. S.; Srinivasan, M. R. Technology of *Khoa*—A Review. *Indian Journal of Dairy Science*, **1979**, *32* (3) 209–216.
66. Ranjeet, K. Studies on the Manufacture of *Basundi* using Conical Process Vat. M. Tech. Thesis, National Dairy Research Institute (Deemed University), Karnal, **2003**, 128.
67. Rao, C. S.; Richard, W. H. Scraped Surface Heat Exchangers. *Critical Reviews in Food Science and Nutrition*, **2006**, *46* (3), 207–219.
68. Saroj, K. Process Mechanization of Rabri Making using Three Stage Thin Film Scraped Surface Heat Exchanger. M. Tech. Thesis, National Dairy Research Institute (Deemed University), Karnal, **2010**, 128.
69. Sawhney, I. K., Kumar, B. Adoption and Popularization of Mechanized *Khoa* Pan. Annual Report, National Dairy Research Institute, Karnal, **1986**, 57.
70. Sawhney, I. K.; Sarma, S. C.; Kumar, B. Development of Village Level *Khoa*-Pan. *Journal of Institution Engineers, India*, **1980**, *61*, 613–618.
71. Sharanabasava, M. R. R. Changes in Moisture and Fat Content of *Gulabjamun* Balls During Sub-Baric Frying and Vacuum Impregnation. *International Journal of Chemical Studies*, **2018**, *6*, 108–1111.
72. Sharma, D. K.; Reuter, H. Ultrafiltration Technique for *Shrikhand* Manufacture. *Indian Journal of Dairy Science*, **1998**, *45* (4), 209–213.
73. Sherawat, K. Design and Development of *Paneer* Making for Domestic Application. M. Tech. Thesis, National Dairy Research Institute, Karnal, **2008**, 125.
74. Singh, A. K. Feasibility Studies on Manufacture of *Basundi* using Three Stage Scraped Surface Heat Exchanger. M. Tech. Thesis, National Dairy Research Institute (Deemed University), Karnal, **2008**, 211.
75. Singh, G. Optimization of Process Parameters for Mechanized Formation of *Khoa-Peda*. M. Tech. Thesis, National Dairy Research Institute (Deemed University), Karnal **2014**, 213.

76. Singh, M. D. Studies on Continuous Acid Coagulation of Buffalo Milk. M. Tech. Thesis, National Dairy Research Institute (Deemed University), Karnal, **1994**, 89.

77. Suryawanshi, A. A.; Ravindra, M. R. Control of Matting Temperature during Pressing Of *Paneer* and Its Effect on Paneer Quality. *Journal of Food Science and Technology*, **2019**, *56* (4), 1715–1722.

78. Talwar, G.; Brar, S. K. Review on Mechanization of Traditional Dairy Products. *Indian Journal of Dairy Science*, **2017**, *70* (1), 1–10.

79. Tellabati, R. Development of Pilot Model for *Basundi* Making and its Performance Evaluation. M. Tech. Thesis, Anand Agricultural University, Anand, **2001**, 211.

80. Upadhyay, J. B.; Bhadania, A. G. Manufacture of *Khoa* Based Sweets and Other Food Products on Scraped Surface Heat Exchanger (SSHE)—An Encouraging Experience. *Indian Dairyman*, **1993**, *45* (6), 224–227.

81. Upadhyay, K. G. Chemical and Microbiological Quality of Shrikhand. M.Sc. Thesis, Gujarat Agricultural, University, S. K. Nagar, Anand, **1974**, 149.

82. Vekariya, Y. V. Studies on Mechanized In-line Production of Burfi. M. Tech. Thesis, National Dairy Research Institute (Deemed University), Karnal, **2011**, 132.

CHAPTER 9

ADVANCES IN FOOD FERMENTATION

BARINDERJEET S. TOOR, ANKITA KATARIA*, AMARJEET KAUR, and
SAVITA SHARMA

ABSTRACT

Food preservation is of prime importance in the food industry where
food fermentation has been playing an important role since ancient times.
Primarily, it was used for the stabilization of perishable agricultural produce
only but eventually, it has been successfully developed as a tool for developing a variety of food products with desired properties. Fermented products
are significant in the human diet which can be attributed to the improved
bioavailability of the nutrients, enhanced digestibility, and desirable organoleptic properties. At the industry level, it is very challenging to control the
fermentation process to achieve the optimum characteristics. Moreover,
longer fermentation times and formation of toxic compounds are the major
concerns. Thus, the scientific community has been striving to overcome these
challenges. Therefore, fermentation coupled with other nonconventional
techniques such as high hydrostatic pressure, pulsed electric field, ultraviolet
radiation, ultrasound, and membrane processing can significantly improve
the overall effectiveness and efficiency.

9.1 INTRODUCTION

Food industry focuses on the preservation of food while maintaining the
nutritional and sensory attributes of the food product. Many processing techniques have been utilized to serve the purpose. In this context, fermentation
has been playing a defined role since ancient times. The technique involves
the metabolic activities of bacteria, yeast, and fungi to bring desirable
changes in the food, for instance, improved bioavailability of the nutrients,
better organoleptic properties, and enhanced digestion.

*Corresponding author. E-mail: ankitakataria92@gmail.com.

The most common species involved in the food fermentations include *Lactobacillus, Acetobacter,* and *Saccharomyces.* Commonly produced products through the utilization of *Lactobacillus* species include sauerkraut, kimchi, curd, and yogurt. *Saccharomyces* is utilized in wine, beer, and bakery products. *Acetobacter* along with *Saccharomyces* use grapes, wine, or apple to produce vinegar or apple cider. Probiotic products are receiving attention gradually due to the health benefits associated with their consumption. *Lactobacillus* and *Bifidobacterium* species are the most common probiotics apart from others like *Pediococcus, Lactococcus,* and *Enterococcus* [13]. Probiotics impart therapeutic and prophylactic properties to the food products as they improve the intestinal microbiota thus benefitting the human health. But, certain parameters such as pH, starter incompatibility, redox potential, and packaging affect the growth of these bacteria [24]. Thus, it is necessary to comprehend their growth stages to maintain their viability during the processing and storage stages for achieving the benefits.

Mostly, fermented products are based on the natural or spontaneous fermentations. Although, this is an inexpensive method but the resulting products lack in uniform quality. Thus, there is a need for advancements, which can be served with the addition of novel starter cultures as well as the utilization of novel processing techniques. Novel starter culture ensures the maximum utilization of substrates at a higher rate resulting in a product with uniform quality and desired sensorial attributes. Nonthermal technologies, the preferred alternatives to the conventional ones, include ultrasonic processing, pulsed electric field (PEF), high-pressure homogenization (HPH), irradiation, and high-pressure processing (HPP), which can be used singly or in combination with other techniques. Ultrasound waves (frequency > 16 kHz) involve the phenomenon of cavitation in which occurs the simultaneous bubble formation and collapse resulting in localized high pressures and temperatures thus generating extreme turbulence and shear energy waves [6, 7]. This can facilitate the inactivation of microorganisms and enzymes at moderate temperatures consequently providing safety and stability to the products [13]. PEF is the process of applying short (very few seconds) but high voltage (15–50 kV/cm) pulses to a commodity present between two electrodes. This technique is based on electroporation, which causes the destruction of undesirable microorganisms, as well as enhances the nutritional quality [1]. High-pressure homogenization includes the depressurization of strongly compressed fluids resulting in high turbulence, intense shearing forces and cavitation, leading to cell rupture, and leakage of intracellular substances. In the irradiation technique, gamma irradiations are made to collide with food products resulting in the production of reactive species, such as free radicals and ions, which contribute to the various biochemical process [60]. High-pressure processing involves the application of high pressure (100–800 MPa) to maintain the microbial quality in the food products.

This chapter focuses on the types of fermentations with major emphasis on the recent advancements in the fermentation technique in different categories of food—grains, dairy, meat, fruits, vegetables, and cocoa—to enhance the properties of the raw materials and the process by application of novel cultures, substrates, and nonthermal technologies.

9.1.1 HISTORY

Since, Neolithic age (nearly 10,000 BC) the perishability of raw material has been tackled using fermentation as a preservation technique [13]. During domestication of animals and plants, fermentation is considered to have begun in India with coagulation of milk to prepare *dahi* during 6000–4000 BC and cheese in Iraq in 7000 BC. Alcoholic and dough fermentation originated during 4000–2000 BC. During the same period, the importance of fermented beverages in terms of nutrition and health started gaining recognition. In the following time, meat sausages, soybean curd, and fermented vegetables were developed. Fermentation of cereals and legumes, establishment of a whisky distillery, and sauerkraut and yoghurt production were carried out till 1500 AD [120]. The principle of fermentation evolved gradually from 1665 when microorganisms were identified by Antonie van Leeuwenhoek and Robert Hooke. This was followed by the designing of "spontaneous generation theory" by Louis Pasteur in 1859 AD. In 1877, Sir John Lister emphasized that milk fermentation was highly carried out by "*Bacterium*" *lactis*, presently known as *Lactococcus lactis* [13]. This industrialization age led to the production of fermented foods commercially using the recognized microorganism, such as lactic acid bacteria (LAB) for meat, vegetable, and dairy products, and yeast for spirit, beer, and wine manufacturing. Defined strains have now replaced the traditional mixed cultures for the preparation of fermented products [120]. Thus, the consistency and quality of the products, which are highly dependent on the performance of the starter, have gradually improved. Since, the proliferation of cultures can be affected by certain factors, such as bacteriophage and pathogens, current investigations are dedicated toward the advanced techniques (novel culture, novel substrate, and novel processing methods) for gradual improvement in the efficiency of fermentation processes.

9.1.2 PRINCIPLE

Fermentation is the technique where the growth and metabolic processes are utilized to preserve and transform food commodities and thus derive energy from the biomolecules for the proliferation of microorganisms [13]. The mechanism includes the activity of enzymes, originated from added starter culture, on the native biomolecules to produce metabolites that can be utilized as substrate for the growth and metabolic activities of other starter cultures. The resulting metabolites contribute to texture, pH, acidity, color, and bioactive composition. Thus impacting the nutritional and structural properties of fermented products. Moreover, these metabolites inhibit the spoilage and pathogenic microflora, consequently enhancing the shelf stability of the perishable commodities [142].

9.2 TYPES OF FERMENTATION

Broadly, industrial fermentation can be divided into two categories, that is, solid-state fermentation and submerged state fermentation. In a solid state, the fermentation is carried out on a solid medium in the absence of free water or very little water. This type of fermentation is applicable in the commercial production of tempeh, and miso from soybean, bread making, mushroom cultivation, cocoa processing, as well as in the industrial production of certain enzymes, such as proteases and amylases by *Aspergillus oryzae*. In contrast, submerged type fermentation includes the growth of microorganisms on the substrate dissolved in a large volume of water [61]. Examples of this type of fermentation include the commercial production of beer, vinegar, cider, wine, soy sauce, kefir, and yoghurt. Table 9.1 states the major comparisons between the two types of fermentations.

Solid-state and submerged-state fermentation can be further divided into aerobic fermentation which occurs in the presence of oxygen and anaerobic fermentation that must be carried out in the absence of oxygen. Aerobic fermentation generally occurs at the starting of the fermentation process for a short period of time and is found to be more intense than other types. Oxygen concentration is a major challenge in aerobic fermentation as oxygen has poor solubility in water; therefore, oxygen transfer rates are kept high to maintain an adequate level of oxygen in the system [10]. Tempeh production by *Rhizopus* species, soy sauce production through koji fermentations, and citric acid production by *A. niger* are a few examples of aerobic fermentation. Examples of anaerobic fermentations include the production of ethanol by *Saccharomyces cerevisiae*, production of sausages, and salami through the activity of *Pediococcus, Micrococcus,* and *Lactobacillus* species and yoghurt production.

TABLE 9.1   Comparison of Solid and Submerged State Fermentation (Adapted from Refs. [61, 90])

Parameter	Solid-State Fermentation	Submerged State Fermentation
Size of bioreactor	Small	Large
Parametric control	Not rigorous apart from oxygen supply, moisture, and heat removal	Highly rigorous
Agitation	May or may not be required	Necessary
Substrate	Carbon, nitrogen, energy, and minerals are provided by only one water-insoluble substrate	Various water-soluble substrate
Concentration of media	Very high	Very low
Inoculum ratio	High	Low
Culture medium flowability	Not free-flowing	Always free-flowing
Phases in the culture system	Three—solid, liquid, gas	Two—liquid, gas
Depth of medium	Usually shallow except for certain fermenters	Variable
Availability of free water	Just adequate to maintain optimum culture viability	Abundant
Nutrient diffusion	Less thereby limiting the culture growth	Optimum due to vigorous mixing
Nutrient distribution	Nutrient concentration gradients are generally observed	Uniform throughout the process
Asepsis of culture system	Not aseptic	Always aseptic
Fungal growth	Deep penetration of the hyphae into the substrate	Growth of mycelial cells as single mycelium or pellets

TABLE 9.1   *(Continued)*

Parameter	Solid-State Fermentation	Submerged State Fermentation
Yeast and bacterial growth	Particle adherence	Uniform distribution
Risk of contamination	Low water availability therefore lower risk	High water availability results in higher risk
Liquid waste	Not produced	Large quantity leading to dumping difficulty
Downstream processing	Cheaper, easier, and less time consuming	Expensive and difficult

9.2.1   SUBMERGED STATE FERMENTATION

As discussed earlier, this type of fermentation is carried out in the system having an abundant amount of water. A controlled atmosphere is a primary requirement to obtain high-quality end product with optimum productivity and efficiency. At an industrial scale, it can be done through either batch operations, as fed-batch process, or in continuous mode. The selection of operation mode ultimately depends upon the type and quality of the end product as well as the scale of the operation.

In batch fermentation, fermenter is filled with nutrient medium and inoculated with a starter culture. The system is kept closed until the completion of fermentation and the final product is harvested after the process. However, in aerobic batch fermentation, oxygen has to be supplied regularly to maintain the optimum concentration. Generally, the process goes up to 3–4 days but sometimes may last for months to obtain the desired results. Most industrial fermentations are conducted in batches. The prime importance of batch fermentation is starter culture which must be kept under controlled conditions. Initially, the starter cultures are grown at lab scale under controlled conditions until it achieves the rapid exponential growth phase and then it is transferred to the fermentation vessel. It is important to mention that the slow-growing culture should be added in large quantities to reduce the fermentation time as longer fermentation time results in low productivity and high cost. At the industry level, production of alcoholic beverages such as beer, whisky, wine, and rum and acidifiers including citric acid, vinegar, and lactic acid is usually achieved through the batch mode of fermentation [90].

In fed-batch or semibatch fermentation, the fermenter is fed with a sterile nutrient medium either periodically or continuously to extend the period of fermentation and the volume of the substrate inside the fermenter increased after every addition. The final product is harvested at the end of each batch. There are few reasons to use this mode instead of the batch system such as a large quantity of substrates in a close vessel can sometimes be toxic for the microbes. For instance, baker's yeast requires glucose for their growth; however, a high concentration of the glucose in the beginning can lead to the production of the ethanol as by-product which can be toxic to the yeast cells. Hence, it is better to add the nutrients at regular intervals to maintain the optimum concentration. Secondly, sometimes few reactants in the reactor increase the viscosity of the solution thus reduces the overall concentration of the dissolved oxygen and the concentration goes further down if these reactants are present in the large amounts thus the death of the microorganisms will be rapid as compared to their growth. On the other hand, semibatch processes maintain the optimum level of oxygen by controlling the solution viscosity. Other advantages of the fed-batch process include the inhibition of catabolite repression and auxotrophic mutants. Examples of this operation include the cultivation of yeast cells through the supply of sugar at regular intervals and commercial production of penicillin [59, 79].

In continuous fermentation, the fermentation vessel is fed with a sterile nutrient medium continuously at a constant rate, and the final product is harvested simultaneously at the same rate; therefore, the net volume inside the vessel remains constant. Technically, continuous fermentations commence as batch processes and feeding of sterile medium begins after the microorganisms reach the specific concentration or maximum population. In certain continuous processes, a small quantity of harvested biomass is added into the fermenter along with the sterile medium to increase the microbial concentration and hence to improve the overall efficiency. Continuous addition of inoculum during processing depends upon the type of mixing and effectiveness of mixing. In some fermenter, for instance, plug fermenters, where long tubes inside fermentation vessel inhibit the back mixing, continuous addition of inoculum is must. In the case of well-mixed fermenters, it is very important to control the feeding rate of the sterile medium as it affects the dilution rate (ratio of feeding rate of medium to the culture volume in the fermenter), which must be kept below the peaked specific growth rate of microbes at particular conditions. Because, if dilution rate exceeds then it will result in washing out of culture from the fermenter. Continuous fermentation can be subdivided into four categories, that is, chemostat, turbidostat, nutristat, and phauxostat. In chemostat type, sterile nutrition medium is added into fermenter at a specific rate and simultaneously the biomass is harvested at the same rate to maintain the net volume constant. In turbidostat, feeding rate of the sterile medium is controlled such that the concentration of the microorganisms remains constant during fermentation. In nutristat, the predecided nutritional concentration is maintained at a constant. When the pH of the nutristat is held constant at a preset value, it becomes a phauxostat. Commercial applications of continuous fermentations include the production of beer and ethanol by *Saccharomyces* species, citric acid production through utilization of *A. niger,* single-cell protein, and apple vinegar (cider) [10].

9.2.1.1 SUBMERGED STATE FERMENTERS

Fermenter is the heart of any fermentation process. There are various types of fermenters working as submerged state fermentations. Table 9.2 shows the major submerge-type fermenters along with their major features.

TABLE 9.2   Major Features of Submerged State fermenters (Adapted from Refs. [10, 61])

Type of Fermenter	Schematic Diagram	Features
Bubble column fermenter	Bubble — Sparger —	• Cylindrical vessel • Height to diameter ratio −4 to 6 • Introduction of air by sparger at the bottom • Agitation is achieved by compressed gas • Not widely used—low efficiency compared to other systems • Unsuitable for media with high number of particulates and viscosity

TABLE 9.2   *(Continued)*

Type of Fermenter	Schematic Diagram	Features
Stirred-tank fermenter	Impellers	• Cylindrical vessel • Height-to-diameter ratio −3 to 5 • Three to four impellers mounted on central shaft to provide mixing • One type of impeller either provide axial flow or radial flow • More than one type of impellers can be used to induce both axial and radial flow simultaneously • Baffles prevent swirl motion→uniform mixing • Generally four baffles are present • Baffle width—9%–11% of the cylindrical vessel diameter
Airlift fermenter	Falling Rising Sparger	• Design—external-loop and internal-loop • Internal-loop—the aerated rising medium and the unaerated falling medium in the same vessel • External-loop—rising and falling media in separate tubes connected at bottom and top • Better solids suspension rate • Comparatively high oxygen transfer rate and heat transfer rate • Unable to handle highly viscous solutions
Fluidized-bed fermenter	Discharge Recycle	• External pump at the bottom produces sufficient viscosity to suspend the solid particles in the solution • Continuous recycling of liquid medium • Efficient particle mixing and uniform temperature gradient • Large upper cross-section area→lowers down the upward velocity→avoid washing out of solid particles

TABLE 9.2   *(Continued)*

Type of Fermenter	Schematic Diagram	Features
Trickle-bed fermenter		• Packed cylindrical shell • Nutrient media sprayed on the top of packing • Support or processing aids—wood shavings, rock particles, and plastic structures • Porous support—allows the movement of nutrient medium and gas • Counter current flow between gas and media • Suitable for low viscosity substrate with dilute suspensions • Major application—vinegar production

9.2.2   SOLID-STATE FERMENTATION

As discussed earlier, solid-state fermentation occurs on a sufficiently moist solid matrix under controlled conditions. This fermentation is best suitable for the cultivation of fungus and yeast as high moisture content is the requirement of bacterial fermentation. Solid-state fermentations have many advantages over submerged state fermentation. It is cost-effective attributed to the use of natural growth medium or low-cost inert substrates. It requires a very small quantity of water, therefore, effluent production is negligible or significantly less thus helps in the reduction of environment pollution concerns. In addition, it avoids the critical operations such as aeration or mixing which further reduce the overall cost. Ease of handling and lesser expensive makes this technique preferable in small industries [150].

There are various factors that need to be considered for solid-state fermentation [77].

9.2.2.1 NATURE OF SUBSTRATE

Nature of the substrate largely depends on the type of the harvested or end product. Natural or agricultural materials are being used as substrate as they are an excellent source of energy and carbon, for instance, the production of tempeh on the cooked soybean or chickpea and cultivation of mushroom on the compost containing wheat straw and horse manure. Various crops including wheat, rice, barley, and maize and by-products of food processing industries including fruit and vegetable pulp, rice bran, oil cakes of soybean, groundnut and coconut, palm kernel, sugarcane and cassava bagasse, saw dust, tamarind seeds, spent grains, coffee pulp, husk, etc., are commonly being utilized as substrate. Besides natural materials, inert substances supplemented with required nutrients and carbon sources can also be employed as the substrate.

9.2.2.2 MOISTURE

Moisture is one of the most critical parameters. Productivity of any fermentation process largely depends on the free water content of the solid matrix or substrate. In solid-state operations, the matrix should have enough water because moisture helps in the diffusion of nutrients throughout the substrate. The presence of low moisture content generally results in the poor and slow diffusion of the nutrients which extends the period of fermentation as well as leads to inefficient processing. In contrast, very high moisture can cause the compression of the solid matrix which again results in the poor reaction rates and low yield.

9.2.2.3 SURFACE AREA

Surface area of the substrate must be large enough, that is, 10^3–10^7 m²/cm³, to ensure the maximum utilization of nutrients present in it. Large surface area also aids in quick diffusion of the moisture and leads to ready growth of the culture.

9.2.2.4 PARTICLE SIZE

Substrate particle size has a significant effect on the aeration as well microbial growth. Substrates with bigger particles offer better aeration properties but demote microbial activity due to lower surface area. On the other hand, small particles offer better surface area therefore promotes microbial activity but impose problems in aeration or respiration. In most of the solid-state operations, the mixture of the different sized particles is being used to counter the problems raised by particles of a single size and to optimize the production.

In addition, there are some other factors such as temperature, pH of media, any pretreatment to substrate, nutrient concentration, carbon source, nature of culture, size of inoculum, incubation time, temperature, and humidity, which needs to be considered during processing.

9.3   APPLICATION IN FOOD PROCESSING INDUSTRY

9.3.1   GRAIN-BASED FERMENTATION

The grain-based products can be majorly classified with respect to the substrates used in the fermentation process. Substrates have varied from wheat, rice, maize, barley, and other conventional grains for a long time to the pseudocereals and super grains like quinoa, amaranth, buckwheat, and oats which are being used currently [41].

9.3.1.1   NOVEL STARTER CULTURE

Bread spoilage is mainly caused by fungi that make it necessary to use certain preservatives like calcium propionate. But, these can be used up to certain limits as higher levels can affect human health. Thus, the shelf life of bread is limited to 2–5 days. LAB act as antifungal starters and can be used in sourdough fermentation. *Lactobacillus plantarum* CRL778 exhibits the highest biopreservation, which is comparable to that of 0.2% calcium propionate. This antifungal effect is due to acetic acid and phenyllactic acid, whereas lactic acid reduces the pH of the dough which does not allow the dissociation of acids and calcium propionate [115]. These undissociated molecules diffuse into the cell and rapidly dissociate to anions and cations due to high pH thus inhibiting the key glycolytic reactions and consequently yield of ATP [54]. A shelf life of 24 days can be achieved by the synergism of 0.4% calcium propionate and *L. plantarum* CRL778 in packaged bread [48]. Antifungal cyclic dipeptides (leucine-proline and phenylalanine-proline) and phenyllactic acid are also produced by *L. plantarum* FST1.7 which improve the shelf life of wheat bread to 10 days when used in combination with 0.3% calcium propionate [35].

Functional foods may be developed using probiotic cultures which, upon acting on a suitable substrate, produce improves technological and nutritional properties. Fermentation of a mixture of oat flour and water (18:82) with *L. plantarum* strains LpB2 and exopolysaccharide producer Lp90 improves the riboflavin content and viscosity. Although, viscosity decreases with shelf life due to the degradation of β-glucan and the exopolysaccharide, the content of vitamin B2 increases during the storage period [128]. Fermentation of legumes like faba bean and mung bean with *L. plantarum* improves the nutritional profile including bioactive peptides, essential amino acids and in vitro protein and starch digestibilities with reduction of antinutritional factors [30, 113, 151]. Horse gram sprout fermentation using *L. plantarum* produces aromatic volatile compounds such as volatile phenols, alcohols, acids, and acid esters, which provide desirable flavor including eugenol which is beneficial for health. The starter also exhibits its probiotic properties by producing short-chain fatty acids [53]. Beer fermented with kefir grains exhibits antioxidant, anti-inflammatory, and antiulcerogenic effects. Moreover, the levels of lipid fractions and transaminases in serum are under control unlike the higher values observed on chronic consumption of ethanol. This is the synergistic effect of polyphenols of the beer and inherent probiotic properties of kefir culture [123].

Contamination of crops by mycotoxins can pose health hazards for humans. Filamentous fungi like *Aspergillus*, *Fusarium*, and *Penicillium* produce different mycotoxins like aflatoxins, citrinin, fumonisin, and zearalenone. Maize crop contamination due to mycotoxins can be overcome during the fermentation process of *ogi* production using lactic acid culture, especially *Lactobacillus paraplantarum*, *Pediococcus acidlactici*, and *Pediococcus pentosaceus*, due to their ability to bind the mycotoxin or degrade it. This ability may be affected by reduction in the microbial count or acidified environment due to which a longer fermentation time decreases the reduction levels of these mycotoxins [96]. A similar effect occurs in *amahewu* (maize-based porridge in South Africa) fermentation by LAB apart from improving its protein digestibility [25]. Similarly, during the fermentation step in beer production, zearalenone, and patulin present in malt are decomposed to less toxic compounds, β-zearalenol and ascladiol, respectively. Also, ochratoxin A adsorbs to the yeast and decomposes to α-ochratoxin A and phenylalanine [68].

Beer fermentation using sourdough yeasts occurs at enhanced rates with increased production of ethanol and volatile compounds, like esters, acids, and alcohols. Thus, craft beer can be produced using sourdough cultures which can improve the organoleptic properties compared to traditional starters [84]. Natural folic acid can be achieved by emulsification of tempeh (soybean fermentation with *R. oligosporus*) and fermented spinach (using kombucha culture) using carboxymethyl cellulose and gelatin in a ratio of 8:0.4 [140]. *Koji* production along with inoculation of halophilic aromatic yeast and *A. oryzae* can produce soy sauce with improved flavor and quality with shortened production time [80].

9.3.1.2   NOVEL INGREDIENT

Bran contains alkylresorcinols which provide health benefits and thus the addition of bran enhances the functionality of bread without affecting the volume and porosity of the crumb [4]. Moreover, fermentation of bran using yeast and/or lactic starter culture provides better bread volume and crumb softness thus improving its palatability [71]. *Monascus* fermentation can be carried out on sorghum grain (whole or hulled) and bran after soaking to produce high ethanol and water-soluble pigments and monacolin K; a highly potential functional component that helps in the regulation of choles-terol [138].

Sourdough fermentation is generally the inoculation of LAB and yeasts. It is considered to be a low glycemic index (≤55) product [36]. Moreover, sourdough fermentation of rye malt using *Lactobacillus reuteri*, a glutamate-accumulating strain, reduces the salt content by 0.5% in bread [157]. Wheat germs too provide saltiness on sourdough fermentation [121]. Sourdough fermentation can improve the resistant starch content of bread attributed to high levels of lactic acid due to which pH is comparably lower than tradi-tional bread since it acts as a prebiotic and improves health due to reduced starch digestibility [16]. It also improves the γ-aminobutyric acid levels by using pseudocereals, that is, amaranth, buckwheat, and quinoa, since they are rich in protein and glutamate is converted to γ-aminobutyric acid by LAB. Celiac patients can consume products prepared from sourdough fermented cereal like wheat, barley, and rye since the process makes them gluten-free by disrupting the protein matrix and thus reducing the risk of irritable bowel syndrome (IBS) [52]. Certain fermentable oligo-, di-, and mono-saccharides and polyphenols, most importantly fructans, present at levels of 1.5%–3.7%

in cereals, are also responsible for IBS which can be reduced by fructophilic LAB strains [50] like *Lactobacillus kunkeii* B23I, PLA21, and PF16 and yeasts like *Kluyveromyces marxianus* [139]. The ratio of polyunsaturated fatty acids (PUFAs) in hempseed can be influenced by sourdough fermentation with *P. pentosaceus* and *P. acidilactici*. This fermentation process can also be used to reduce the antinutritional factors like phytic acid, polyphenols, and protease inhibitors especially in legumes, and improve the in vitro protein digestibility, as well as the essential amino acid content. Sourdough fermentation can also be applied on by-products like bran and germ and enables the positive effect on nutrition like antioxidant activity, bioavailability, lipase inactivation, etc., and sensory and textural properties [51]. Pasta, for the millennials, can be prepared using sourdough fermented flour especially using nonwheat flours due to the multiple nutritional and textural benefits [89].

9.3.1.3 NOVEL TECHNOLOGY

Standardization and optimization of the quality of the fermented product are mandatory. Fermentation process control requires the frequent analysis of different parameters, such as temperature, pH, and soluble solids content (SSC). These are generally determined chemically and/or involve destructive techniques. Moreover, they are time-consuming and require skilled labor. This leads to inconsistency in the batches of beer especially for small-scale industries which produce small quantities per batch. Thus, real-time evaluation of all the quality parameters is necessary to achieve consistent quality. Devices on the principle of Fourier-transform near infrared (FT-NIR) can be used in the range of 10,500–5500/cm to assess the pH and °Brix for filtered beers and more accurately, the biomass [56]. On a similar basis, visible-NIR spectroscopy when are operated in the range of 450–980 nm can be used for evaluating pH and SSC especially for filtered craft beers, and for nonfiltered beer after calibration with a large dataset. It also identifies the time to halt the process of fermentation [49]. Instead of SSC, determination of all the components of wort, which include glucose, fructose, sucrose, maltose, maltotriose and dextrins, and ethanol throughout the fermentation processes improves the assessment validity. FT-IR in the region of 1200–950/cm reveals the changes in these compounds during the process of fermentation and thus elaborates the biotransformation of sugars to ethanol [57]. Low-frequency impedance spectroscopy may also be used to monitor the proliferation and viability of yeast cells as well as their physiological states related to cell respiratory conditions [136, 148]. These novel techniques can be used for on-line monitoring of the brewing industries.

Cereal products with novel textures may be produced on the application of high pressure (50–250 MPa at 20 °C for 2 min) to yeasted wheat doughs. The microflora viability is adequate for fermentation of the dough and the product is hard, adhesive, and less sticky with brownish color due to changes in protein [8].

Disruption of structures can assist in accelerating the release of components thus improving the rate of fermentation of the substrates. Ultrasound, before or after liquefaction, assists in the release of glucose from cornmeal thus enhancing the yield of ethanol upon fermentation by *S. cerevisiae* var. *ellipsoideus*. This is due to the degradation of starch granules and consequent glucose release which facilitates the hydrolytic action of the enzyme and thus fermentation, thereby reducing the process time. Similarly, high-pressure application on a mixture of barley malt and water causes gelatinization of starch due to granule disruption thus accelerating the beer production process [112]. Ultrasonication during the preparation of bread dough can disrupt the starch fraction and uniformly distribute the fat molecules. This enhances the saccharification process to produce fermentable sugars thus enhancing the yeast activity and consequently the structural and mechanical characteristics [9].

9.3.2   DAIRY-BASED FERMENTATION

Fermentation of milk using different cultures and processing techniques has been followed for a long time. The objectives of milk fermentation changed from shelf-life extension, eventually to achieving diverse organoleptic, and physical properties with a specific texture, rheology, and microstructure and, health and functional benefits by the utilization of functional ingredients especially probiotics and prebiotics. Thus, various ingredients (whey protein concentrates and isolates, casein, prebiotics, enzymes-like transglutaminase), starter cultures (conventional, probiotics, nonconventional), or some novel techniques are applied to improve both product quality and functionality [87].

9.3.2.1   NOVEL STARTER CULTURE

Lactobacillus and *Bifidobacterium* species are the most common probiotics apart from others like *Pediococcus, Lactococcus,* and *Enterococcus.*

Generally, probiotic bacteria are used along with the conventional starters of the product since their growth in milk is slow due to the absence of proteolytic enzymes due to which they do not affect the organoleptic and rheological properties [82]. Also, utilization of growth-promoting food ingredients (prebiotics), antioxidants, cell-protective agents, microencapsulation, and oxygen barrier packaging materials can improve the viability of probiotic bacteria [146]. Certain strains of cultures such as *B. longum* and *Streptococcus thermophilus* produce exopolysaccharides which improve the viscosity, thereby the physicochemical and rheological properties of products like low-fat yogurt [114, 116]. Also, exopolysaccharides can act as prebiotics, improve the viability of starter cultures against pathogenic bacteria, phage attack, osmotic stress and desiccation and exhibit immunostimulatory, antiulcer and anticholesterolemic properties [125, 126].

Certain biogenic amines like tyramine may be produced in cheeses due to the conventional cheese microflora or due to the amino acids released during ripening which act as substrates for biogenic amines. Bacteriophage Q69 has the ability to reduce the LAB which produces tyramine, thus reducing its net content in cheese, without affecting those which are required for the complete fermentation of cheese. Since phages have a narrow range of hosts, a combination of phages is required for the reduction of different biogenic amines in cheeses and other dairy products [63].

Certain wild lactococci strains prove to be of high technological and functional potential. *L. laudensis* and *L. hircilactis* isolated from cow and goat milk, respectively, in Italy, exhibit high antioxidant activity and produce sufficient amounts of ethanol, diacetyl, acetoin, and acetic acid in the final product. Moreover, *L. hircilactis* produces high quantities of folate in the fermented product compared to raw milk [143]. Kefir grains contain a multitude of microorganisms. Brazilian kefir grains have *L. lactis* ssp. *cremoris* MRS47 which improves the content of PUFA and reduces the saturated fatty acid content. It is due to an increase in Δ9-desaturase ability consequently increasing the production of conjugated linoleic acid in fermented milk [149].

9.3.2.2   PREBIOTICS

Prebiotics like inulin and lactulose are indigestible ingredients that essentially improve the activity and/or stimulate the growth of the colonic bacteria, thus benefitting human health [37]. More specifically, inulin increases the bifidobacterial count (bifidogenic effect) for all age-groups, thereby improving calcium absorption [86], bowel habits [83], weight management by a beneficial effect on satiety [19], glucose tolerance in patients of diabetes, and resistance against infections [34], reducing the serum lipids, plasma triglycerides and cholesterol, and stimulating the immune system. Technologically, it is used to replace sugar and fat and provide texture (e.g.,, long-chain inulin exhibits improves creaminess in yoghurt) especially to low-fat dairy products. The dietary fiber and bifidogenic effects of inulin are observed when it is present in the range from 3 to 6 g per 100 g/mL [104] and 3 to 8 g per serving [33]. Significant augmentation in the growth and viability of probiotics like *B. lactis*, *Lactobacillus acidophilus*, and *Lactobacillus rhamnosus* has been noticed for nonfat fermented milk supplemented inulin at low amounts. Lactulose is an isomer of lactose and contains fructose instead of glucose. It affects the acidification and postacidification processes, growth, viability, and metabolic activity of *L. acidophilus*, *Lactobacillus bulgaricus*, *L. rhamnosus,* and *B. lactis* with *Streptococcus thermophilus* in skim milk [98].

9.3.2.3   TRANSGLUTAMINASE

Application of the enzyme transglutaminase (TGase) on milk proteins modifies the functional properties like solubility, emulsification, hydration, gelation, heat stability, rennetability, and rheology by forming covalent crosslinks in the protein polymers. This is due to the intra- or inter-molecular bonding by catalyzing the transfer of acyl group between γ-carboxyamide groups of peptide-bound glutamine residues and the ε-amino groups of lysine residues [87]. Preferentially, caseins are acted upon by TGase compared to whey proteins since structural modification is required in the latter to contribute in crosslinks with caseins or within their fractions [66, 67] due to which treatment of milk proteins by TGase leads to improves structural stabilization and modification thereby enhancing the physicochemical, rheological, and textural properties especially in low-fat yogurt. TGase concentrations may vary from 9–14 U/g (glutathione absent [100]) to 0.6–1.0 U/g (glutathione present [12]) to enhance the viscosity in set nonfat yoghurt. For stirred low-fat probiotic yoghurt, low concentrations of TGase (0.01%–0.05% w/w) exhibit better physical and textural characteristics. Utilization of TGase (0.01%) and whey protein concentrate (0.3%) does not improve the textural properties due to the prevalence of cross-linking of casein over whey protein aggregation consequently leading to a gel matrix with a nonuniform network leading to product that is lumpy and coarse [88].

9.3.2.4 NOVEL TECHNOLOGIES

Firm gels with significantly low syneresis are obtained on presonication of micellar casein solution followed by the addition of gelation agents like tetrasodium pyrophosphate [22]. Application of ultrasound on milk produces fat droplets of smaller size and if carried out prior to inoculation at a high amplitude level, significantly enhances the viscosity, water holding capacity, hardness and cohesiveness, and reduced syneresis [133] probably due to whey protein denaturation [44]. Moreover, the activity of the starter culture can be enhanced thereby reducing the fermentation time on the application of sonication before the fermentation [91].

Milk protein structures are affected due to the pulsed electric treatment depending upon the intensity, temperature, and duration which then impacts the functional properties like thermal stability, gel strength, gelation rate, emulsion stability, rennetability, and hydrophobicity of casein and whey proteins [15]. Yoghurt-like product prepared from milk processed using both thermal and PEF (60 °C for 30 s and 30 kV/cm for 32 µs) has comparable organoleptic properties with reduced growth of molds, yeasts, and mesophilic aerobic organisms [153].

High-pressure homogenization results in partial whey protein denaturation and uniform size of fat globules, and inactivation of microflora and enzymes [65]. Firm gels, lower syneresis, and titration acidity are obtained in yogurts prepared from HPH (200–300 MPa) processed milk containing skim milk powder [132] but may negatively affect the flavor [131].

High-pressure processing modifies the milk proteins to produce a compact network and inactivates the spoilage and pathogenic organisms thus providing better quality and nutritional properties to yogurt. Combination of HPP (676 MPa for 5 min) and thermal treatments (85 °C for 30 min) to milk leads to better physicochemical and functional properties especially water holding capacity [107] due to whey protein denaturation and casein micelle aggregation thus reducing syneresis and improving the yield stress [62].

9.3.3 MEAT-BASED FERMENTATION

Meat and meat products are rich in fat, proteins, essential amino acids, vitamins, minerals, and other bioactive components [11]. The major issue is the safety of the food which mandates the development of those products which encourage both health and nutritional benefits [58, 147]. These include products that contain functional components and less nitrates/nitrites, fat, cholesterol, and salt [156].

Meat fermentation involves the action of enzymes that are present endogenously or produced microbially. These cause biochemical, physical, and microbial changes such as solubilization and gelation of sarcoplasmic and myofibrillar proteins, acidification due to carbohydrate catabolism, lipid and protein degradation, nitrite and nitrate reduction, dehydration and nitroso-myoglobin formation [144]. Natural fermentation of meat produces highly acceptable organoleptic properties but affects the final quality due to inconsistency and the presence of pathogenic or spoilage microorganisms which may be hetero-fermentative or produce biogenic amines thus leading to questionable safety of the product [47]. Thus, defined starters mainly LAB, coagulase-negative staphylococci, catalase-positive bacteria like *Staphylococcus* and *Kocuria*, yeasts like *Debaromyces,* and molds like *Penicillum* are commonly used since the final products exhibit acceptable organoleptic properties [95, 147].

9.3.3.1   STARTER CULTURE

The starter culture highly affects the overall quality of the final product. The microbial safety can be enhanced by applying starters that produce bacteriocins that inhibit the pathogenic microbes like *Staphylococcus aureus*, *Salmonella enteritidis*, and *Listeria monocytogenes*. *L. acidophilus* produces a bacteriocin and thus increases the safety and stability of goat minced meat when fermentation is carried out at 26 °C for 7 days [94]. Similarly, *Lactobacillus sakei* improves the quality of fermented sausages, due to the production of bacteriocin [155], control of spoilage and pathogenic microorganisms, and reduction in malondialdehyde and nitrite along with improvement in sensory properties [46], even without the culture activity being affected in the presence of mesenterocin Y [155]. When used along with *Staph. carnosus* and *Staph. xylosus*, it reduces the accumulation of biogenic amines by 80%–90% in sausages which have been fermented naturally [14]. Fermentation of mutton sausages using *Lactobacillus pentosus, Lactobacillus pentosaceus,* and *Staph. carnosus* improves the nutritional content due to the reduction of saturated fatty acids and elevation of monounsaturated and PUFAs [159]. Sausages fermented with yeast cultures like *Debaromyces hansenii* exhibit improved sensory properties due to high proteolytic and lipolytic activity and consequent production of volatile compounds like 3-methylbutanal,

3-methylbutanol, and 2-propanone which improve the flavor and aroma profiles [5].

Novel starters like probiotics are used due to their health-benefitting properties. Since meat product preparation involves negligible or no heat processing, they act as perfect carriers for probiotics [72]. Certain *Lactobacillus* strains have been isolated from fermented sausages which show better growth with various prebiotics and higher adherence to the intestinal human Caco-2 cell lines [108].

9.3.3.2 NOVEL TECHNOLOGIES

Safety of fermented meat products can also be ensured by the application of thermal and nonthermal technologies such as microwave (MW) and radio frequency, ohmic heating, HPP, PEF, pulsed ultraviolet light, ozonation, irradiation, and ultrasonication.

HPP application after-ripening of fermented sausages (especially low-acid) has proved to improve their quality by reducing the pathogenic counts. Treatment of ripened sausage a 600 MPa for 5 min immediately decreases the counts of *Enterobacteriaceae* to less than 1 log cfu/g, *Staph. aureus* and *Escherichia faecium* CTC8005 by 1 log cfu/g and *L. monocytogenes* by 2 log cfu/g [127]. HHP treatment thus lowers the addition of sodium chloride and sodium nitrite for the production of low-sodium meat products [99]. Lower pressures can be used when combined with a carbon dioxide packaging environment since the cell membrane damage caused by HPP allows the penetration of higher amounts of CO_2 thus enhancing the lethal effect [3]. HPP also inhibits the diamine (putrescine and cadaverine) producing microbes without affecting the activity of the starter [135].

Application of pulsed ultraviolet light of 11.9 J/cm^2 on dry-fermented sausage reduces the counts of *L. monocytogenes* and *Salmonella enterica* by 1.5 and 1.8 cfu/cm^2, respectively, without affecting the organoleptic characteristics of the product [45].

Decontamination of frozen meat has also been achieved by irradiation of 2 to 4 kGy, which reduces the counts of *Escherichia coli* O157:H7 by 5 log but higher radiation is required for *L. monocytogenes*. The application of gamma radiation effectively reduces the amounts of all biogenic amines during storage and can be used effectively as a preservation treatment for fermented meat products [119].

9.3.4   FRUITS- AND VEGETABLES-BASED FERMENTATION

9.3.4.1   WINE

Wine is the one of largely consumed alcoholic beverage produced through a complex biochemical process in which yeasts converts the grapes sugars into alcohol, that is, ethanol. Earlier, grape fermentation was carried out naturally by wild yeasts but the inferior quality of the end product forced the manufacturers to utilize the specific strains of yeast, that is, *S. cerevisiae*. The wine-making process generally includes four to five basic steps such as crushing, pressing, fermentation, clarification, and ageing. The resulted product is expected to have characteristics aromatic properties and color values attributed to the presence of various compounds including anthocyanins, esters, etc.

Saccharomyces species are the dominant population in the natural and inoculated grapes fermentation during winemaking. However, sometimes the biological activities of unwanted yeasts during prefermentation or fermentation stages resulted in the production of certain metabolites (acetic acid, 2,3-butanediol, diacetyl, etc.) which can have a detrimental effect on the overall quality of the wine. Generally, manufactures use sulfur dioxide to counter the presence of undesirable yeast, as sulfur dioxide has antiseptic effects. However, the consumption of sulfur dioxide can impose some serious health issues. Therefore, some new approaches, for instance, non-*Saccharomyces* yeasts, can be used to avoid the interference of undesirable yeasts. Earlier, non-*Saccharomyces* yeasts were classified as spoilage yeasts due to their ability to produce such metabolites which have a negative effect on the sensorial properties of the wine. However, their use along with *S. cerevisiae* can have a synergistic effect on the quality of the wine. *Tetrapisispora phaffii* is one of the examples of non-*Saccharomyces* yeasts which has the ability to produce KpKt toxins that can kill undesirable apiculate yeasts (*Kloeckera* and *Hanseniaspora*). KpKt toxins are stable over the period of 13–14 days at pH range of 3–5 and at a temperature below 40 °C and their antimicrobial activity is somewhere equivalent to sulfur dioxide. Similarly, *Pichia anomala* and *Kluyveromyces wickerhamii* produce toxins, that is, Pikt and KwKt, which can inhibit the detrimental activity of *Dekkera* or *Brettanomyces* yeasts and are stable over the period of 10 days at low pH [28, 31].

Non-*Saccharomyces* yeasts, when using as pure culture or as mixed culture can reduce the ethanol concentration leading to the production of low alcohol wines, which is the demand of the present consumer. Yeasts like

Pichia subpelliculosa and *Williopsis saturnus* can be used with *S. cerevisiae* to produce quality wines with an ethanol concentration of 3% v/v. The mechanism behind the low ethanol concentration includes the utilization of available sugar for respiration purposes rather than fermentation [118]. Non-*Saccharomyces* yeasts like *Candida pulcherrima* and *Hansenula anomala* contribute to the aromatic properties of the wine through the production of esters or by secreting hydrolytic enzymes (glucosidase), which release the volatile compounds bounded to the sugars. Non-*Saccharomyces* yeast can also enhance glycerol production thus contribute to smoothness and sweetness. However, high glycerol concentrations are associated with the high level of acetic acid, which is undesirable. Non-*Saccharomyces* yeasts can be beneficial in terms of microbial activity and sensory properties, but keeping in mind that every strain is not beneficial, only tested strains should be used [85, 101].

Efficiency and yield are two important factors in any fermentation system that must be considered before the actual process. Cell immobilization techniques can be successfully implemented to improve certain factors. In wine production, yeast cell can be immobilized on the bacterial cellulose through adsorption-incubation technique. Certain practices can improve the ethanol production rate by 19.6%–116% and sugar uptake rate by 16.4%–91.8% as compared to traditional processing. Thus, by improving the sugar uptake rate and ethanol production rate, the overall process time can be reduced [145].

9.3.4.1.1 Novel Technologies in Wine Production

PEF can successfully inhibit the activities of undesirable non-*Saccharomyces* yeasts through the electroporation mechanism. In red wine, application of 24 and 31 kV/cm at 30 °C can reduce the *Lactobacillus delbrueckii* ssp. *bulgaricus* population from 6.14 ± 0.48 log cfu/ml to 2.68 ± 0.45 and 2.04 ± 0.19 log cfu/ml, respectively. Similarly, the combination of 31 kV/cm and 30 °C can reduce the *E. coli* O157:H7 population from 4.97 ± 0.08 log cfu/ml to 1.26 ± 0.12 log cfu/ml and *Candida lipolytica* population from 5.30 ± 0.24 log cfu/mL to 0.00 ± 0.00 log cfu/ml [1]. PEF application of 5–10 kV/cm before the fermentation stage can significantly improves the total phenol content as well as color intensity which is attributed to the ability of PEF to induce the mass transfer from the skin and seeds to juice. However, no significant change occurs in titratable acidity, pH, color, °Brix, anthocyanin content, antioxidant capacity, and sensory properties after the processing at 31 kV/cm and $40 \pm 2°C$. PEF can also be used in combination with enzymatic treatment because such treatments can increase the anthocyanin content, total phenolic content, and color intensity by 26%, 11%, and 28%, respectively [29, 117].

High-pressure processing or high hydrostatic pressure processing is becoming popular in the beverage industry, especially in fruit juices. In winemaking process, it is being utilized to serve many benefits. One of the main objectives of HPP is to reduce or eliminate the utilization of sulfur dioxide. Exposure to 200–500 MPa for 1–20 min can successfully inactivate the bacteria, yeast, and fungi without affecting the sensory properties of the wine. Some specific examples include the 8 and 6 log reduction of bacterial (*Acetobacter aceti, Acetobacter pasteurianus,* and *Lactobacillus plantarum*) and yeast (*Brettanomyces bruxellensis* and *S. cerevisiae*) population in wine at 400–500 MPa for 5–15 min [17]. On the other hand, the high-pressure values above 500 MPa can significantly change the chemical composition and sensory properties of the wine. For instance, the application of 650 MPa for 1 to 2 h can reduce the total phenols by 4%–5%, anthocyanin content by 5%–6%, and flavonols content by 10%–12%. There are various factors, which can be responsible for the reduction of total phenol content, such as transfer of thermal energy between the pressure vessel and food particles and accumulation of highly reactive species. During aging also, total phenols and anthocyanins decrease attributed to the precipitation, condensation, and polymerization reactions [130, 141].

Power ultrasound or low-frequency treatments have a significant impact on wine microbiology, physicochemical composition, and fermentation period. Power ultrasound application of 20 kHz in the range of 330–360 W s/m^3 can reduce the fermentation time by shortening the lag phase of yeast by almost 17%. Such a phenomenon happens, when ultrasounds increase the availability of nutrients in the matrix resulting in the acceleration of uptake rate. Sometimes, larger ultrasound energies, such as more than 850 W s/m^3, negatively affect the growth of wine yeast, thus increase the overall processing time [69]. This generally happens due to the cavitation phenomenon which tends to inactivate the yeast cell. Power ultrasound significantly contributes to the elimination of chemical preservatives and thermal treatments as these treatments have a detrimental effect on the overall quality of the wine. Ultrasound treatment of 24 kHz for 20 min is capable to reduce the stability of undesirable yeasts (*Hanseniaspora uvarum*). Similar instances include 89.1%–99.7% reduction of *Dekkera* or *Brettanomyces* yeasts at 24 kHz and 400 W s/m^3 [55, 81]. Besides power ultra sound waves, high-frequency ultrasound waves can also be utilized in winemaking processing. During the fermentation process, the velocity of high-frequency waves can estimate the concentration of sugars and alcohol in the fermentation vessel due to the positive correlation between such parameters. Similar type of correlation can also advise the concentration of organic acids such as malic acid and lactic acid produced by LAB. Yeast cell concentration in suspension can be measured at higher frequencies, that is, above 15 MHz [40, 93].

MW is the electromagnet waves, which have a potential application in enhancing the quality of grape wine. MW treatment induces the extraction of phenolic compounds into juice, thus improves the anthocyanin content, total phenols, tannins content, and antioxidant activity by 2- to 3-folds. MWs can enhance the rate of fermentation which is attributed to their ability to rapidly degrade the undesirable yeast in grapes, which shortens the lag phase and accelerate alcohol fermentation. MW processing has ability to maintain the microbial quality of the food matrix. Application of MWs for 3 min can successfully reduce the total yeast content by 36%–38%, *Brettanomyces* population by 35%–67%, and approximately 90%–100% in case of acetic acid and LAB. MWs can also be used to sanitize fermentation vessels and wine barrels [20, 21, 29].

Pretreatment of grapes with gamma radiations has a significant impact on the total phenolic content, antioxidant activity, and organoleptic properties. The application of irradiation (1.5– 2 kGy) before fermentation, improves the antioxidant activity by 19%–37%, total phenols by 18%–31% and total anthocyanin content by 47%–77%. This technique also contributes to fruity flavor of the wine without inducing the major changes in aromatic constituents. No other changes are generally observed for irradiation [60].

9.3.4.2   CIDER

Cider is an alcoholic beverage, traditionally produced by the natural fermentation of apple juices or must. *S. cerevisiae* and *S. bayanus* are the dominant species in the fermentation of apple. One of the important steps in the commercial production of cider is pasteurization, which is carried out with the objective to eliminate pathogens. However, thermal processing has a negative effect on the organoleptic and nutritional quality. Recently, a combined application of microfiltration and ultraviolet radiation in cider achieves more than 5 log reduction of pathogens like *E. coli*, *Alicyclobacillus acidoterrestri*, and *Cryptosporidium parvum* without affecting the

nutritional and sensory characteristics. The mechanism involves the elimination of microbial cell and suspended solids by microfiltration, which allows
the ultraviolet treatment at lower doses to achieve desired results [158].

9.3.4.3   VINEGAR

Vinegar is the resulted product of alcoholic fermentation and acetic acid
fermentation. *Saccharomyces* species are utilized for alcoholic fermentation while *Acetobacter* species convert the available alcohol to acetic acid.
Based on the raw material, vinegar can be classified into various categories
including wine vinegar, cereal vinegar, balsamic vinegar, cane vinegar, fruit
vinegar, etc. The demand for fruits vinegar is increasing gradually due to the
high concentrations of phenolic compounds, anthocyanins, and tannins as
well as the high antioxidant activities of fruit vinegars. Fruits such as apple,
pomegranate, and citrus are being utilized to serve the purpose. Utilization of
pineapple waste for vinegar production and glasswort vinegars are the recent
advancements in this context. Glasswort vinegars are getting attention due to
their antioxidant-rich profile [26, 122].

9.3.4.4   PROBIOTIC PRODUCTS

At present, most of the probiotics are dairy based due to the great compat-
ibility with culture and carrier. But gradually, the trend is shifting to fruits
and vegetables-based probiotics. The reason behind this shift may be the
allergy issues related to the consumption of dairy products or may be due to
the nutrient and polyphenols-rich profile of fruits and vegetables. The main
challenge for the development of fruits and vegetables as probiotic foods,
is the low pH, which is not favorable for the viability of many beneficial
microorganisms. Even in this scenario, there are many examples that show
the potential of fruits and vegetables. For instance, cashew apple juice
significantly promotes the growth of *Lactobacillus casei* during the 1–2 days
of fermentation at 30 °C and 6.4 pH after inoculating with same culture.
Also, cashew apple juice able to maintain a sufficient cell count (8 log cfu/
mL) during the storage period of 42 days at 4 °C [109]. Similarly, sugar cane
juices can promote the growth of certain probiotic cultures (*Lactobacillus
acidophilus*) during the 24 h fermentation at 37 °C and maintain the population (4.0×10^8 cfu/mL) during 3 weeks storage at 3–4 °C [73]. In some case,
fruit juices promotes and maintain the culture of one particular strain but fail

to do the same with other. It can happen due to the fact that every strain has their own favorable conditions to grow and survive. For example, peach juice promotes the growth of *Lactobacillus delbrueckii* at 30 °C and maintains the cell count during low temperature (4 °C) storage of 4 weeks, but fails to do the same with *Lactobacillus casei* [102].

Mixed juices made from vegetables like bitter gourd, carrot, and bottle gourd have the ability to induce the growth of LAB like *L. plantarum, L. acidophilus,* and *P. pantosaceus.* During the fermentation process of 72 h at 30 °C, the LAB utilizes the available nutrients and grows up to the required microbial count (>10^6 cfu/mL). Vegetable juices like potato juice and cabbage juice also show the same intent in promoting the growth of LAB during processing [74, 134].

As we discussed above that acidic conditions and pH can inhibit the growth of probiotic cultures. In this context, encapsulation of culture can be the potential solution to protect it from the unfavorable environment. Encapsulation of *S. cerevisiae* with the solution of xanthan gum, inulin, and alginate allows the growth of the culture in the acidic environment of the berry juice, and even at cold storage temperatures the cell integrity remains intact [43]. Novel technologies like sonication can also be collaborate with the fermentation process for the development of probiotic drinks. It (sonication) improves the availability of nutrients in the pineapple juice and promotes the growth of *L. casei* during fermentation at pH 5.8 and 31 °C, thus contributing in development of high quality prebiotic drinks [32].

9.3.4.5   SAUERKRAUT

Sauerkraut is a traditional fermented product, produced by the natural lactic acid fermentation of cabbage in the presence of salt. Various organisms are involved in the natural fermentation, which carried out their activity in different phases. Efforts are being made to modify the traditional methods to produce better quality product with higher efficiencies. In this context, starter cultures like *L. casei* are being used for cabbage fermentation. *L. casei* initiated fermentations show better retention of vitamin C content (44.64 ± 2.12 mg/kg) as compared to natural fermentations (31.04 ± 1.23 mg/kg) and high fermentation rates depicts better efficiency. Also, the final product shows the higher concentration of phenolic compounds due to the enzymatic hydrolysis of the cell wall resulting in the production of glycosides and esters [38, 97]. It is important to mention that salt concentration also affects the fermentation process, such as the higher concentrations favor the growth of certain yeasts, which can be responsible for the off-flavors in the product; therefore, the salt concentration should be considered before actual processing [152].

In few instances, inoculation with some specific strains can produce such compounds which can be beneficial for human gut health. *Leuconostoc citreum* KACC 91035 has such abilities to produce isomaltooligosaccharides during sauerkraut fermentation in the presence of sugars. Isomaltooligosaccharides are the nondigestible carbohydrates that can be utilized by the gut microflora for beneficial purpose [27].

9.3.4.6   KIMCHI

Traditionally, kimchi is produced by the natural fermentation of vegetables like cabbage and radish. Like sauerkraut, the major challenge is to control the fermentation to develop a uniform quality product. Starter culture fermentations and application of novel techniques are getting attention, as they can develop superior quality kimchi. Different cultures can be utilized to serve different functions during processing, such as *L. plantarum* presence accelerates the rate of process as well as improves the functionality in terms of better antioxidant activities and nutritional profile in the product [78]. Certain bacterial strains like *L. citreum* act as antimicrobial agents and control the microbial diversity during the fermentation. The antimicrobial activity of the *L. citreum* is attributed to the production of bacteriocins (kimchicin GJ7) during lactic acid fermentation. Bacteriocins (kimchicin GJ7) have the ability to reduce the viability of pathogens like *Staph. aureus*, *Salmonella typhi*, and *E. coli* O157:H7, from 5.52 log cfu/mL to 2.52 cfu/mL after 48 h of fermentation. In addition, *L. citreum* driven fermentation contributes to the overall sensory quality of the kimchi through the production of mannitol [23, 137].

The microbial population during kimchi fermentation can also be controlled by the addition of antimicrobial agents. In this context, allyl isothiocyanate can serve the benefit by inhibiting the growth of undesirable microbial species. However, the higher concentration of these agents in the food matrix can provide off-flavor; therefore, the level of addition needs to be considered before addition. Other options include the encapsulation of antimicrobial agents, which can prevent the dispersion of off-flavors [75].

9.3.5 COCOA-BASED FERMENTATION

Fermentation of cocoa beans is the most important process in the manufacturing of chocolate. Traditional fermentation of cocoa includes three stages which take 4–7 days to develop the desirable characteristics. The first stage of fermentation is dominant by yeasts including *Saccharomyces, Pichia, Hanseniaspora,* and *Kluyveromyces*, which produce ethanol and induce the hydrolytic degradation of pectin [39, 92]. The second stage of fermentation is dominant by LAB which takes 36 h to reach maximum concentration. LAB including *L. fermentum* is involved in the production of few volatile compounds like acetoin, 2,3-butanediol, and diacetyl [2]. However, certain studies deny the necessity of LAB in the cocoa bean fermentation [64]. In the last stage, acetic acid bacteria utilize the available ethanol for their growth and produce acetic acid. The penetration of acetic acid and lactic acid into cocoa bean along with high temperature and low pH [42, 110] destroys the seed embryo and give rise to the various biochemical reaction which provides the characteristics properties to the final product [18, 42, 103].

The extreme conditions such as high temperature, extreme pH values, as well as production of certain metabolites, affect the fermentation process resulting in quality degradation. In addition, the traditional fermentation process takes a very long time, that is, 4–7 days, and also the quality of the end product is inconsistent. Therefore, it is very important to consider these factors before commencing the fermentation process for commercial application. Computer vision methods can be used to assess the fermentation process [106] and classify the cocoa beans based on their fermentation levels (unfermented, partially fermented, or fermented) [154]. Such systems utilize artificial vision, electronic nose, and electronic tongue.

Extreme conditions such as high temperatures and ethanol content can be overcome by using different yeast starters such as *Pichia kudriavzeii*. These starter cultures can also metabolize the citric acid in the cocoa pulp [111, 129] thereby reducing the acidity and equilibrating the pH to 5.0–5.5, which is suitable for the action of endogenous enzymatic reactions, such as proteolysis. *P. kudriavzeii* is an aromatic yeast and produces acetaldehyde (59.57–78.52 µL/L) and ethyl acetate (22.32–62.48 µL/L) which contribute unique fruity flavors to the fermented product. Elevated concentrations of ethanol (8.96–9.37 mg/g) can be achieved as compared to the natural fermentation (0.875 mg/g) after 120 h, which acts as a source of carbon for acetic acid-producing bacteria. The aroma formation due to the biochemical reactions occurring during the roasting process is thus highly activated due to the ethanol and acetic acid [110, 124].

Surface sterile cocoa beans when incubated with media containing specific amounts of acetic acid (may also contain ethanol and pectinase) at definite time-temperature combination for 5 days (30 °C/24 h β→35 °C/24 h→45 °C/24 h→50 °C/24 h→50 °C/24 h), exhibit the optimum combination of free amino acids (15.4–25 mg/g FFDM), reducing sugars (16.42–28 mg/g FFDM), phenolic compounds (2.2–5.1 mg/g FFDM), and acetic acid (8.1–14.6 mg/g FFDM). Thus standardized fermentation is a reproducible process that shows the potential to exhibit high chocolate flavors, limited acidification and astringency, and unaffected maillard reaction to obtain high-quality beans even without the addition of starter in the medium [70].

As we know that the traditional cocoa fermentation time generally ranges from 4 to 7 days. This can be reduced to 48 h when the beans are fermented in a controlled environment in two stages, that is, prefermentation under anaerobic conditions (24 h) and fermentation under aerobic conditions (24 h). Kefir grains, which are rich in LAB and yeasts such as *S. cerevisiae* and *Candida,* are used as starter culture wherein the former act during the prefermentation stage and the latter enhances the fermentation process rate [105].

As mentioned earlier, traditional fermentation is carried out in baskets or banana leaves which provide inadequate thermal insulation and product mixing. Local materials such as coconut fibers can be used in a cylindro-rotative structure integrated with a 360° axis of rotation which facilitates proper positioning during the different fermentation phases. Such a fermenter confines the heat and enhances the movement of the beans thereby homogenizing the temperature and increasing the efficacy (83.23%) of the process [76].

9.4   SUMMARY

Fermentation is used as a technology for food processing since time immemorial. The process involves the metabolic activity of bacteria, yeast, and fungi on food products resulting in the production of bioactive compounds that have a beneficial effect on the human body. With the population explosion, consumer demands for fermented foods are increasing gradually, which is difficult to fulfill using conventional technologies. It has become necessary to move toward advanced techniques which ensure a rapid supply of fermented foods with excellent nutrition composition, uniform quality and consistency, and desired organoleptic properties. The escalating competitiveness has led to the emergence of modern techniques and innovations with the help of scientific research which includes the application of novel cultures or ingredients or techniques. These advancements can successfully counter the hurdles faced during application of the conventional techniques. However, the major challenge is the scaling-up of the advanced technologies without the loss of characteristic attributes related to the fermented products.

KEYWORDS

- **alcoholic fermentation**
- **fermentation**
- **fermenters**
- **lactic fermentation**
- **nonthermal technologie**
- **probiotics**
- **solid and submerged state**

REFERENCES

1. Abca, E. E.; Evrendilek, G. A. Processing of Red Wine by Pulsed Electric Fields with Respect to Quality Parameters. *Journal of Food Processing and Preservation*, **2014**, *39*, 758–767.
2. Adler, P.; Bolten, C. J.; Dohnt, K.; Hansen, C. E.; Wittmann, C. Core Fluxome and Metafluxome of Lactic Acid Bacteria Under Simulated Cocoa Pulp Fermentation Conditions. *Applied and Environmental Microbiology*, **2013**, *79*, 5670–5681.
3. Al-Nehlawi, A.; Guri, S.; Guamis, B; Saldo, J. Synergistic Effect of Carbon Dioxide Atmospheres and High Hydrostatic Pressure to Reduce Spoilage Bacteria on Poultry Sausages. *LWT—Food Science and Technology*, **2014**, *58*, 404–411.
4. Andersson, A. A.; Landberg, R.; Söderman, T.; Hedkvist, S.; Katina, K.; Juvonen, R.; Holopainen, U.; Lehtinen, P.; Åman, P. Effects of Alkylresorcinols on Volume and Structure of Yeast-leavened Bread. *Journal of the Science of Food and Agriculture*, **2011**, *91* (2), 226–232.
5. Andrade, M. J.; Córdoba, J. J.; Casado, E. M.; Córdoba, M. G.; Rodríguez, M. Effect of Selected Strains of *Debaryomyces hansenii* on the Volatile Compound Production of Dry Fermented Sausage "Salchichón." *Meat Science*, **2010**, *85*, 256–264.

6. Anese, M.; De Bonis, M. V.; Mirolo, G.; Ruocco, G. Effect of Low Frequency, High Power Pool Ultrasonics on Viscosity of Fluid Food: Modeling and Experimental Validation. *Journal of Food Engineering*, **2013**, *119* (3), 627–632.

7. Barbosa-Canóvas, G. V.; Rodríguez, J. J. Update on Non-thermal Food Processing Technologies: Pulsed Electric Field, High Hydrostatic Pressure, Irradiation and Ultrasound. *Food Australia*, **2002**, *54*, 513–518.

8. Bárcenas, M. E.; Altamirano-Fortoul, R; Rosell, C. M. Effect of High Pressure Processing on Wheat Dough and Bread Characteristics. *LWT—Food Science and Technology*, **2010**, *43* (1), 12–19.

9. Berezina, N. A.; Komolikov, A. S.; Galagan, T. V.; Rumyanceva, V. V.; Nikitin, I. A.; Zavalishin, I. V. Investigation of Ultrasonic Dough Processing Influence on Bread Quality. In: *Paper Presented at International Scientific and Practical Conference "Agro-SMART—Smart Solutions for Agriculture" (Agro-SMART 2018)*; Paris: Atlantis Press, **2018**.

10. Bhatia, S. C. (Ed.) *Food Biotechnology*. Woodhead Publishing India Pvt Ltd, New Delhi: CRC Press, **2016**, 411.

11. Biesalski, H. K. Meat as a Component of a Healthy Diet—Are There Any Risks or Benefits if Meat is avoided in the Diet? *Meat science*, **2005**, *70*, 509–524.

12. Bönisch, M. P.; Huss, M.; Weitl, K.; Kulozik, U. Transglutaminase Cross-linking of Milk Proteins and Impact on Yoghurt Gel Properties. *International Dairy Journal*, **2007**, *17* (11), 1360–1371.

13. Bourdichon, F.; Casaregola, S.; Farrokh, C.; Frisvad, J. C.; Gerds, M. L.; Hammes, W. P.; Harnett, J.; Huys, G.; Laulund, S.; Ouwehand, A.; Powell, I. B.; Prajapati, J. B.; Seto, Y.; Schure, E, T.; Boven, A. V.; Vankerckhoven, V.; Zgoda, A.; Tuijtelaars, S.; Hansen, E. B. Food Fermentations: Microorganisms with Technological Beneficial Use. *International Journal of Food Microbiology*, **2012**, *154*, 87–97.

14. Bover-Cid, S.; Izquierdo-Pulido, M.; Vidal-Carou, M. C. Mixed Starter Cultures to Control Biogenic Amine Production in Dry Fermented Sausages. *Journal of Food Protection*, **2000**, *63*, 1556–1562.

15. Buckow, R.; Chandry, P. S.; Ng, S. Y.; McAuley, C. M.; Swanson, B. G. Opportunities and Challenges in Pulsed Electric Field Processing of Dairy Products. *International Dairy Journal*, **2014**, *34* (2), 199–212.

16. Buddrick, O.; Jones, O. A.; Hughes, J. G.; Small, D. M. The Effect of Fermentation and Addition of Vegetable Oil on Resistant Starch Formation in Wholegrain Breads. *Food Chemistry*, **2015**, *180*, 181–185.

17. Buzrul, S. High Hydrostatic Pressure Treatment of Beer and Wine: A Review. *Innovative Food Science and Emerging Technologies*, **2012**, *13*, 1–12.

18. Camu, N.; González, A.; De Winter, T.; Van Schoor, A.; De Bruyne, K.; Vandamme, P.; Takrama, J. S.; Addo, S. K.; De Vuyst, L. Influence of Turning and Environmental Contamination on the Dynamics of Populations of Lactic Acid and Acetic Acid Bacteria Involved in Spontaneous Cocoa Bean Heap Fermentation in Ghana. *Applied and Environmental Microbiology*, **2008**, *74*, 86–98.

19. Cani, P.; Joly, E.; Horsmans, Y.; Delzenne, N. M. Oligofructose promotes Satiety in Healthy Human: A Pilot Study. *European Journal of Clinical Nutrition*, **2006**, *60*, 567–572.

20. Carew, A. Here's A Hot Idea: Straight Out of the Microwave. *Australian and New Zealand Grape grower and Winemaker*, **2013**, *58*, 62–63.

21. Carew, A. L.; Gill, W.; Close, D. C.; Dambergs, R. G. Microwave Maceration with Early Press off Improves Phenolics and Fermentation Kinetics in Pinot Noir. *American Journal of Enology and Viticulture*, **2014**, *65*, 401–406.

22. Chandrapala, J.; Zisu, B.; Kentish, S.; Ashokkumar, M. Influence of Ultrasound on Chemically Induced Gelation of Micellar Casein Systems. *Journal of Dairy Research*, **2013**, *80*, 138–143.

23. Chang, J.; Chang, H. Growth Inhibition of Foodborne Pathogens by Kimchi Prepared with Bacteriocin-Producing Starter Culture. *Journal of Food Science*, **2011**, *76*, 72–78.

24. Charalampopoulos, D.; Rastall, R. A. (Eds.) *Prebiotics and Probiotics Science and Technology.* Volume 1; New York, NY: Springer, **2009**, 1262.

25. Chelule, P. K.; Mbongwa, H. P.; Carries, S.; Gqaleni, N. Lactic acid fermentation improves the Quality of *Amahewu*, a Traditional South African Maize-based Porridge. *Food Chemistry*, **2010**, *122*, 656–661.

26. Cho, H. D.; Lee, J. H.; Jeong, J. H.; Kim, J. Y.; Yee, S. T.; Park, S. K.; Lee, M. K.; Seo, K. I. Production of Novel Vinegar having Antioxidant and Anti-fatigue Activities from *Salicornia herbacea* L. *Journal of the Science of Food and Agriculture*, **2016**, *96*, 1085–1092.

27. Cho, S.; Shin, S.-Y.; Lee, S.; Li, L.; Moon, J. S.; Kim, D.-J.; Im, W.-T.; Han, N. Simple Synthesis of Isomaltooligosaccharides during Sauerkraut Fermentation by Addition of *Leuconostoc* Starter and Sugars. *Food Science and Biotechnology*, **2015**, *24*, 1443–1446.

28. Ciani, M.; Comitini, F. Non-*Saccharomyces* Wine Yeasts have a Promising Role in Biotechnological Approaches to Winemaking. *Annals of Microbiology*, **2011**, *61*, 25–32.

29. Clodoveo, M. L.; Dipalmo, T.; Rizzello, C. G.; Corbo, F; Crupi, P. Emerging Technology to Develop Novel Red Winemaking Practices: An Overview. *Innovative Food Science and Emerging Technologies*, **2016**, *38*, 41–56.

30. Coda, R.; Melama, L.; Rizzello, C.G.; Curiel, J.A.; Sibakov, J.; Holopainen, U.; Pulkkinen, M.; Sozer, N. Effect of Air Classification and Fermentation by *Lactobacillus plantarum* VTT E-133328 on Faba bean (*Vicia faba* L.) Flour Nutritional Properties. *International Journal of Food Microbiology*, **2015**, *193*, 34–42.

31. Comitini, F.; Ciani, M. The Zymocidial Activity of *Tetrapisispora phaffii* in the Control of *Hanseniaspora uvarum* during the Early Stages of Winemaking. *Letters in Applied Microbiology*, **2010**, *50*, 50–56.

32. Costa, M. G.; Fonteles, T. V.; Jesus, A. L.; Rodrigues, S. Sonicated Pineapple Juice as Substrate for *L. casei* Cultivation for Probiotic Beverage Development: Process Optimisation and Product Stability. *Food Chemistry*, **2013**, *139*, 261–266.

33. Coussement, P. A. Inulin and Oligofructose: Safe Intakes and Legal Status. *The Journal of Nutrition*, **1999**, *129* (7), 1412S-1417S.

34. Cummings, J. H.; Christie, S.; Cole, T. J. A Study of Fructo-oligosaccharides in the Prevention of Travellers' Diarrhoea. *Alimentary Pharmacology & Therapeutics*, **2001**, *15* (8), 1139–1145.

35. Dal Bello, F.; Clarke, C. I.; Ryan, L. A. M.; Ulmer, H.; Schober, T. J..; Ström, K.; Sjögren, J.; Van Sinderen, D.; Schnürer, J.; Arendt. E. K. Improvement of the Quality and Shelf life of Wheat Bread by Fermentation with the Antifungal strain *Lactobacillus plantarum* FST 1.7. *Journal of Cereal Science*, **2007**, *45*, 309–318.

36. D'Alessandro, A.; De Pergola, G. Mediterranean Diet Pyramid: A Proposal for Italian People. *Nutrients*, **2014**, *6* (10), 4302–4316.

37. Debon, J.; Prudêncio, E. S.; Petrus, J. C. C. Rheological and Physico-chemical Characterization of Prebiotic Microfiltered Fermented Milk. *Journal of Food Engineering*, **2010**, *99* (2), 128–135.

38. Du, R.; Song, G.; Zhao, D.; Sun, J.; Ping, W.; Ge, J. *Lactobacillus casei* Starter Culture Improves Vitamin Content, Increases Acidity and Decreases Nitrite Concentration during Sauerkraut Fermentation. *International Journal of Food Science & Technology*, **2018**, *53*, 1925–1931.

39. Dujon, B. A.; Louis, E. J. Genome Diversity and Evolution in the Budding Yeasts (*Saccharomycotina*). *Genetics*, **2017**, *206*, 717–750.

40. Elvira, L.; Vera, P.; Cañadas, F. J.; Shukla, S. K.; Montero, F. Concentration Measurement of Yeast Suspensions Using High Frequency Ultrasound Backscattering. *Ultrasonics*, **2016**, *64*, 151–161.

41. Emerald, M.; Rajauria, G.; Kumar, V. Novel Fermented Grain-Based Products. In: *Novel Food Fermentation Technologies*; Ojha, K. S. and Tiwari, B. K. (Eds.); Switzerland: Springer International Publishing, **2016**, 263-278.

42. Figueroa-Hernández, C.; Mota-Gutierrez, J.; Ferrocino, I.; Hernández-Estradac, Z. J.; González-Ríos, O.; Cocolin, L.; Suárez-Quiroz, M. L. The Challenges and Perspectives of the Selection of Starter Cultures for Fermented Cocoa Beans. *International Journal of Food Microbiology*, **2019**, *301*, 41–50.

43. Fratianni, F.; Cardinale, F.; Russo, I.; Iuliano, C.; Tremonte, P.; Coppola, R.; Nazzaro, F. Ability of Synbiotic Encapsulated *Saccharomyces cerevisiae boulardii* to Grow in Berry Juice and to Survive Under Simulated Gastrointestinal Conditions. *Journal of microencapsulation*, **2014**, *31*, 299–305.

44. Frydenberg, R. P.; Hammershøj, M.; Andersen, U.; Greve, M. T.; Wiking, L. Protein Denaturation of Whey Protein Isolates (WPIs) Induced by High Intensity Ultrasound during Heat Gelation. *Food Chemistry*, **2016**, *192*, 415–423.

45. Ganan, M.; Hierro, E.; Hospital, X. F.; Barroso, E; Fernández, M. Use of Pulsed Light to Increase the Safety of Ready-To-Eat Cured Meat Products. *Food Control*, **2013**, *32*, 512–517.

46. Gao, Y.; Li, D.; Liu, X. Bacteriocin-Producing *Lactobacillus sakei* C2 as Starter Culture in Fermented Sausages. *Food Control*, **2014**, *35*, 1–6.

47. García-Varona, M.; Santos, E. M.; Jaime, I.; Rovira, J. Characterisation of Micrococcaceae Isolated from Different Varieties of Chorizo. *International Journal of Food Microbiology*, **2000**, *54*, 189–195.

48. Gerez, C. L.; Torino, M. I.; Obregozo, M. D.; De Valdez, G. F. A Ready-to-use Antifungal Starter Culture improves the Shelf Life of Packaged Bread. *Journal of Food Protection*, **2010**, *73* (4), 758–762.

49. Giovenzana, V.; Beghi, R.; Guidetti, R. Rapid Evaluation of Craft Beer Quality during Fermentation process by vis/NIR Spectroscopy. *Journal of Food Engineering*, **2014**, *142*, 80–86.

50. Gobbetti, M.; De Angelis, M.; Di Cagno, R.; Calasso, M.; Archetti, G.; Rizzello, C. G. Novel Insights on the Functional/Nutritional Features of the Sourdough Fermentation. *International Journal of Food Microbiology*, **2019a**, *302*, 103–113.

51. Gobbetti, M.; De Angelis, M.; Di Cagno, R.; Polo, A.; Rizzello, C. G. The Sourdough Fermentation is the Powerful Process to Exploit the Potential of Legumes, Pseudo-cereals and Milling By-products in Baking Industry. *Critical Reviews in Food Science and Nutrition*, **2019b**, *60*, 1–16.

52. Gobbetti, M.; Rizzello, C. G.; Di Cagno, R.; De Angelis, M. How the Sourdough may affect the Functional Features of Leavened Baked Goods. *Food Microbiology*, **2014**, *37*, 30–40.

53. Goswami, R. P.; Jayaprakasha, G. K.; Shetty, K.; Patil, B. S. *Lactobacillus plantarum* and Natural Fermentation-mediated Biotransformation of Flavor and Aromatic compounds in Horse gram Sprouts. *Process Biochemistry*, **2018**, *66*, 7–18.

54. Gould, G. W. Methods for Preservation and Extension of Shelf Life. *International Journal of Food Microbiology*, **1996**, *33*, 51–64.

55. Gracin, L.; Jambrak, A. R.; Juretić, H.; Dobrović, S.; Barukčić, I.; Grozdanović, M., Smoljanić, G. Influence of High Power Ultrasound on *Brettanomyces* and Lactic Acid Bacteria in Wine in Continuous Flow Treatment. *Applied Acoustics*, **2015**, *103*, 143–147.

56. Grassi, S.; Amigo, J. M.; Lyndgaard, C. B.; Foschino, R.; Casiraghi, E. Beer Fermentation: Monitoring of Process Parameters by FT-NIR and Multivariate Data Analysis. *Food Chemistry*, **2014a**, *155*, 279–286.

57. Grassi, S.; Amigo, J. M.; Lyndgaard, C. B.; Foschino, R.; Casiraghi, E. Assessment of the Sugars and Ethanol Development in Beer Fermentation with FT-IR and Multivariate Curve Resolution Models. *Food Research International*, **2014b**, *62*, 602–608.

58. Grasso, S.; Brunton, N. P.; Lyng, J. G.; Lalor, F.; Monahan, F. J. Healthy Processed Meat Products–Regulatory, Reformulation and Consumer Challenges. *Trends in Food Science & Technology*, **2014**, *39*, 4–17.

59. Gupta, S. K.; Dangi, A. K.; Smita, M.; Dwivedi, S.; Shukla, P. Effectual Bioprocess Development for Protein Production. In: *Applied Microbiology and Bioengineering: An Interdisciplinary Approach*; Shukla, P. (Ed.); San Diego, CA: Academic Press, **2018**, 203–227.

60. Gupta, S.; Padole, R.; Variyar, P. S.; Sharma, A. Influence of Radiation Processing of Grapes on Wine Quality. *Radiation Physics and Chemistry*, **2015**, *111*, 46–56.

61. Harada, Y.; Sakata, K.; Sato, S.; Takayama, S. Fermentation Pilot Plant. In: *Fermentation and Biochemical Engineering Handbook*. 3rd ed. Todaro, C. L. and Vogel, H. C. (Eds.); Alpharetta, GA: Elsevier Science Ltd, **2014**, 3–15.

62. Harte, F.; Luedecke, L.; Swanson, B.; Barbosa-Canovas, G. V. Low-Fat Set Yogurt Made from Milk Subjected to Combinations of High Hydrostatic Pressure and Thermal Processing. *Journal of Dairy Science*, **2003**, *86*, 1074–1082.

63. Hill, D.; Sugrue, I.; Arendt, E.; Hill, C.; Stanton, C; Ross, R. P. Recent Advances in Microbial Fermentation for Dairy and Health. *F1000Research*, **2017**, *6*, 751.

64. Ho, V. T. T.; Zhao, J.; Fleet, G. The Effect of Lactic Acid Bacteria on Cocoa Bean Fermentation. *International Journal of Food Microbiology*, **2015**, *205*, 54–67.

65. Huppertz, T. Homogenization of Milk: High-Pressure Homogenizers. In: *Encyclopedia of Dairy Sciences*. 2nd ed. Fuquay, P. F.; Fox, P. L. H. and McSweeney, J. W. (Eds.); London: Elsevier Academic, **2011**, 755–760.

66. Iličić, M. D.; Carić, M. Đ.; Milanović, S. D.; Dokić, L. P.; Đurić, M. S.; Bošnjak, G. S.; Duraković, K. G. Viscosity Changes of Probiotic Yoghurt with Transglutaminase during Storage. *Structure*, **2008**, *3*, 4.

67. Iličić, M.; Milanović, S.; Carić, M. Rheology and Texture of Fermented Milk Products. In: *Rheology: Principal, Application and Environmental Impacts*; Karpushkin, E. (Ed.); New York, NY: Nova Science, **2015**, 27–64.

68. Inoue, T.; Nagatomi, Y.; Uyama, A.; Mochizuki, N. Fate of Mycotoxins during Beer Brewing and Fermentation. *Bioscience, Biotechnology, and Biochemistry*, **2013**, *77* (7), 130027-1-6.

69. Jomdecha, C.; Prateepasen, A. Effects of Pulse Ultrasonic Irradiation on the Lag Phase of *Saccharomyces cerevisiae* Growth. *Letters in Applied Microbiology*, **2011**, *52*, 62–69.

70. Kadow, D.; Niemenak, N.; Rohn, S.; Lieberei, R. Fermentation-like Incubation of Cocoa Seeds (*Theobroma cacao* L.)—Reconstruction and Guidance of the Fermentation Process. *LWT—Food Science and Technology*, **2015**, *62*, 357–361.

71. Katina, K.; Laitila, A.; Juvonen, R.; Liukkonen, K. H.; Kariluoto, S.; Piironen, V.; Landberg, R.; Åman, P.; Poutanen, K. Bran Fermentation as a Means to Enhance Technological Properties and Bioactivity of Rye. *Food Microbiology*, **2007**, *24* (2), 175–186.

72. Khan, M. I.; Arshad, M. S.; Anjum, F. M.; Sameen, A.; Gill, W. T. Meat as a Functional Food with Special Reference to Probiotic Sausages. *Food Research International*, **2011**, *44*, 3125–3133.

73. Khatoon, N.; Gupta, R. K. Probiotics Beverages of Sweet Lime and Sugarcane Juices and its Physiochemical, Microbiological & Shelf-life Studies. *Journal of Pharmacognosy and Phytochemistry*, **2015**, *4*, 25–34.

74. Kim, N.; Jang, H. L.; Yoon, K. Potato Juice Fermented with *Lactobacillus casei* as a Probiotic Functional Beverage. *Food Science and Biotechnology*, **2012**, *21*, 1301–1307.

75. Ko, J.; Kim, W. Y.; Park, H. J. Effects of Microencapsulated Allyl Isothiocyanate (AITC) on the Extension of the Shelf-Life of Kimchi. *International Journal of Food Microbiology*, **2011**, *153*, 92–98.

76. Koffi, A. S.; Yao, N.; Bastide, P.; Bruneau, D.; Kadjo, D. Homogenization of Cocoa Beans Fermentation to Upgrade Quality Using an Original Improved Fermenter. *International Journal of Food Science and Nutrition Engineering*, **2017**, *11*, 558–563.

77. Krishna, C. Solid-State Fermentation Systems—An Overview. *Critical Review in Biotechnology*, **2005**, *25*, 1–30.

78. Lee, M.-E.; Jang, J.-Y.; Lee, J.; Park, H. W.; Choi, H. J.; Kim, T.-W. Starter Cultures for Kimchi Fermentation. *Journal of Microbiology and Biotechnology*, **2015**, *25*, 559–568.

79. Lim, H.; Shin, H. Introduction to Fed-Batch Cultures. In: *Fed-Batch Cultures; Principles and Applications of Semi-Batch Bioreactors*; Lim, H. and Shin, H. (Eds.); Cambridge: Cambridge University Press, **2013**, 1–18.

80. Liu, J.; Li, D.; Hu, Y.; Wang, C.; Gao, B.; Xu, N. Effect of a Halophilic Aromatic Yeast together with *Aspergillus oryzae* in *Koji* making on the Volatile Compounds and Quality of Soy Sauce Moromi. *International Journal of Food Science & Technology*, **2015**, *50* (6), 1352–1358.

81. Luo, H.; Schmid, F.; Grbin, P. R.; Jiranek, V. Viability of Common Wine Spoilage Organisms after Exposure to High Power Ultrasonics. *Ultrasonics Sonochemistry*, **2012**, *19*, 415–420.

82. Marafon, A. P.; Sumi, A.; Alcântara, M. R.; Tamime, A. Y.; De Oliveira, M. N. Optimization of the Rheological Properties of Probiotic Yoghurts Supplemented with Milk Proteins. *LWT—Food Science and Technology*, **2011**, *44* (2), 511–519.

83. Marteau, P.; Jacobs, H.; Cazaubiel, M.; Signoret, C.; Prevel, J. M.; Housez, B. Effects of Chicory Inulin in Constipated Elderly People: A Double-Blind Controlled Trial. *International Journal of Food Sciences and Nutrition*, **2011**, *62* (2), 164–170.

84. Mascia, I.; Fadda, C.; Dostálek, P.; Karabín, M.; Zara, G.; Budroni, M.; Del Caro, A. Is it Possible to Create an Innovative Craft Durum Wheat Beer with Sourdough Yeasts? A case study. *Journal of the Institute of Brewing*, **2015**, *121* (2), 283–286.

85. Mateo, J. J.; Maicas, S. Application of Non-*Saccharomyces* Yeasts to Wine-Making Process. *Fermentation*, **2016**, *2*, 1–13.

86. Meyer, D.; Stasse-Wolthuis, M. Inulin and Bone Health. *Current Topics in Nutraceutical Research*, **2006**, *4*, 211–225.

87. Milanović, S. D.; Hrnjez, D. V.; Iličić, M. D.; Kanurić, K. G.; Vukić, V. R. Novel Fermented Dairy Products. In: *Novel Food Fermentation Technologies*; Ojha, K. S. and Tiwari, B. K. (Eds.); Switzerland: Springer International Publishing, **2016**, 165–202.

88. Milanović, S.; Iličić, M.; Djurić, M.; Carić, M. Effect of Transglutaminase and Whey Protein Concentrate on Textural Characteristics of Low Fat Probiotic Yoghurt. *Milchwissenschaft*, **2009**, *64* (4), 388–392.

89. Montemurro, M.; Coda, R.; Rizzello, C.G. Recent Advances in the Use of Sourdough Biotechnology in Pasta Making. *Foods*, **2019**, *8* (4), 129.

90. Najafpour, G. D. Application of Fermentation Processes. In: *Biochemical Engineering and Biotechnology*. 2nd ed. Najafpour, G. D. (Ed.); Amsterdam: Elsevier Science Ltd, **2007**, 252–262.

91. Nguyen, N. H.; Anema, S. G. Effect of Ultrasonication on the Properties of Skim Milk used in the Formation of Acid Gels. *Innovative Food Science & Emerging Technologies*, **2010**, *11* (4), 616–622.

92. Nielsen, D. S.; Hønholt, S.; Tano-Debrah, K.; Jespersen, L. Yeast Populations Associated with Ghanaian Cocoa Fermentations Analysed Using Denaturing Gradient Gel Electrophoresis (DGGE). *Yeast*, **2005**, *22*, 271–284.

93. Novoa-Díaz, D.; Rodríguez-Nogales, J.; Fernández-Fernández, E.; Vila-Crespo, J.; García-Álvarez, J.; Amer, M.; Chávez, D.; Juan, A.; Turó, A.; Garcia-Hernandez, M.; Salazar, J. Ultrasonic Monitoring of Malolactic Fermentation in Red Wines. *Ultrasonics*, **2014**, *54*, 1575–1580.

94. Ogunbanwo, S. T.; Okanlawon, B. M. Biopreservative activities of *Lactobacillus acidophilus* U1 during Fermentation of Fresh Minced Goat Meat. *Journal of Applied Biosciences*, **2008**, *12*, 650–656.

95. Ojha, K. S.; Kerry, J. P.; Duffy, G.; Beresford, T.; Tiwari, B. K. Technological Advances for Enhancing Quality and Safety of Fermented Meat Products. *Trends in Food Science & Technology*, **2015**, *44*, 105–116.

96. Okeke, C. A.; Ezekiel, C. N.; Nwangburuka, C. C.; Sulyok, M.; Ezeamagu, C. O.; Adeleke, R. A.; Dike, S. K.; Krska, R. Bacterial Diversity and Mycotoxin Reduction During Maize Fermentation (Steeping) for *Ogi* Production. *Frontiers in Microbiology*, **2015**, *6*, 1402.

97. Olatunji, T.; Chakraborty, S.; Tripathi, M.; Kotwaliwale, N.; Chandra, P. Quality Characteristics of Sauerkraut Fermented by using a *Lactobacillus paracasei* Starter Culture Grown in Tofu Whey. *Food Science and Technology International*, **2017**, *24*, 187–197.

98. Oliveira, R. P. D. S.; Florence, A. C. R.; Perego, P.; De Oliveira, M. N.; Converti, A. Use of Lactulose as Prebiotic and its Influence on the Growth, Acidification Profile and Viable Counts of Different Probiotics in Fermented Skim Milk. *International Journal of Food Microbiology*, **2011**, *145* (1), 22–27.

99. Omer, M. K.; Alvseike, O.; Holck, A.; Axelsson, L.; Prieto, M.; Skjerve, E.; Heir, E. Application of High Pressure Processing to Reduce Verotoxigenic *E. coli* in Two Types of Dry-Fermented Sausage. *Meat Science*, **2010**, *86*, 1005–1009.

100. Ozer, B.; Kirmaci, H. A.; Oztekin, S.; Hayaloglu, A.; Atamer, M. Incorporation of Microbial Transglutaminase into Non-Fat Yogurt Production. *International Dairy Journal*, **2007**, *17* (3), 199–207.

101. Padilla, B.; Gil, J. V.; Manzanares, P. Past and Future of Non-*Saccharomyces* Yeasts: From Spoilage Microorganisms to Biotechnological Tools For Improving Wine Aroma Complexity. *Frontiers in Microbiology*, **2016**, *7*, 411.

102. Pakbin, B.; Razavi, S.; Mahmoudi, R.; Gajarbeygi, P. Producing Probiotic Peach Juice. *Journal of Biotechnology and Health Science*, **2014**, *1*, 1–5.

103. Papalexandratou, Z.; Lefeber, T.; Bahrim, B.; Lee, O. S.; Daniel, H.-M.; De Vuyst, L. *Hanseniaspora opuntiae, Saccharomyces cerevisiae, Lactobacillus fermentum*, and *Acetobacter pasteurianus* Predominate During Well-Performed Malaysian Cocoa Bean Box Fermentations, Underlining the Importance of these Microbial Species for a Successful Cocoa. *Food Microbiology*, **2013**, *35*, 73–85.

104. Reuterswärd, A. L. The New EC Regulation on Nutrition and Health Claims on Foods. *Scandinavian Journal of Food and Nutrition*, **2007**, *51*, 100–106.

105. Parra, P.; Castillo, O.; Maldonado, P. Alternative Method for the Fermentation of Cocoa Beans. In: *IEEE International Conference on Automation/23rd Congress of the Chilean Association of Automatic Control*, **2018**, 1–6.

106. Parra, P.; Negrete, T.; Llaguno, J.; Vega, N. Computer Vision Methods in the Process of Fermentation of the Cocoa Bean. In: *IEEE Third Ecuador Technical Chapters Meeting (ETCM)*; Cuenca, **2018**, 1–6.

107. Penna, A.; Rao-Gurram, S.; Barbosa-Canovas, G. V. Effect of Milk Treatment on Acidification, Physicochemical Characteristics, and Probiotic Cell Counts in Low Fat Yogurt. *Milchwissenschaft*, **2007**, *62*, 48–52.

108. Pennacchia, C.; Vaughan, E. E.; Villani, F. Potential Probiotic Lactobacillus Strains from Fermented Sausages: Further Investigations on their Probiotic Properties. *Meat Science*, **2006**, *73*, 90–101.

109. Pereira, A. L.; Maciel. T.; Rodrigues, S. Probiotic Beverage from Cashew Apple Juice Fermented with *Lactobacillus casei*. *Food Research International*, **2011**, *44*, 1276–1283.

110. Pereira, G. V. M.; Alvarez, J. P.; Neto, Dã. Pedro. de. C.; Soccol, V. T.; Tanobe, V. O. A.; Rogez, H.; Góes-Neto, A.; Soccol, C. R. Great Intraspecies Diversity of *Pichia Kudriavzevii* in Cocoa Fermentation Highlights the Importance of Yeast Strain Selection for Flavour Modulation of Cocoa Beans. *LWT—Food Science and Technology*, **2017**, *84*, 290–297.

111. Pereira, G. V. M.; Miguel, M. G. C. P.; Ramos, C. L.; Schwan, R. F. Microbiological and Physicochemical Characterization of Small-Scale Cocoa Fermentations and Screening of Yeast and Bacterial Strains to Develop a Defined Starter Culture. *Applied Environmental Microbiology*, **2012**, *78*, 5395–5405.

112. Pérez-Lamela, C.; Ledward, D. A.; Reed, R. J. R.; Simal-Gándara, J. Application of High-Pressure Treatment in the Mashing of White Malt in the Elaboration Process of Beer. *Journal of the Science of Food and Agriculture*, **2002**, *82* (3), 258–262.

113. Petruláková, M.; Valík, Ľ. Legumes as Potential Plants for Probiotic Strain *Lactobacillus rhamnosus* GG. *Acta Universitatis Agriculturae et Silviculturae Mendelianae Brunensis*, **2015**, *63* (5), 1505–1511.

114. Petry, S.; Furlan, S.; Waghorne, E.; Saulnier, L.; Cerning, J.; Maguin, E. Comparison of the Thickening Properties of Four *Lactobacillus delbrueckii* subsp. *bulgaricus* strains and Physicochemical Characterization of their Exopolysaccharides. *FEMS Microbiology Letters*, **2003**, *221* (2), 285–291.

115. Piper, P.; Ortiz Calderon, C.; Hatzixanthis, K.; Mollapour, M. Weak Acid Adaptation: The Stress Response that Confers Yeast with Resistance to Organic Acid Food Preservatives. *Microbiology*, **2001**, *147*, 2635–2642.

116. Prasanna, P. H. P.; Grandison, A. S.; Charalampopoulos, D. Microbiological, Chemical and Rheological Properties of Low Fat Set Yoghurt produced with Exopolysaccharide (EPS) producing *Bifidobacterium* strains. *Food Research International*, **2013**, *51* (1), 15–22.

117. Puértolas, E.; Saldaña, G.; Condón, S.; Alvarez, I.; Raso, J. A Comparison of the Effect of Macerating Enzymes and Pulsed Electric Fields Technology on Phenolic Content and Color of Red Wine. *Journal of Food Science*, **2009**, *74*, 647–652.

118. Quiros, M.; Rojas, V.; Gonzalez, R.; Morales, P. Selection of Non-*Saccharomyces* Yeast Strains for Reducing Alcohol Levels in Wine by Sugar Respiration. *International Journal of Food Microbiology*, **2014**, *181*, 85–91.

119. Rabie, M. A.; Siliha, H.; El-Saidy, S.; El-Badawy, A. A; Malcata, F. X. Effects of γ-Irradiation upon Biogenic Amine Formation in Egyptian Ripened Sausages during Storage. *Innovative Food Science & Emerging Technologies*, **2010**, *11*, 661–665.

120. Ray, R. C.; Joshi, V. K. Fermented foods: Past, Present and Future. In: *Microorganisms and Fermentation of Traditional Foods*; Ramesh C. R. and Montet, D. (Eds.); New York, NY: CRC Press, **2014**, 1–36.

121. Rizzello, C. G.; Nionelli, L.; Coda, R.; Di Cagno, R.; Gobbetti, M. Use of Sourdough Fermented Wheat Germ for Enhancing the Nutritional, Texture and Sensory Characteristics of the White Bread. *European Food Research and Technology*, **2010**, *230* (4), 645–654.

122. Roda, A.; Faveri, D. M.; Giacosa, S.; Dordoni, R.; Lambri, M. Effect of Pre-treatments on the Saccharification of Pineapple Waste as a Potential Source for Vinegar Production. *Journal of Cleaner Production*, **2016**, *112*, 4477–4484.

123. Rodrigues, K. L.; Araújo, T. H.; Schneedorf, J. M.; De Souza Ferreira, C.; Moraes, G. D. O. I.; Coimbra, R. S.; Rodrigues, M. R. A Novel Beer fermented by Kefir enhances Anti-inflammatory and Anti-ulcerogenic Activities found Isolated in its Constituents. *Journal of Functional Foods*, **2016**, *21*, 58–69.

124. Rodriguez-Campos, J.; Escalona-Buendía, H. B.; Orozco-Avila, I.; Lugo-Cervantes, E.; Jaramillo-Flores, M. E. Dynamics of Volatile and Non-Volatile Compounds in Cocoa (*Theobroma Cacao* L.) During Fermentation and Drying Processes using Principal Components Analysis. *Food Research International*, **2011**, *44*, 250–258.

125. Ruas-Madiedo, P.; Hugenholtz, J.; Zoon, P. An Overview of the Functionality of Exopolysaccharides produced by Lactic Acid Bacteria. *International Dairy Journal*, **2002**, *12* (2–3), 163–171.

126. Ruas-Madiedo, P.; Moreno, J. A.; Salazar, N.; Delgado, S.; Mayo, B.; Margolles, A.; Clara, G. Screening of Exopolysaccharide-producing *Lactobacillus* and *Bifidobacterium* strains Isolated from the Human Intestinal Microbiota. *Applied and Environmental Microbiology*, **2007**, *73* (13), 4385–4388.

127. Rubio, R.; Bover-Cid, S.; Martin, B.; Garriga, M.; Aymerich, T. Assessment of Safe *Enterococci* as Bioprotective Cultures in Low-Acid Fermented Sausages Combined with High Hydrostatic Pressure. *Food Microbiology*, **2013**, *33*, 158–165.

128. Russo, P.; De Chiara, M. L. V.; Capozzi, V.; Arena, M. P.; Amodio, M. L.; Rascón, A.; Dueñas, M. T.; López, P.; Spano, G. *Lactobacillus plantarum* strains for Multifunctional Oat-based Foods. *LWT—Food Science and Technology*, **2016**, *68*, 288–294.

129. Samagaci, L.; Ouattara, H.; Niamké, S.; Lemaire, M. *Pichia kudrazevii* and *Candida nitrativorans* are the Most Well-Adapted and Relevant Yeast Species Fermenting Cocoa in Agneby-Tiassa, a Local Ivorian Cocoa Producing Region. *Food Research International*, **2016**, *89*, 773–780.

130. Santos, M. D.; Nunes, C.; Rocha, M. A.; Rodrigues, A.; Rocha, S. M.; Saraiva, J. A.; Coimbra, M. A. High Pressure Treatments Accelerate Changes in Volatile Composition of Sulphur Dioxide-Free Wine During Bottle Storage. *Food Chemistry*, **2015**, *188*, 406–414.

131. Serra, M.; Trujillo, A. J.; Guamis, B.; Ferragut, V. Flavour Profiles and Survival of Starter Cultures of Yoghurt Produced from High-Pressure Homogenized Milk. *International Dairy Journal*, **2009**, *19*, 100–106.

132. Serra, M.; Trujillo, A. J.; Quevedo, J. M.; Guamis, B.; Ferragut, V. Acid Coagulation Properties and Suitability For Yogurt Production of Cows' Milk Treated by High-Pressure Homogenisation. *International Dairy Journal*, **2007**, *17*, 782–790.

133. Sfakianakis, P.; Tzia, C. Flavor and Sensory Characteristics of Yogurt Derived from Milk Treated by High Intensity Ultrasound. In: *Nutrition, Functional and Sensory Properties of Foods;* Ho, C. T.; Mussinan, C.; Shahidi, F. and Contis, E. T. (Eds); Camridge: RSC Publishing;, **2013**, 92–97.

134. Sharma, V.; Mishra, H. N. Fermentation of Vegetable Juice Mixture by Probiotic Lactic Acid Bacteria. *Nutrafoods*, **2013**, *12*, 17- 22.

135. Simon-Sarkadi, L.; Pásztor-Huszár, K.; Dalmadi, I.; Kiskó, G. Effect of High Hydrostatic Pressure Processing on Biogenic Amine Content of Sausage During Storage. *Food Research International*, **2012**, *47*, 380–384.

136. Slouka, C.; Brunauer, G.; Kopp, J.; Strahammer, M.; Fricke, J.; Fleig, J.; Herwig, C. Low-frequency Electrochemical Impedance Spectroscopy as a Monitoring Tool for Yeast Growth in Industrial Brewing Processes. *Chemosensors*, **2017**, *5* (3), 24.

137. Song, H. M.; Cho, R. K.; Hae, C. C. Heterofermentative Lactic Acid Bacteria as a Starter Culture to Control Kimchi Fermentation. *LWT—Food Science and Technology*, **2018**, *88*, 181–188.

138. Srianta, I. *Monascus*-Fermented Sorghum: Pigments and Monacolin K produced by *Monascus purpureus* on Whole grain, Dehulled grain and Bran substrates. *International Food Research Journal*, **2015**, *22* (1), 377–382.

139. Struyf, N.; Laurent, J.; Verspreet, J.; Verstrepen, K. J.; Courtin, C.M. *Saccharomyces cerevisiae* and *Kluyveromyces marxianus* cocultures allow Reduction of Fermentable Oligo-, Di-, and Monosaccharides and Polyols levels in Whole Wheat Bread. *Journal of Agricultural and Food Chemistry*, **2017**, *65* (39), 8704–8713.

140. Susilowati, A.; Aspiyanto, A.; Lotulung, P. D.; Maryati, Y. Formulation of Emulsion of Soybean (*Glycinia soyae* L.) Tempeh and Fermented Spinach (*Amaranthus* sp.) Using Combination of Gelatin and CMC as Thickener. *Indonesian Journal of Pharmaceutical Science and Technology*, **2019**, *6* (3), 95–103.

141. Tao, Y.; Sun, D.; Górecki, A. R.; Błaszczak, W.; Lamparski, G.; Amarowicz, R.; Fornal, J.; Jeliński, T. Effects of High Hydrostatic Pressure Processing on the Physicochemical and Sensorial Properties of a Red Wine. *Innovative Food Science and Emerging Technologies*, **2012**, *16*, 409–416.

142. Terefe, N. S. Emerging Trends and Opportunities in Food Fermentation. In: *Reference Module in Food Science*; Trinetta, V. (Ed.); Amsterdam: Elsevier, **2016**.

143. Tidona, F.; Meucci, A.; Povolo, M.; Pelizzola, V.; Zago, M.; Contarini, G.; Carminati, D; Giraffa, G. Applicability of *Lactococcus hircilactis* and *Lactococcus laudensis* as Dairy Cultures. *International Journal of Food Microbiology*, **2018**, *271*, 1–7.

144. Toldrá, F.; Hui, Y. H.; Astiasarán, I.; Sebranek, J.; Talon, R. (Eds.) *Handbook of Fermented Meat and Poultry*. Hoboken, NJ: John Wiley & Sons, **2014**, 528.

145. Ton, N. M. N.; Le, V. V. M. Application of Immobilized Yeast in Bacterial Cellulose to the Repeated Batch Fermentation in Wine-Making. *International Food Research Journal*, **2011**, *18*, 983–987.

146. Tripathi, M. K.; Giri, S. K. Probiotic Functional Foods: Survival of Probiotics during Processing and Storage. *Journal of Functional Foods*, **2014**, *9*, 225–241.

147. Työppönen, S.; Petäjä, E.; Mattila-Sandholm, T. Bioprotectives and Probiotics for Dry Sausages. *International Journal of Food Microbiology*, **2003**, *83*, 233–244.

148. Van Wyk, S.; Silva, F. V. Impedance Technology reduces the Enumeration Time of *Brettanomyces* Yeast during Beer Fermentation. *Biotechnology Journal*, **2016**, *11* (12), 1667–1672.

149. Vieira, C. P.; Cabral, C. C.; da Costa Lima, B. R.; Paschoalin, V. M. F.; Leandro, K. C.; Conte-Junior, C. A. *Lactococcus lactis* ssp. *cremoris* MRS47, a Potential Probiotic Strain Isolated From Kefir Grains, Increases Cis-9, Trans-11-CLA and PUFA Contents in Fermented Milk. *Journal of Functional Foods*, **2017**, *31*, 172–178.

150. Wang, L.; Yang, S. T. Solid State Fermentation and Its Applications. In: *Bioprocessing for Value-Added Products from Renewable Resources*; Yang, S. T. (Ed.); Amsterdam: Elsevier, **2007**, 465–489.

151. Wu, H.; Rui, X.; Li, W.; Chen, X.; Jiang, M.; Dong, M. Mung bean (*Vigna radiata*) as Probiotic food through Fermentation with *Lactobacillus plantarum* B1–6. *LWT—Food Science and Technology*, **2015**, *63* (1), 445–451.

152. Xiaozhe, Y.; Wenzhong, H.; Aili, J.; Zhilong, X.; Yaru, J.; Yuge, G.; Sarengaowa, X. Y. Effect of Salt Concentration on Quality of Chinese Northeast Sauerkraut Fermented by *Leuconostoc mesenteroides* and *Lactobacillus plantarum*. *Food Bioscience*, **2019**, *30*, 100421.

153. Yeom, H. W.; Evrendilek, G. A.; Jin, Z. T.; Zhang, Q. H. Processing of Yogurt-based Products with Pulsed Electric Fields: Microbial, Sensory and Physical Evaluations. *Journal of Food Processing and Preservation*, **2004**, *28* (3), 161–178.

154. Yro, A.; N'zi, C. E.; Kpalma, K. Cocoa Beans Fermentation Degree Assessment for Quality Control Using Machine Vision and Multiclass SVM Classifier. *International Journal of Innovation and Applied Studies*, **2018**, *24*, 1711–1717.

155. Zdolec, N.; Hadžiosmanović, M.; Kozačinski, L.; Cvrtila, Ž.; Filipović, I.; Škrivanko, M. Influence of Protective Cultures on Listeria Monocytogenes in Fermented Sausages: A Review. *Archiv für Lebensmittelhygiene*, **2008**, *59*, 60–64.

156. Zhang, W.; Xiao, S.; Samaraweera, H.; Lee, E. J.; Ahn, D. U. Improving Functional Value of Meat Products. *Meat Science*, **2010**, *86*, 15–31.

157. Zhao, C. J.; Kinner, M.; Wismer, W.; Gänzle, M. G. Effect of Glutamate Accumulation during Sourdough Fermentation with *Lactobacillus reuteri* on the Taste of Bread and Sodium-reduced Bread. *Cereal Chemistry*, **2015**, *92*, 224–230.

158. Zhao, D.; Barrientos, J. U.; Wang, Q.; Markland, S. M.; Churey, J. J.; Padilla-Zakour, O. I.; Worobo, R. W.; Kniel, K. E.; Moraru, C. I. Efficient Reduction of Pathogenic and

Spoilage Microorganisms from Apple Cider by Combining Microfiltration with UV Treatment. *Journal of Food Protection*, **2015**, *78*, 716–722.

159. Zhao, L.; Jin, Y.; Ma, C.; Song, H.; Li, H.; Wang, Z.; Xiao, S. Physico-Chemical Characteristics and Free Fatty Acid Composition of Dry Fermented Mutton Sausages as Affected by the use of Various Combinations of Starter Cultures and Spices. *Meat Science*, **2011**, *88*, 761–766.

CHAPTER 10

PHYSICOCHEMICAL AND THERMAL PROPERTIES OF BAKERY PRODUCTS

SUDHARSHAN R. RAVULA, DIVYASREE AREPALLY*, A.K. DATTA, and T.K. GOSWAMI

ABSTRACT

In the baking process, mode of radiation is the predominant heat transfer accounting for 50%–80% of total heat compared to convection and conduction modes. In the past two decades, mathematical modeling and simulations in food processing have become principal tools for comprehension and predicting the transport phenomena. Hence, the knowledge of physicothermal properties of foods is needed in order to design and select the efficient processing equipment, to estimate the process time, and to control the operating costs. Moreover, the fortification of bakery products with different fiber and protein-rich sources is currently being investigated to improve the nutritional and functional properties. This chapter focuses on physical, chemical, thermal properties, and fortification of bakery products: the various measurement techniques; predictive models; and computer tools for the estimation of thermophysical properties.

10.1 INTRODUCTION

The interest in bakery products is growing day by day because of their nutritional properties and feasibility of their use in the feeding programs and in disaster situations, such as earthquakes. Bakery products are produced on small-scale or in local bakeries serving local consumers to larger-scale bakeries serving the entire country. Among the unit operations to manufacture bakery products, the baking process is an energy-consuming process.

*Corresponding author. E-mail: divyasreearepally@gmail.com

It transforms semisolid dough to eatable product under the influence of heat through intricate physical and biochemical reactions, such as protein denaturation, water evaporation, starch gelatinization, browning, Mallard reactions, and dough deformation [14].

Bakery products are produced by the straight-dough method, sponge and dough method, and Chorleywood method. During baking, the dough pieces are often spread, the extent of which depends on the components of the recipe, flour, sugar, fat characteristics, dough preparation, and hot-oven conditions. In general, large commercial bakeries use less energy per unit mass of the product compared to small bakeries due to efficiency. Bakery products are commonly made from cereals. Cereal and sprouted grains play a significance role in the production of functional bakery products. The nutritional value of bakery products can be increased by fortification and supplementation with a wide variety of protein-rich cereals and pulses.

Heat, mass, and momentum transfer occur simultaneously during the baking process. Among three modes of heat transfer (convection, conduction, and radiation) during the baking process, radiation is the most predominant mode of heat transfer. The heat transfer is through convection and radiation to the surface of the product from the medium and interior walls of the oven, respectively; and by conduction from the baking pan/mold to the product and within the product. Simultaneously, moisture diffuses inside out of the product.

The temperature and moisture distribution play a significance role to assess the quality of the product. In order to predict these spatial distributions using suitable heat and mass transfer diffusion equations, knowledge of physical and thermal properties is a must. As the baking process proceeds, the temperature/moisture (particularly crust for bread and cake)/and volume changes resulting changes in properties of the product. The physicothermal properties are helpful for analyzing the heat transfer and for designing process equipment. Recently, numerical modeling and simulation models have been successfully applied to understand the physical phenomena during the processing. The predicted thermophysical properties of bakery products are not always readily available.

Baik et al. [7] and Rask [36] studied the properties of bakery products with their predicted models. The thermal properties of most water-rich foods are temperature-dependent and also depend on the food composition. Several research studies have reported measured values of properties and few data on predicted models. However, the properties vary with product formulations and processing conditions.

This chapter focuses on (1) physical, chemical, thermal properties, and fortification of bakery products; (2) the various measurement techniques; and (3) predictive models and computer tools for the estimation of thermo-physical properties.

Appendix A.I indicates all equations to estimate the thermal and physical properties of bakery products.

10.2 MEASUREMENT TECHNIQUES FOR THERMAL AND PHYSICAL PROPERTIES

10.2.1 DENSITY

10.2.1.1 SEED REPLACEMENT METHOD

This method is used to determine the density of irregular materials, for example, bakery products in which deformation occurs. The rape and mustard seeds are most commonly used for baked products. To determine the bulk density of seeds, a glass container is filled with known volume uniformly with seeds. The seeds and sample materials are placed together in a container and then the weight of the whole material is measured. Total volume of the sample is then calculated using Eq. (A.1). After knowing the volume, the apparent density (ρ_b) is calculated from Eq. (A.2).

All equations to estimate the thermal and physical properties of bakery products are indicated in Appendix A.I in this chapter.

10.2.1.2 PYCNOMETRY

This method is mostly applicable to liquid or semisolid samples. The mass of the sample is weighed. The same sample is kept in a cylinder or container of known volume [41]. Density is calculated based on data obtained for mass and volume.

10.2.2 VOLUME MEASUREMENT

20.2.2.1 SOLID DISPLACEMENT METHOD

Water displacement method (WPM) is not an accurate method for the measurement of volume of porous materials, such as bread, cake, etc. The

WPM has been modified with seeds (e.g., rape seeds). This solid displacement method is mainly useful to determine the volume of irregular baked products. Prior to the test, density (ρ_b) of the seed is determined. The mass of the sample (W_s) and the container or cylinder (W_c) is measured. The sample is filled with seeds in a cylinder. The whole weight is taken ($W_{c+seed+s}$). Knowing the mass of the container (W_c), the sample volume (V_s) can be determined with Eq. (A.3).

However, the solid displacement method is less accurate because the bread or cake may be compressed when seeds are filled. Under such situations, computer vision technique is very useful. For example, in the case of bread loaf and bread, 3D scanner and computer program MeshLab can be used.

10.2.3   SPECIFIC HEAT (C_p) MEASUREMENTS

10.2.3.1   METHOD OF MIXING

Budžaki and Šeruga [10] described the determination of C_p of dough with an indirect method that is described here. A known quantity of water is taken at low temperatures in the calorimeter. The temperature change in water in the calorimeter is determined. In this method, heat losses are neglected. Then, the heat capacity is estimated from Eq. (A.4).

From Eq. (A.4), the average value of H_f is obtained. The samples are prepared in the homogeneous, filled, vacuumed, and packed in polyethylene pouches to prevent direct exposure to the heating medium. At room temperature, packed samples are placed in the heating medium in the flask calorimeter. The thermal energy is then released from the material to the calorimeter. Once the equilibrium state has reached, the C_p of dough can be calculated. Moreover, during calculations, the heat capacity of the polyethylene pouch is ignored. The energy balance equation is used to determine the C_p of the dough (Eq. (A.5)).

From Eq. (A.5), the average values of specific heat can be obtained over a range of temperatures. Budžaki and Šeruga [10] reported that the C_p of unleavened dough with indirect mixing method varied significantly with the temperature at different levels of water content. Hwang and Hayakawa [20] measured the C_p of biscuit with an indirect mixing method at a temperature >100 °C. The C_p for biscuit varied from 1875 to 1942.7 J/(kg K) for a moisture content ranging from 3.15% to 3.87%. Using the mixing method, Kulacki and Kennedy [25] found that the uncertainty in C_p of cookie-dough was 5%; and the C_p was increased with an increase in temperature.

10.2.3.2 DIFFERENTIAL SCANNING CALORIMETER

Differential scanning calorimeter (DSC) is a thermo-analytical technique [34] to measure the specific heat of a homogeneous product over a wide range of temperatures [15]. This method requires an only small amount of sample for measurement and therefore heterogeneous samples require large number of replications. Nowadays, a modulated differential scanning calorimeter (MDSC) technique, which provides the more accurate value of C_p, has been successfully used.

In MDSC method, the product is exposed to a linear heating, which is provided with temperature modulation that results in a cyclic heating profile. This temperature modulation separates the total heat flow into its reversing (C_p related) and nonreversing (kinetic) components. Therefore, MDSC gives more accurate C_p values compared to conventional DSC. The main variables considered for the measurement of C_p with MDSC are heating speed, modulation period, and amplitude.

Christenson et al. [16] determined the C_p of muffin, biscuit, and cupcake for a wide range of temperatures (20–85 °C) and moisture (0%–60%) using DSC. Similarly, Baik et al. [6] determined that C_p of baked samples with MDSC at specific baking times (0–19.5 min); and they observed that the C_p of cake batter was increased after 13 min of baking time, followed by decrease toward the end of the baking process.

10.2.4 THERMAL CONDUCTIVITY

The steady-state and transient techniques are mostly used to determine the thermal conductivity of food. In the steady-state method, it takes more time (>12 h) for the thermal conductivity analysis, depending on size of the sample. Since longer exposure time to high temperatures, the steady-state methods provide poor testing results due to migration of moisture and changes in property of the product. Therefore, this method is not frequently used for the determination of thermal conductivity (k) of bakery products. Due to the major drawbacks of steady-state technique, the transient or unsteady-state methods are more suitable and recommended. The transient methods provide better results and quicker measurements [26].

10.2.4.1   STEADY-STATE METHODS

In this method, the heat flux is estimated either by employing the electric resistance or induction coils. Based on Newton's law of cooling, heat transfer coefficient is estimated. The temperature of the surface is computed by taking the average surface temperature of the system. No mass transfer is assumed. In this process, it takes more time for reaching the thermal equilibrium.

10.2.4.1.1   Guarded Hot Plate

It is a simple method that can easily handle experimental conditions. The guarded hot plate technique is known for determining thermal conductivity (k) of biological materials using Fourier's equation based on the steady-state conditions. According to ASTM Standards [5], the test sample is placed between the hot plate (which is maintained at a specific temperature) and cold plates (which are maintained at lower temperatures by controlling with water). The heat is then transferred from the hot plate to the cold plate through the sample. The k of test sample is then estimated by determining the heat flux across the tested material for a resulting temperature gradient. This method poses the problem of escaping heat from the edges of slab. If edges are insulted, then it may give two-dimensional temperature profile, resulting errors in the measurements. These losses can be prevented by guard-rings or heaters. The thermal conductivity is calculated with Fourier's equation (A.6).

Based on experimental studies, Kulacki and Kennedy [25] estimated the uncertainty in the k of dough to be 7.4% for the AACC-formula biscuit dough and 9% for hard sweet biscuit dough using the single-plate method. They concluded that the k of biscuit dough was moderately temperature dependent.

10.2.4.1.2   Differential Scanning Calorimeter

Buhari and Singh [11] used the DSC method to determine the thermal conductivity of food. This method can overcome the disadvantages that are posed in steady-state methods. The advantages of this method are fast measurements in 600–900 s, small sample size, and no water loss. Moreover, DSC technique is reliable, precise, and suitable for low and high moisture foods.

In this technique, the probe is placed into the selected food. The temperature of the DSC heating pan is maintained at 40 °C. The initial temperature of food is registered after 5 min. The pan temperature is then immediately raised by 10 °C. The final temperature of the sample is recorded, once the new steady state has reached after 10–15 min. Thermal conductivity is then calculated with Eq. (A.7).

10.2.4.1.3 Capped Column Test

The capped column test (CCT) method was designed by Zhou et al. [48] for the determination of thermal conductivity. The authors tested potato, bread, and bread dough. In their experiment, the constant heat flux and no moisture loss were taken into account. The constant heat flux was given by providing the circulation of hot and cold water at constant temperatures on both ends of the cylindrical apparatus. To lower the heat losses, the tested material was enclosed in a polystyrene foam. To stop the effects of gravity-induced moisture migration, the CCT device was operated in the horizontal plane. Once the system reached the steady state, temperature and moisture distributions were measured. This temperature distribution was determined by T-type thermocouples and registered using the HP-3497A data acquisition system along with the height of the test sample and water stream. The moisture was determined by the standard oven method. For moisture measurement, the sample was cut into various sections of equal thicknesses at the end of the method. The thermal conductivity was estimated by using the temperature and moisture gradient data that were used simultaneously in the heat and mass transfer equations.

10.2.4.2 TRANSIENT METHODS

10.2.4.2.1 Line Heat Source Probe

This method is commonly used for the determination of k of biological materials. This method has been found to be simple, fast, and reliable. It takes a shorter time for measurements and requires relatively smaller samples. A sensor is kept into the material, which is assumed to be an infinite body. The thermal conductivity is determined from the observed temperature rise in the sensor due to the heating power. The thermal conductivity is estimated from of slope of a straight line between log time versus temperature as mentioned in Eq. (A.8).

10.2.5 THERMAL DIFFUSIVITY

Thermal diffusivity of any food material can be estimated either with an analytical method or probe method. In the analytical method, the properties (viz., mass density, thermal conductivity, and specific heat) must be known from experimental studies. This method is widely used for the estimation of diffusivity using Eq. (A.9).

In the probe method, k and of material are measured simultaneously. This method involves attaching one more temperature sensor at a particular distance (r_d) from the probe heater. This distance should be maintained as defined in Eq. (A.10).

Nevertheless, this method is not used for porous material (e.g., bread, cake, muffin, etc.), where significant deformation occurs. In a research study by Kulacki and Kennedy [25], the thermal diffusivity of biscuit dough ranged from 0.8×10^{-9} to 1.2×10^{-9} m²/s at the variation of moisture from 4.1% to 8.5%, and density from 1.25 to 1.28 g/cm³. Sweat [42] found the thermal diffusivity of cake from the relationship between k, C_p, and. The thermal diffusivity was increased from 1.09×10^{-7} to 1.43×10^{-7} m²/s during the baking process of a cake.

10.3 THEORETICAL MODELS

There are different theoretical models for thermal properties that widely use computer simulations to understand heat and mass mechanisms. The most common models used are parallel model, series model, random model, Maxwell–Eucken model, and Levy model, etc.

In the parallel model, the effective thermal properties are expressed in terms of the weighted arithmetic mean of conductivities of each phase of a component. A general expression for the thermal properties of foods can be expressed as shown in Eqs. (A.11)–(A.13).

Moreover, the effective heat capacity of food can be estimated by summing the volume heat capacities of each phase as suggested by Nicolas

[31] in Eq. (A.14). The effective thermal conductivity of bakery products uses the Law of Addition of thermal resistances for a multiphase system in series (Eq. (A.15)).

The series and parallel models (A.11)–(A.15) are analogies to Ohm's law describing that the flow of current is related to the resistance and voltage. However, the parallel model is mostly used for liquid food.

10.3.1 RANDOM MODEL

This model predicts values that are in "between" the series and parallel model. For multiphase porous materials, thermal conductivity was observed in between series and parallel models [47]. It is higher than that of a series model. The random model for multiphase porous materials can be expressed as a weighted geometric of the means of the component phase (Eq. (A.16)).

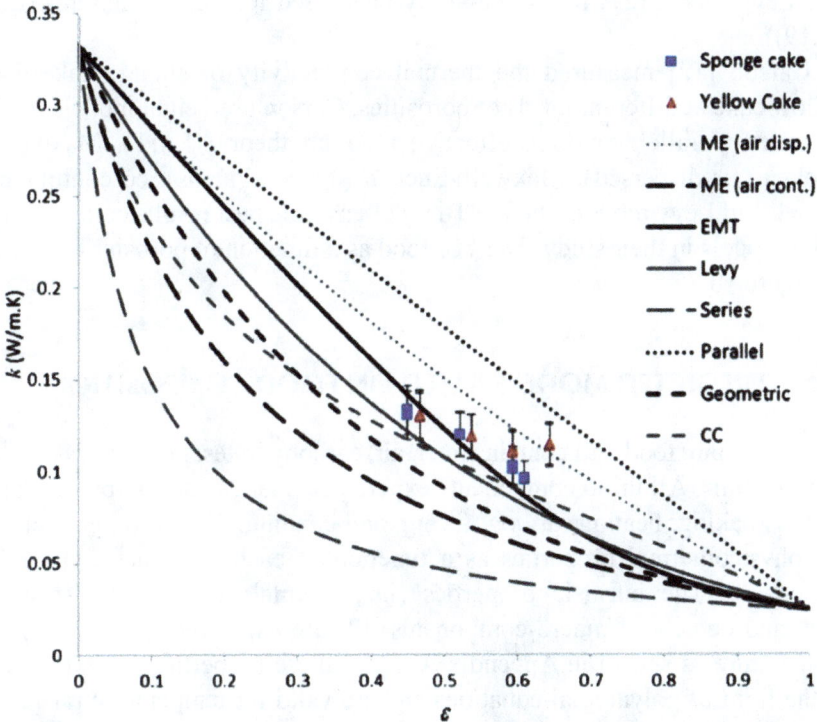

FIGURE 10.1 Thermal conductivity as a function of porosity plotted for eight effective thermal conductivity models [12].

Zhang and Datta [47] have used the Eq. (A.16) to obtain the k_{eff} of bread by considering the multiphase system. Other researchers have also used this equation for different products, for example, fresh seafoods [35].

10.3.2   MAXWELL–EUCKEN MODEL

The Maxwell model was derived on the basis of the random distribution of discontinuous spheres in a continuous medium. This equation provides the estimation of thermal conductivity if air is dispersed (Eq. (A.17)) and continuous (Eq. (A.18)).

a.   Maxwell model, if air forms the dispersed phase.

10.3.3   LEVY MODEL

The Levy model [27] has successfully been used for binary mixtures (Eq. (A.19)).

Carson [12] measured the thermal conductivity of sponge cake and yellow cake at different levels of porosities. Carson used eight models, such as series, parallel, random, effective medium theory (EMT), Maxwell–Eucken (air dispersed), Maxwell–Eucken (air continuous), co-continuous model, and Levy model. The EMT model gave the best results compared to other models in their study. The k of food as a function of porosity is shown in Figure 10.1.

10.4   PREDICTED MODELS BASED ON FOOD COMPOSITION

In general, any food will contain water, air, carbohydrates, proteins, fats, ash, and vitamins. All these components experience changes during processing, such as baking, heating, drying, frying, and canning. Therefore, predicting the physicothermal properties as a function of each component phase is reasonable. The intrinsic properties (i.e., thermal conductivity, specific heat, and density of macro-components) [7] are expressed as functions of temperature as shown in Appendix A.II. All these properties are expressed in the form of polynomial equations and are valid for temperature range of −40–150 °C.

10.5 PREDICTED THERMOPHYSICAL PROPERTIES DURING BAKING

Although predicting the physicothermal properties as a function of each component phase is reasonable, yet it is necessary to develop the thermophysical properties to achieve precise simulated results taking into account the phase change. Purlis and Salvadori [32] developed effective thermal properties (density, thermal conductivity, and specific heat) equations of bread including phase transition to accomplish the simulated results with experimental data. According to the moving boundary problem, the effective thermophysical properties are precisely stated including the phase change taking place during the processing of a product. The effective properties (density, thermal conductivity, and specific heat) were determined for dough/crumb and crust of bakery products. For these properties, a smoothed heavy-side function or piece-wise function is taken into account with continuous derivative/first derivative to include the phase change. The phase change with parameters are $T_f = 100$ °C and = 0.5 °C. Moreover, the delta-type function $d(T - T_f)$ term must be included to simulate the enthalpy jump, and it is stated by the sum of two smoothed heavy-side functions with different signs.

The suitable effective thermophysical properties with phase transition for the simulation of bread, cake, and biscuit during baking can be estimated using Eqs. (A.22)–(A.32).

Appendix A.III represents the physicothermal properties of bakery products using empirical equations.

10.6 SOFTWARE-BASED MODELS

10.6.1 ARTIFICIAL NEURAL NETWORK

The prediction models of thermal conductivity can be categorized into theoretical models, regression models, and distribution models. Generally, relationship between the k of the food and its components is assumed as linear. However, these types of linear relationships have been mentioned in the published literature and these are restricted to a limited range of moisture content, temperature, density, and porosity. They also differ in the nature of food material.

However, the nonlinear equations cover a wide range of variables. In such cases, artificial neural network (ANN) is an excellent tool to predict the properties (e.g., as shown in Figure 10.2) for handling highly nonlinear relationships among the parameters. This ANN technique has been widely accepted for estimating and predicting the properties and processing of related parameters over wide range of data.

Sablani et al. [39] developed an ANN model for computing the k of different bakery products for a wide range of temperature, water content, and density. They observed that the thermal conductivity values are obtained for all bakery products with a relative error of 10% and standard error of 0.003 W/(m K). Moreover, a good predicted values of thermal conductivity were observed for products with conventional ANN model with MRE > 15%.

10.6.2 INVERSE TECHNIQUE

For heat transfer modeling, all thermal properties can be determined by direct measurement methods; however, these measurements are sometimes difficult, or even impossible (e.g., effective thermal conductivity of bread by probe method). Under such situations, satisfactory results can be achieved using the inverse method with simple instruments.

The research studies on inverse heat transfer methods in food processing are very limited. Inverse methods can be used to determine the unknown parameters (such as heat flux, heat transfer coefficient, thermal conductivity, and specific heat), and boundary conditions [30] by utilizing the temperature measurement within the food material. In the direct heat problem, the temperature distributions inside the food material are computed with different discretization techniques, such as finite element method, finite volume method, and finite difference method when other parameters are specified.

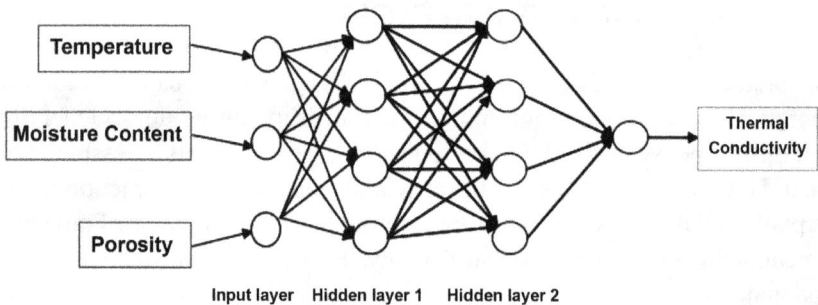

Input layer   Hidden layer 1   Hidden layer 2

FIGURE 10.2   Prediction of thermal conductivity using ANN.

Recently for a baking process, Monteau [30] determined thermal conductivity of sandwich bread by inverse method during the convective cooling process under natural or forced convection. Also in another process (e.g., dying, heating, freezing), thawing process of different food materials have successfully used inverse method for the determination of thermal conductivity, specific heat, and thermal diffusivity [28, 29, 49].

10.7 CHEMICAL PROPERTIES, HEALTH BENEFITS, AND FORTIFICATION OF BAKERY PRODUCTS

Bakery items are cereal-based products. Wheat is the major ingredient in the production of bakery products. Wheat has many beneficial components for its use in bakery products [37], such as:

1. complex carbohydrates,
2. dietary fiber,
3. low-fat content (without containing cholesterol),
4. minerals, especially calcium, phosphorus, iron, and potassium, and
5. vitamin B.

Nevertheless, wheat has some unique properties, for instance, gluten in wheat allows the formation of bread with good volume and uniform cell structure. Cereal and sprouted grains play significance role in the production of functional bakery products. Sprouted grains are a good source of ferulic acid, glutathione, and plant sterols and also have vitamins (E, B1, B2, and B3) and minerals (P, K, Mg, Ca, Zn, and Mn) [33]. Since the last decade, extensive work has been carried out to produce bakery products using cereals (such as rice, corn, oat, other different flours (buckwheat, amaranth), dietary fibers, and protein sources of flour, corn germ and rice bran, etc.

Ktenioudaki and Gallagher [24] reported the health benefits of dietary fiber, such as blood glucose and cholesterol attenuation, protection against cardiovascular diseases, regulation of intestinal function and promotion of gut health, protection against colon cancer, and risk of colorectal cancer. Nevertheless, high intake of dietary fiber was associated with a reduced risk of colorectal cancer. However, there are major challenges to obtain good quality of end-product with addition of fiber in bakery products. For example, they produce low loaf volume, crumb hardness, bitter flavor, and dark color. Bread is a high-energy food that contains more fat and carbohydrates with low amounts of protein, vitamins, and minerals [4].

Ameh et al. [4] determined the physicochemical and sensory properties of wheat bread with supplementation of rice bran @95/5, 90/10, and 85/15. The crude protein, crude fat, crude fiber, and ash of the composite bread loaves were increased significantly from 12.04% to 13.10%, 1.57% to 3.77%, 1.76% to 2.91%, and 1.46% to 2.41%, respectively. Further, there was a decrease in carbohydrate content with an increase in supplementation level from 62.10% to 54.14%. The thiamine and Niacin were significantly increased from 0.15 to 0.47 mg/100 g, and 3.31 to 4.04 mg/100 g, respectively; however, there was no significant increase in Riboflavin. Iron, potassium, calcium, and magnesium (mg/100 g) were increased significantly with an increase in supplementation level from 9.32 to 20.52, 80.74 to 188.20, 81.31 to 130.70, and 13.65 to 132.22, respectively.

Bhol et al. [9] incorporated the pomegranate whole fruit bagasse powder into the bread @5 g and 15 g. They concluded that mineral content was significantly improved. Eshak [17] evaluated the physicochemical and sensory properties of flatbread with partial replacement of banana peels BP @5% and 10%. The protein, fiber, and K, Na, Ca, Fe, Mg, and Zn contents of BP bread were higher compared to the control bread.

Also, Giami et al. [19] found that the calcium, sodium, potassium, and phosphorus contents were significantly increased with increased level of defatted fluted seed (FPF) flour of cookies (0%–25%); and also protein content was significantly higher than the conventional cookies. Ismail et al. [21] supplemented the cookies with pomegranate peel. They observed that levels of Ca, K, Fe, and Zn were significantly increased in the supplemented cookies. Moreover, significant reduction in protein content and negligible content in water content was observed in supplemented cookies.

Adeola and Ohizua [2] prepared the biscuits with supplementation of flour of unripe cooking banana, pigeon pea, and sweet potato. They found that the crude protein, crude fiber, and total ash content of biscuits were increased with an increased amount of pigeon pea flour. Moreover, such biscuit samples had more energy content and were rich in Mg with a favorable Na/K ratio (<1.0) and antinutritional contents within the acceptable levels.

Shalaby et al. [40] reported that storage time had no significant effect on pH of bread crumbs and it had significant effect on soluble starch in bread crumb. Recently, Ade et al. [1] prepared the bread and cookies with blends of wheat flour (WF) and African yam bean water-extractable proteins (AYBWEP) at proportion of 100/0, 95/5, 90/10, 85/15, and 80/20. They observed that the pH of bread and cookie samples was decreased with increased level of AYBWEP blend from 5% to 20%. Cookies, biscuits, and crackers are mostly made from patent flours with low gluten. They are low in protein content than the bread. Therefore, these products represent the largest categories of snack foods [13].

There are several advantages with protein fortification of cookies, biscuits, and crackers [13], such as (1) high-level consumption, (2) good eating qualities, (3) acceptable among varied sections of the population, and (4) longer shelf life. These facts make their large-scale production and distribution possible.

In the last decade, bakery items have been produced with different products, such as nuts, pea, orange, sugar beet, peach, mango, potato, and apple, as a source of fiber. The major significance of dietary fiber is to increase stool bulk, to avoid constipation, to facilitate movement of food along the gastrointestinal tract, to aid in proper digestion of nutrients, and to lower blood cholesterol levels by lowering the absorption of fats and cholesterol [13].

Another interesting area is the addition of marine functional ingredients in the production of bakery products (bread, biscuit, and crackers). Generally, marine food contains omega-3 fatty acids, chitin and chitosan, algal ingredients, antioxidants, minerals, and vitamins like calcium, carotenoids, and bioactive compounds. However, there are many challenges associated with the incorporation of seafoods, which may be polluted with industrial waste material and heavy metals etc. As a result, sensory and physicochemical properties of the product may change [23].

Rosell and Garzon [37] defined fortification as "the incorporation of one or more vitamins and/or minerals to a food, regardless of its usual content in the food, in order to prevent or correct a demonstrated deficiency of one or more vitamins and/or minerals in the population or specific population groups or to improve the nutritional status of the population and dietary intakes of vitamins or minerals due to changes in dietary habits, and that addition must be based on generally accepted scientific knowledge of the role of vitamins and minerals in health." To control nutrient deficiencies, proper fortification should be used. In the past, food fortification has been used to prevent micronutrient malnutrition, but today, food fortification is used to improve health and wellness. Fortification of foods and ingredients can prevent deficiency of micronutrients, diseases; can improve nutritional status and promote overall health.

Fortification may be the easiest, cheapest, and best way to reduce nutrient deficiency problems. In general, during the milling of cereals, some of the nutrients (protein, minerals, and vitamins) may be lost. These can be restored with fortification to improve the intake level of nutrients for the specific population [37]. If the food is consumed with low and without fiber, it may lead to health problems. Therefore, high-fiber foods (fruits and vegetables, legumes, and cereal grains) are now being advised in the treatment and controlling of various health disorders. For instance, bakery products (bread) are considered the best vehicle to enhance the dietary fiber content. In order to enhance the dietary fiber content in bakery products (e.g., bread, biscuit), several studies have been attempted to utilize bran of wheat bran, corn, oat, soybean, triticale bran potato peels.

Furthermore, many factors involved in the improvement of nutritional composition of bakery products are wheat quality for the purpose of milling, rate of extraction in higher level, air classification of flours, and supplementing the refined flours with nonwheat protein sources (eggs, milk, and milk-based products), and non-WFs (legumes, tubers, oilseeds, protein concentrates, plant leaves, and microorganisms). Although, egg and milk products have shown good nutritional and functional properties, yet they lack dietary fiber.

The vegetable protein sources are cheap and attractive supplements for bakery products [13]. Nevertheless, rheological and sensory properties of the bakery products are adversely affected by the addition of vegetable protein sources and also these lead to increase in acrylamide. Al-Dmoor [3] reviewed the ingredients and fortification of the flatbread. He stated that it was beneficial with fortification using thiamin, calcium and ascorbic acid in bread-making flour @0.64 mg/100 g flour, 211 mg/100 g flour, and 15 to 100 ppm, respectively.

10.8 SUMMARY

The thermophysical properties of bakery products depend on the food ingredients. However, the properties vary with product formulations and processing conditions. Recently, numerical modeling and simulation have been successfully applied to understand the physical phenomena during processing. There are different theoretical, ANN, inverse, and empirical models to estimate thermal properties. The nutritional value of bakery products can be improved by fortification and supplementation with a wide variety of protein-rich cereals and pulses.

KEYWORDS

- **bakery products**
- **cereals**
- **fortification**
- **pulses**
- **supplementation**

REFERENCES

1. Ade, I.C.; Ingbian, E.K.; Abu, J.O. Physical, Chemical and Sensory Properties of Baked Products from Blends of Wheat and African Yam Bean (*Sphenostylis stenocarpa*) Water-Extractable Proteins. *Nigerian Food Journal*, **2012**, *30* (1), 109–115.

2. Adeola, A.A.; Ohizua, E.R. Physical, Chemical, and Sensory Properties of Biscuits Prepared from Flour Blends of Unripe Cooking Banana, Pigeon Pea, and Sweet Potato. *Food Science & Nutrition*, **2018**, *6* (3), 532–540.

3. Al-Dmoor, H.M. Flat Bread: Ingredients and Fortification. *Quality Assurance and Safety of Crops and Foods*, **2012**, *4* (1), 2–8.

4. Ameh, M.O.; Gernah, D.I.; Igbabul, B.D. Physicochemical and Sensory Evaluation of Wheat Bread Supplemented with Stabilized Undefatted Rice Bran. *Food and Nutrition Sciences*, **2013**, *4* (9), 43–48.

5. ASTM. Standard method of test for thermal conductivity of materials by means of the guarded hot plate. In: *ASTM Standards, Part 3*; West Conshohocken, PA: American Society for Testing and Materials (ASTM), **1955**, 1084.

6. Baik, O.D.; Sablani, S.S.; Marcotte, M.; Castaigne, F. Modeling the Thermal Properties of Cup Cake During Baking. *Journal of Food Science*, **1999**, *64* (2), 295–299.

7. Baik, O.D.; Marcotte, M.; Sablani, S.S.; Castaigne, F. Thermal and Physical Properties of Bakery Products. *Critical Reviews in Food Science and Nutrition*, **2001**, *41* (5), 321–352.

8. Bakshi, A.S.; Yoon, J. Thermophysical Properties of Bread Rolls during Baking. *Lebensmittel Wissenschaft Technologie* (Food Science Technology), **1984**, *17* (2), 90–93.

9. Bhol, S.; Lanka, D.; Bosco, S.J.D. Quality Characteristics and Antioxidant Properties of Breads Incorporated with Pomegranate Whole Fruit Bagasse. *Journal of Food Science and Technology*, **2016**, *53* (3), 1717–1721.

10. Budžaki, S.; Šeruga, B. Specific Heat and Thermal Conductivity of the Croatian Unleavened Dough. *International Journal of Food Properties*, **2015**, *18* (10), 2300–2311.

11. Buhri, A.B.; Singh, R.P. Measurement of Food Thermal Conductivity Using Differential Scanning Calorimetry. *Journal of Food science*, **1993**, *58* (5), 1145–1147.

12. Carson, J.K. Measurement and Modelling of Thermal Conductivity of Sponge and Yellow Cakes as a Function of Porosity. *International Journal of Food Properties*, **2014**, *17* (6), 1254–1263.

13. Chavan, J.K.; Kadam, S.S.; Reddy, N.R. Nutritional Enrichment of Bakery Products by Supplementation with Non-Wheat Flours. *Critical Reviews in Food Science & Nutrition*, **1993**, *33* (3), 189–226.

14. Chevallier, S.; Colonna, P.; Della Valle, G.; Lourdin, D. Contribution of Major Ingredients during Baking of Biscuit Dough Systems. *Journal of Cereal Science*, **2000**, *31* (3), 241–252.

15. Choi, Y.; Okos, M.R. Thermal Properties of Liquid Foods—Review. Paper No. 83–6516 Presented at the Winter Meeting of ASAE, Chicago, IL: ASABE, **1983**, 11.

16. Christenson, M.E.; Tong, C.H.; Lund, D.B. Physical Properties of Baked Products as Functions of Moisture and Temperature. *Journal of Food Processing and Preservation*, **1989**, *13* (3), 201–217.

17. Eshak, N.S. Sensory Evaluation and Nutritional Value of Flat Bread Supplemented with Banana Peels as a Natural Source of Dietary Fiber. *Annals of Agricultural Sciences*, **2016**, *61* (2), 229–235.

18. Ferrari, E.; Marai, S.V.; Guidetti, R.; Piazza, L. Modelling of Heat and Moisture Transfer Phenomena during Dry Biscuit Baking by Using Finite Element Method. *International Journal of Food Engineering*, **2012**, *8* (3), Published Online July 23. https://doi.org/10.1515/1556–3758.2326.

19. Giami, S.Y.; Achinewhu, S.C.; Ibaakee, C. The Quality and Sensory Attributes of Cookies Supplemented with Fluted Pumpkin (*Telfairia occidentalis*) Seed Flour. *International Journal of Food Science & Technology*, **2005**, *40* (6), 613–620.

20. Hwang, M.P.; Hayakawa, K.I. Specific Heat Calorimeter for Foods. *Journal of Food Science*, **1979**, *44* (2), 435–448.

21. Ismail, T.; Akhtar, S. Effect of Pomegranate Peel Supplementation on Nutritional, Organoleptic and Stability Properties of Cookies. *International Journal of Food Sciences and Nutrition*, **2014**, *65* (6), 661–666.

22. Johnsson, C.; Skjöldebrand, C. Thermal Properties of Bread during Baking. In: *Engineering of Food*; McKenna, B. M., Ed.; New York, NY: Elsevier Applied Science Publishers, **1984**, volume *1*; 333–341.

23. Kadam, S.U.; Prabhasankar, P. Marine Foods as Functional Ingredients in Bakery and Pasta Products. *Food Research International*, **2010**, *43* (8), 1975–1980.

24. Ktenioudaki, A.; Gallagher, E. Recent Advances in the Development of High-Fiber Baked Products. *Trends in Food Science & Technology*, **2012**, *28* (1), 4–14.

25. Kulacki, F.A.; Kennedy, S.C. Measurement of the Thermo-physical Properties of Common Cookie Dough. *Journal of Food Science*, **1978**, *43* (2), 380–384.

26. Kustermann, M.; Scherer, R.; Kutzbach, H.D.. Thermal Conductivity and Diffusivity of Shelled Corn and Grain. *Journal of Food Process Engineering*, **1981**, *4* (3), 137–153.

27. Levy, F.L. Modified Maxwell-Eucken Equation for Calculating the Thermal Conductivity of Two-Component Solutions or Mixtures. *International Journal of Refrigeration*, **1981**, *4* (4), 223–225.

28. Martins, R.C.; Silva, C.L.M. Inverse Problem Methodology for Thermal-physical Properties Estimation of Frozen Green Beans. *Journal of Food Engineering*, **2004**, *63* (4), 383–392.

29. Mendonça, S.L.; Celso-Filho, R.B.; Da-Silva, Z.E. Transient Conduction in Spherical Fruits: Method to Estimate Thermal Conductivity and Volumetric Thermal Capacity. *Journal of Food Engineering*, **2005**, *67* (3), 261–266.

30. Monteau, J.Y. Estimation of Thermal Conductivity of Sandwich Bread Using an Inverse Method. *Journal of Food Engineering*, **2008**, *85* (1), 132–140.

31. Nicolas, V.; Glouannec, P.; Ploteau, J.P.; Salagnac, P. Experiment and Multiphysics Simulation of Dough Baking by Convection, Infrared Radiation and Direct Conduction. *International Journal of Thermal Sciences*, **2017**, *115*, 65–78.

32. Purlis, E.; Salvadori, V.O. Bread Baking as Moving Boundary Problem, Part 1: Mathematical Modelling. *Journal of Food Engineering*, **2009**, *91* (3), 428–433.

33. Rahaie, S.; Gharibzahedi, S.M.T.; Razavi, S.H.; Jafari, S.M. Recent Developments on New Formulations Based on Nutrient-Dense Ingredients for the Production of Healthy-Functional Bread: Review. *Journal of Food Science and Technology*, **2014**, *51* (11), 2896–2906.

34. Rahman, M.S. *Food Properties Handbook*. 2nd ed.; Boca Raton, FL: CRC Press, **2009**, 856.

35. Rahman, M.S.; Driscoll, R.H. Thermal Conductivity of Sea foods: Calamari, Octopus and Prawn. *Food Australia*, **1991**, *43* (8), 356–361.

36. Rask, C. Thermal Properties of Dough and Bakery Products: Review. *Journal of Food Engineering*, **1989**, *9*, 167–193.

37. Rosell, C.M.; Garzon, R. Chemical Composition of Bakery Products. *Handbook of Food Chemistry*, **2014**, *2014*, 1–28.

38. Rubio, A.R.I.; Sweat, V.E. Measurement and Modeling Thermal Conductivity of Baked Products. Presented at Annual Meeting of ASAE, Chicago, IL, **1990**, 6.

39. Sablani, S.S.; Baik, O.D.; Marcotte, M. Neural Networks for Predicting Thermal Conductivity of Bakery Products. *Journal of Food Engineering*, **2002**, *52* (3), 299–304.

40. Shalaby, M.T.; Abou-Raya, M.A.; EL-gammal, R.E.; Al–Janabi, H.A.A. Effect of Storage on Some Physical and Chemical Properties of Bakery Bread. *Journal of Food and Dairy Science Mansoura University*, **2014**, *5* (12), 891–904.

41. Shugar, G.J.; Ballinger, J.T. (Eds.). Chemical Technicians' Ready Reference Handbook. 5th ed.; New York, NY: McGraw Hill Education, **2011**, 704.

42. Sweat, V.E.. Experimental Measurement of the Thermal Conductivity of Yellow Cake. In: *Proceedings of the 13rd International Conference on Thermal Conductivity*; Lake Ozark, MO, **1973**, 6.

43. Tadano, T.; Tou, K.; Yasuda, H. Study on Effective Thermal Conductivity of White Bread. *Bulletin of the College of Agriculture and Veterinary Medicine-Nihon University (Japan)*; Nihon, **1990**, 31.

44. Ureta, M.M.; Goñi, S.M.; Salvadori, V.O.; Olivera, D.F. Energy Requirements During Sponge Cake Baking: Experimental and Simulated Approach. *Applied Thermal Engineering*, **2017**, *115*, 637–643.

45. Zanoni, B.; Petronio, M. Effect of Moisture and Temperature on the Specific Heat of Bread. *Italian Journal of Food Science (Italy)*, **1991**, 12. http://agris.fao.org/agris-search/search.do?recordID=IT9261311.

46. Zanoni, B.; Peri, C.; Gianotti, R. Determination of the Thermal Diffusivity of Bread as Function of Porosity. *Journal of Food Engineering*, **1995**, *26* (4), 497–510.

47. Zhang, J.; Datta, A.K. Mathematical Modelling of Bread Baking Process. *Journal of Food Engineering*, **2006**, *75* (1), 78–89.

48. Zhou, L.; Puri, V.M. Measurement of Coefficients for Simultaneous Heat and Mass Transfer in Food Products. *Drying Technology*, **1994**, *12*, 607–627.

49. Zueco, J.; Alhama, F.; Gonzalez-Fernandez, C.F. Inverse Determination of Specific Heat of Foods. *Journal of Food Engineering*, **2004**, *64* (3), 347–353.

APPENDIX A.I   Equations for Estimation of Thermal and Physical Properties of Bakery Products

Parameter and method of determination	Eq.

Direct Measurement Methods:

Density

Seed Replacement Method

$$V_s = \frac{\left[W_{c+s} - \left(W_{c+s+\text{sample}} - W_s \right) \right]}{\rho_s}$$

(A.1)

where W_{c+s} = weight of (container + seeds); $W_{c+s+\text{sample}}$ = weight of (container + seeds + sample); W = weight of the sample; V_s = volume of the sample.

$$\rho_b = \frac{W}{V_s}$$

(A.2)

Volume Measurement

Solid Displacement Method

$$V_s = V_{\text{container}} - \frac{\left[W_{(c+\text{seed}+s)} - W_c - W_s \right]}{\rho_b}$$

(A.3)

Specific Heat (C_p) Measurements

Method of Mixing: Heat Capacity

$$H_f = \frac{C_w M_{cw} \left(T_{eq} - T_{cw} \right) - C_w M_{hw} \left(T_{hw} - T_{eq} \right)}{T_{hw} - T_{eq}}$$

(A.4)

The C_p of the dough

$$C_p = \frac{\left(H_f + C_w M_{cw} \right) \left(T_{eq} - T_{cw} \right) - \left(T_d - T_{eq} \right)}{M_d \left(T_{hw} - T_{eq} \right)} C_{pw}$$

(A.5)

where H_f and C_w = specific heats of calorimeter and water (kJ/(kg °C)), respectively; T_{cw}, T_{hw}, T_d, and T_{eq} are temperatures of cold water, hot water, dough and equilibrium of mixture, respectively; and M_d = mass.

Thermal Conductivity (k)

Fourier's equation:

$$q = \frac{kA(T_1 - T_2)}{x}$$

(A.6)

where q is the heat in J/s, x is the sample thickness, T_1 and T_2 are the temperatures.

Parameter and method of determination	Eq.

Differential Scanning Calorimeter (DSC)

$$k = \frac{L\Delta Q}{A(\Delta T_2 - \Delta T_1)}$$ (A.7)

where L = length of the product, ΔQ = heat required to maintain pan temperature, A = area of the product, ΔT_1 and ΔT_2 = difference between initial and final temperatures.

Line Heat Source Probe

$$k = \frac{Q}{4\pi}\left[\frac{\ln\left(\frac{t_2}{t_1}\right)}{(T_2 - T_1)}\right]$$ (A.8)

where k = thermal conductivity; t_1, t_2 = initial and final time of measurements; T_1 and T_2 = temperatures at time t_1 and t_2, respectively; q = heat flux generated by the probe heater.

Thermal Diffusivity (α): Analytical Method

$$k = \frac{Q}{4\pi}\left[\frac{\ln\left(\frac{t_2}{t_1}\right)}{(T_2 - T_1)}\right]$$ (A.9)

where k = thermal conductivity; t_1, t_2 = initial and final time of measurements; T_1 and T_2 = temperatures at time t_1 and t_2, respectively; q = heat flux generated by the probe heater.

$$\alpha = \frac{k}{\rho C_p}$$ (A.10)

Probe Method:

$$0.32\sqrt{\alpha\tau} < r_d < 6.2\sqrt{\alpha\tau}$$

where r_d = distance from the probe heater. The probe method is mostly used for liquids (e.g., cake batter) and nonporous soft solids (e.g., dough).

Parameter and method of determination	Eq.
Theoretical (Mathematical) Methods	

$$c_p = \sum \varepsilon_i C_{pi}$$

(A.11)

$$k = \sum \varepsilon_i k_i$$

(A.12)

$$\rho = \frac{1}{\sum \left(\varepsilon_i / \rho_i \right)}$$

(A.13)

Series Method (Law of Addition of thermal resistances)

$$\rho C_p = \rho_s^a C_{ps} + \rho_w^a C_{pw} + \rho_g^a C_{pg}$$

(A.14)

Parallel Method

$$\frac{1}{k_{eff}} = \sum_{i=1}^{n} \frac{\varepsilon_i}{k_i}$$

(A.15)

where ε_i= volume fraction of the ith phase; K_i = conductivity of ith phase; k_{eff} = effective thermal conductivity of the ith phase.

Random model

$$k_{eff} = k_1^{\varepsilon_1} k_2^{\varepsilon_2} k_3^{\varepsilon_3} \cdots k_n^{\varepsilon_n}$$

Maxwell model, if air is dispersed:

(A.16)

$$k_{eff} = k_c \left[\frac{k_c + 2k_a - 2\varepsilon \left(k_c - k_a \right)}{k_c + 2k_a + \varepsilon \left(k_c - k_a \right)} \right]$$

(A.17)

Maxwell model, if air is continuous:

$$k_{eff} = k_a \left[\frac{2k_a + k_c - 2\left(1 - \varepsilon\right)\left(k_a - k_c \right)}{2k_a + k_c + \left(1 + \varepsilon\right)\left(k_a - k_c \right)} \right]$$

Levy model for binary mixtures

(A.18)

$$k_{eff} = k_a \left[\frac{2k_a + k_c - 2F\left(k_a - k_c \right)}{2k_a + k_c + F\left(k_a - k_c \right)} \right]$$

(A.19)

Parameter and method of determination	Eq.
where	

$$F = \frac{2/E - 1 + 2(1-\varepsilon)\sqrt{\left[2/E - 1 + 2(1-\varepsilon)\right]^2 - 8(1-\varepsilon)/E}}{2}$$ (A.20)

$$E = \frac{\left(k_c - k_a\right)^2}{\left(k_c + k_a\right)^2 + k_c k_a / 2}$$ (A.21)

Predicted Thermophysical Properties During Baking
(Empirical Equations)

Bread

Specific heat

$$C_p(T,M) = C_p^*(T,M) + \lambda_v M \delta(T - T_f, \Delta T)$$ (A.22)

$$C_p^*(T,M) = C_{p,s}(T) + MC_{p,w}(T)$$ (A.23)

$$C_{p,s}(T) = 5T + 25$$
$$C_{p,w} = 5207 - 7.317T + 1.35 \times 10^{-2} T^2$$ (A.24)

Thermal Conductivity

$$\begin{cases} 0.2 + 0.9/1[1 + \exp(-0.1(T - 353.16))] & \text{If } T \le T_f \Delta T \\ 0.21 & \text{If } T \ge T_f \Delta T \end{cases}$$ (A.25)

Density

$$\begin{cases} 180.6 & \text{If } T \le T_f \Delta T \\ 321.3 & \text{If } T \ge T_f \Delta T \end{cases}$$ (A.26)

where M is moisture content.

Cake [44]

Density, $\rho(T)$

$$\begin{cases} 1013 - 6.13T & \text{If } T \le 100 \\ 400 & \text{If } T \ge 100 \end{cases}$$ (A.27)

Parameter and method of determination	Eq.

Thermal Conductivity, $k(T)$

$$\begin{cases} 0.18\times10^{-2}T+0.2 & \text{If } T\leq100 \\ 0.2 & \text{If } T\geq100 \end{cases}$$

(A.28)

Biscuit [18]

Density, $\rho(T, c)$

$$\frac{c}{\rho_0}\rho_w+\left(1-\frac{c}{\rho_0}\right)R$$

(A.29)

Thermal Conductivity, $k(T, c)$

$$\begin{cases} \dfrac{c}{\rho_0}\rho_w+\left(1-\dfrac{c}{\rho_0}\right)R & \text{If } T<100\ °C>c_{fc} \\ 0.06 & \text{If } T>100\ °C>c_{fc} \end{cases}$$

(A.30)

Specific heat, $C_p\ (T, c)$

$$\begin{cases} \dfrac{c}{\rho_0}\rho_w+\left(1-\dfrac{c}{\rho_0}\right)R & \text{If } T\leq T_f-\Delta T \\[2mm] \dfrac{c}{\rho_0}\rho_w+\left(1-\dfrac{c}{\rho_0}\right)R+\lambda G\left(\dfrac{c}{\rho_o}\right)+A\dfrac{\lambda}{T_e}\cdot\dfrac{c}{\rho_o} & \text{If } T_f-\Delta T<T\leq T_f+\Delta T \\[2mm] \dfrac{c}{\rho_0}\rho_w+\left(1-\dfrac{c}{\rho_0}\right)R+\dfrac{\lambda}{T_e}\dfrac{c}{\rho_o} & \text{If } T>T_f+\Delta T \end{cases}$$

(A.31)

(A.32)

Note: Equations in this appendix are referred in the text of this chapter.

APPENDIX: A.II   Properties of Macro-components (expressed in °C) in Bakery Products

Density (kg/m³)	Thermal Conductivity (W/(m °C))
Ash	
$2.4328\times10^3-0.28063T$	$3.2962\times10^2+1.4011\times10^{-3}T-2.907\times10^{-3}T^2$
Carbohydrate	
$1.599\times10^2-0.3104\ T$	$2.014\times10^{-1}+1.39\times10^{-3}T-4.33\times10^{-6}\ T^2$
CO_2	
$\dfrac{PM_{CO_2}}{RT}$; *Note:* T is in K	$4.066238\times10^{-5}+1.34528\times10^{-7}T$; Note: T is in K.

Density (kg/m³)	Thermal Conductivity (W/(m °C))
Fat	
$1.599 \times 10^2 - 0.3104T$	$1.807 \times 10^{-1} + 2.760 \times 10^{-3}T - 1.775 \times 10^{-7}\,T^2$
Liquid vapor	
$\dfrac{PM_w}{RT}$; *Note*: T is in K	$2.0995 \times 10^{-5} + 1.34528 \times 10^{-7}T$ Note: T is in K
Liquid water	
$9.972 \times 10^2 + 3.14 \times 10^{-3}T$ $- 3.76 \times 10^{-3}T^2$	$5.7109 \times 10^2 + 1.7625 \times 10^{-3}T - 6.7036 \times 10^{-3}T^2$
Protein	
$1.33 \times 10^3 - 0.5184T$	$1.78 \times 10^{-1} + 1.196 \times 10^{-3}T - 2.718 \times 10^{-6}\,T^2$
Specific heat (kJ/(kg °C))	
Ash	$1.0926 + 1.8896 \times 10^{-3}T - 3.682 \times 10^{-6}\,T^2$
Carbohydrate	$1.5488 + 1.9625 \times 10^{-3}T - 5.94 \times 10^{-6}\,T^2$
CO_2	$\left(\begin{array}{l} 3.2868 \ + \ 5.12014 \times 10^{-3}T + 2.2352 \times 10^{-6}T^2 \\ -3.3522 \times 10^{-10}T^3 \end{array} \right) R$ *Note*: T is in K
Fat	$1.9842 + 1.473 \times 10^{-3}T - 4.8008 \times 10^{-6}\,T^2$
Liquid vapor	$\left(\begin{array}{l} 3.49708 + 1.5226033 \times 10^{-3}T + 2.2301684 \times 10^{-8}T^2 \\ -5.9706577 \times 10^{-11}T^3 \end{array} \right) R$ *Note*: T is in K
Liquid water	$4.1762 + 9.0862 \times 10^{-5}T - 5.4731 \times 10^{-6}\,T^2$
Protein	$2.0082 + 1.2089 \times 10^{-3}T - 1.313 \times 10^{-6}\,T^2$

Legend:
P = pressure, Pa;
M = molecular weight (kg/mol);
R = universal gas constant—8.314 J/(mol.K)

APPENDIX: A.III   Predicted Models of Bakery Products

Properties			Ref.
Density (kg/m³)	**Thermal Conductivity (W/ (m °C))**	**Specific heat (kJ/(kg °C))**	
Bread			
$225 \times 10^{-0.0095T}$	$0.6792 - 5.51M +$ $0.0020\rho + 9M^2$ $- 0.0024M\rho$	$3056M + 1130.44$	[8]
–	–	**Bread crumb:** $C_{pb} = MC_{pw} + (1-M)C_{pd}$; $C_{pd} = 1.60T + 1373$; $C_p = 1.60(1-M)$ $T + MC_{pw} + 1373(1-M)$	[22]
–	–	**Bread crust:** $C_{pd} = 2.62T + 1263$ $Cp = 2.62(1-M)T +$ $MC_{pw} + 1263(1-M)$	[22]
Crumb: $979 - 9.90\varepsilon$	**Crumb:** $0.768 - 5.0 \times 10^{-3}\varepsilon$	–	[46]
Crust: $895 - 9.0\varepsilon$	**Crust:** $0.398 - 3.1 \times 10^{-3}\varepsilon$	–	[46]
–	–	$C_{pb} = MC_{pw} + (1-M)C_{pd}$	[45]
–	$\{0.0502 (1+0.002T)\} \times$ $\{[1+11.4 (1+0.0385T)V_w]\}$	–	[43]
–	$0.0598 + 0.0001270\rho$	–	[38]
	$\ln k = 6.03 - 17.7M - 0.061\ T +$ $0.065\ MT + 1.0 \times 10^{-4}T^2$; $\ln k = -4.12 - 17.8M$ $+ 0.0031T + 0.065MT$	**Bread solid:** $C_{pd} = [0.098 + 0.0049T] \times 10^3$	[16]
Cake/muffin			
–	$0.0844 + 0.0000892\rho$	–	[38]
–	$0.00263T - 0.831$ $M - 0.000910\rho$ $+0.00422\ M\rho$	$0.00263T - 0.831M -$ $0.000910\ \rho$ $+0.00422\ M\rho$	[6]
–	–	$[1.0 - 0.5 (1-M)](C_{p,w})$	[42]
–	–	$(0.40 + 0.0039T)10^3$	[7]

Properties			Ref.
Density (kg/m³)	Thermal Conductivity (W/(m °C))	Specific heat (kJ/(kg °C))	
–	$\ln k = -48.0-10.9M + 0.272T + 0.053MT-4.1\times10^{-4}T^2$	–	[16]
–	$\ln k = -7.79-7.80M +0.015\ T+0.043MT$	–	[7]
Biscuit			
–	$\text{Ln}\ k = -15.8-7.90M+ 0.072T +0.038MT-9.7\times10^{-5}T^2;$ $\ln k = -5.95-8.61M +0.0098T+0.041MT$	$C_{pd} =[\ 0.8+0.003T]\times10^3,$ for 331–358 K; $C_{pd} =[1.17+ 0.003T]\times10^3,$ for 303–331 K	[16]

Legend: T, temperature; M, moisture content; ε, porosity; ρ, density (kg/m³).

APPENDIX AII Properties of Macrocomponents (expressed in °C) in Bakery Products

Density (kg/m³)	Thermal Conductivity (W/(m °C))
Ash	
$2.4328 \times10^3 -0.28063T$	$3.2962\times10^2 + 1.4011\times10^{-3}T - 2.907\times10^{-3}T^2$
Carbohydrate	
$1.599\times10^2 -0.3104\ T$	$2.014\times10^{-1}+ 1.39\times10^{-3}T -4.33\times10^{-6}\ T^2$
CO$_2$	
$\dfrac{PM_{CO_2}}{RT}$; Note; T is in K	$4.066238\times10^{-5} + 1.34528\times10^{-7}T;$ Note: T is in K
Fat	
$1.599\times10^2 - 0.3104T$	$1.807\times10^{-1}+2.760\times10^{-3}T-1.775\times10^{-7}\ T^2$
Liquid vapor	
$\dfrac{PM_w}{RT}$; *Note*: T is in K	$2.0995\times10^{-5} + 1.34528\times10^{-7}T$ Note: T is in K
Liquid water	
$9.972\times10^2 + 3.14\times10^{-3}T -3.76\times10^{-3}T^2$	$5.7109\times10^2 + 1.7625\times10^{-3}T - 6.7036\times10^{-3}T^2$
Protein	
$1.33\times10^3 -0.5184T$	$1.78\times10^{-1} + 1.196\times10^{-3}T - 2.718\times10^{-6}\ T^2$

Density (kg/m³)	Thermal Conductivity (W/(m °C))
Specific heat (kJ/(kg °C))	
Ash	$1.0926 + 1.8896 \times 10^{-3}T - 3.682 \times 10^{-6}\ T^2$
Carbohydrate	$1.5488 + 1.9625 \times 10^{-3}T - 5.94 \times 10^{-6}\ T^2$
CO_2	$\left(\begin{array}{l}3.2868\ +\ 5.12014 \times 10^{-3}T + 2.2352 \times 10^{-6}T^2 \\ -3.3522 \times 10^{-10}T^3\end{array}\right)R$ Note: T is in K
Fat	$1.9842 + 1.473 \times 10^{-3}T - 4.8008 \times 10^{-6}\ T^2$
Liquid vapor	$\left(\begin{array}{l}3.49708 + 1.5226033 \times 10^{-3}\ T + 2.2301684 \times 10^{-8}T^2 \\ -5.9706577 \times 10^{-11}T^3\end{array}\right)R$ Note: T is in K
Liquid water	$4.1762 + 9.0862 \times 10^{-5}T - 5.4731 \times 10^{-6}\ T^2$
Protein	$2.0082 + 1.2089 \times 10^{-3}T - 1.313 \times 10^{-6}\ T^2$

Legend:

P = pressure, Pa;

M = molecular weight (kg/mol);

R = universal gas constant—8.314 J/(mol K)

APPENDIX A.III Predicted Models of Bakery Products

Properties			Ref.
Density (kg/m³)	Thermal Conductivity (W/(m °C))	Specific heat (kJ/(kg °C))	
Bread			
$225 \times 10^{-0.0095T}$	$0.6792 - 5.51M +$ $0.0020\rho + 9M^2 - 0.0024M\rho$	$3056M + 1130.44$	[8]
–	–	**Bread crumb:** $C_{pb} = MC_{pw} + (1-M)C_{pd}$; $C_{pd} = 1.60T + 1373$; $C_p = 1.60(1-M)T + MC_{pw} + 1373(1-M)$	[22]
–	–	**Bread crust:** $C_{pd} = 2.62T + 1263$ $C_p = 2.62(1-M)T + MC_{pw} + 1263(1-M)$	[22]

Properties			Ref.
Density (kg/m³)	Thermal Conductivity (W/(m °C))	Specific heat (kJ/(kg °C))	
Crumb:	**Crumb:**	–	[46]
979–9.90	$0.768-5.0\times10^{-3}$		
Crust:	**Crust:**	–	[46]
895–9.0	$0.398-3.1\times10^{-3}$		
–	–	$C_{pb} = MC_{pw} + (1-M)C_{pd}$	[45]
–	$\{0.0502\,(1+0.002T)\}\times$	–	[43]
	$\{[1+11.4\,(1+0.0385T)V_w]\}$		
–	$0.0598 + 0.0001270$	–	[38]
	$\ln k = 6.03-17.7M - 0.061\,T +$	**Bread solid:**	[16]
	$0.065MT + 1.0\times10^{-4}T^2$;	$C_{pd} = [0.098+0.0049T]\times10^3$	
	$\ln k = -4.12-17.8M$		
	$+ 0.0031T + 0.065MT$		
Cake/muffin			
–	$0.0844 + 0.0000892$	–	[38]
–	$0.00263T-0.831\,M- 0.000910$	$0.00263T-0.831M- 0.000910$	[6]
	$+0.00422\,M$	ρ	
		$+0.00422\,M$	
–	–	$[1.0-0.5\,(1-M)](C_{p,w})$	[42]
–	–	$(0.40+0.0039T)10^3$	[7]
–	$\ln k = -48.0-10.9M$	–	[16]
	$+ 0.272T +$		
	$0.053MT-4.1\times10^{-4}T^2$		
–	$\ln k = -7.79-7.80\,M$	–	[7]
	$+0.015\,T+0.043MT$		
Biscuit			
–	$\ln k = -15.8-7.90M+ 0.072T$	$C_{pd}=[\,0.8+0.003T]\times10^3$,	[16]
	$+0.038MT-9.7\times10^{-5}T^2$;	for 331–358 K;	
	$\ln k = -5.95-8.61M$	$C_{pd}=[1.17+ 0.003T]\times10^3$,	
	$+0.0098T+0.041MT$	for 303–331 K	

T, temperature; M, moisture content; ε, porosity; ρ, density (kg/m³).

INDEX

β

β-casein, 173
β-cyclodextrin, 74
β-lactoglobulin, 173
β-sheet, 83

A

Acetic acid, 214, 219, 224, 225, 227, 228, 231, 232
Acetobacter, 204, 226, 228
Acidification, 220, 222, 232
Acoustic
  drying, 151, 152
  streaming, 70, 83, 87, 152
  waves, 70, 151, 152
Actomyosin, 84, 85
Additive manufacturing (AM), 102–104, 109
Aerobic
  fermentation, 206
  mesophilic bacteria, 125
African yam bean water-extractable proteins (AYBWEP), 258
Air-liquid interface, 164, 168
Albumen, 22
Alcoholic fermentation, 228, 233
Algae elimination, 128
Algal biomass, 105
Alicyclobacillus acidoterrestris, 52, 121
Allergenicity, 60
Amino acids, 20, 42, 55, 215, 219, 221, 232
Anaerobic
  digestion, 72
  fermentation, 206
Anionic polysaccharide, 175
Anthocyanin, 54–56, 73, 75, 224–228
Antimicrobial
  activity, 120, 224, 230
  agent, 120, 230
  potential, 28
Antioxidant, 73, 219, 259
  activity, 122, 217, 219, 227

capacity, 23, 225
rich profile, 228
Apparent density (ρ_b), 247
APV baker unit, 50
Aromatic
  compounds, 23, 73
  constituents, 227
  yeast, 216, 231
Artificial neural network (ANN), 255, 256, 260
Ascorbic acid, 49, 55, 78, 83, 145, 260
Asepsis, 207
Aseptic packaging plant, 50
Aspergillus
  ochraceus, 124
  oryzae, 206
  parasiticus, 25
Atmospheric freeze-drying (AFD), 152–155
Auto-cleaning, 86

B

Bacillus atrophaeus, 52
Bacteriocins, 222, 230
Bacteriophage, 205
Bacterium, 205
Bakery products, 8, 12, 86, 204, 245–247, 249, 253, 255–257, 259–261
Basundi, 183, 188, 195
Batch wash ozone sanitation system (BWOSS), 127
Billiard ball model, 5
Binder jetting, 103, 104
Bioactive
  components, 3, 72, 221
  compounds, 23, 74, 75, 78, 232, 259
  degradation, 80
  ingredients, 90
Biochemicals, 86
Biocompounds, 158
Bio-films, 127
Biofuels, 86
Biogenic amines, 219, 222, 223

Biological oxygen demand (BOD), 119, 120
Biomaterials, 106
Bioprinting, 106, 107
Bioreactor, 207
Biowaste valorization, 70, 93
Blackcurrant pulp, 175
Blanching, 3, 4, 37, 44, 55, 56, 84, 119,
    143–145
Blowing agents, 172
Boiling point, 74, 117
Botrytis cinerea, 26
Brettanomyces, 25, 224, 226, 227
    bruxellenxis, 25
    yeasts, 224, 226
Bubbles, 73, 82, 83, 166, 168–170, 173, 178
Buffalo milk (BM), 184, 185, 187–189
Burfi, 183, 185, 186, 195

C

Calcium, 84, 105, 214, 215, 219, 257–260
Calorimeter, 248
Cane vinegar, 228
Capacitance, 10
Capped column test (CCT), 251
Caramelization, 20
Carbohydrate, 72, 230, 254, 257
    catabolism, 222
    foods, 8
Carbon
    dioxide, 19, 28, 81, 82, 223
    monoxide (CO), 124
Carboxyl
    groups, 85
    methylcellulose, 174, 175
Carotenoids, 55, 59, 75, 80, 259
Cartesian coordinate system, 103
Casein micelle aggregation, 221
Catalytic activity, 81
Cavitation, 23, 38, 70, 72–74, 79–83, 87,
    89, 93, 152, 204, 226
    threshold, 74
Cell
    disintegration, 86
    immobilization techniques, 225
    matrix, 3
    membrane, 42, 81, 223
    rupture, 204
    wall, 52, 80, 81, 229

Cellular macromolecules, 130
Cellulose, 56, 216, 225
Cereals, 105, 113, 128, 183, 205, 217, 246,
    257, 259–261
Chemical oxygen demand (COD), 119, 120,
    128
Chemo nuclear, 114
Chimaphilin, 23
Chitosan, 24, 259
Chlorophyll, 53, 58
    degradation, 53
Chlorpyrifos, 129
Chorleywood method, 246
Chroma, 53
Chronic diseases, 128
Coagulase-negative staphylococci, 222
Cocoa bean fermentation, 231
Complex relative conductivities, 41
Computational
    fluid dynamics, 141, 148
    modeling, 28, 29
Conductors, 7
Consumer acceptability, 60, 177
Contemporary electrothermal techniques, 39
Conventional
    baking, 17–19
    cutting, 86
    disinfectants, 127
    drying, 16, 20, 80, 141, 177
    extraction, 23, 24, 74
    food, 71, 91
        processing technology, 71
    heating, 3, 9, 11, 17, 18, 20, 21, 41, 47,
        52, 58
    immersion, 129
    method, 22, 23, 38–40, 51, 54–57, 72, 74,
        158, 188, 195
    pasteurization, 22
    processing techniques, 101
    solvent, 75
    technique, 54, 58
    thermal processing techniques, 12, 39
    water treatment, 128
Corona wind drying, 149, 150
Cotyledon, 86
Crab meat imitation, 54
Critical control point (CCP), 69, 90, 91
Cryptosporidium parvum, 227
Crystallization, 71, 82, 85, 146, 192

Customary hypochloric acid, 128
Cyclic heating profile, 249
Cylindrical electrodes, 49

D

Dairy
  based fermentation
    prebiotics, 219
    transglutaminase, 220
  industry, 57, 126, 181, 197
Death kinetics, 26, 52, 60
Deep eutectic
  mixtures, 72
  solvent (DES), 75
Dehulling, 86
Dehydration, 4, 11, 16, 17, 27, 47, 53, 60, 79, 80, 146, 147, 222
Dehydrator, 157
Denaturation, 17, 24, 86, 154, 168, 172, 173, 221, 246
Deoxynivalenol (DON), 121, 122, 129, 130
Design parameters, 60
Detente instantanee controlee (DIC), 144
Dichlorvos, 129
Dielectric, 7
  barrier discharge, 25
  component, 7
  constant, 7, 8, 10, 22
  drying, 12
  heating, 3–6, 9, 11–13, 16–20, 22, 23, 25, 26, 28, 29, 39
    dipolar rotation, 6
    ionic conduction, 5
    operation principles, 5
    recent trends and future advances, 26
  loss factor, 8, 10
  property, 8, 9, 84
Dietary fiber, 220, 257, 259, 260
Difenoconazole, 129
Differential scanning calorimeter (DSC), 249–251
Dipolar
  molecules, 6
  rotation, 6
Dipole rotation, 5
Direct-resistance heating, 47
Disinfectant, 113, 118, 121, 127, 130, 131
Disordered kinetic energy, 5

Dissipation, 6, 7, 10, 11
Dough deformation, 246
Drying methods, 27, 141, 150, 157, 165

E

Effective
  medium theory (EMT), 254
  thermal conductivity, 253, 256
Electric
  field, 4–6, 8, 41, 42, 51–57, 59, 149, 150
  parameters, 52
Electrical
  conductance, 57
  conductivity, 41–48, 53, 60
  energy, 5, 17, 40–42, 86, 195, 196
  impedance spectroscopy, 38
  resistance, 41, 46
Electrochemical reactions, 46, 50, 60
Electrode, 4, 5, 19, 41–43, 45, 46, 48–50, 53, 114, 149, 150, 204
  product junction, 46
Electrohydrodynamic drying (EHD), 149, 150
Electrolysis, 39, 45, 46
Electrolyte ions, 42
Electromagnetic
  energy, 11, 24, 38, 144
  radiation, 4, 7–10
  spectrum, 11, 12, 16
  waves, 7, 9, 16, 156
Electro-osmosis, 42
Electroporation, 42, 43, 51, 204, 225
Ellagitannins, 78
Emulsification, 71, 78, 87–89, 93, 173, 216, 220
Emulsion, 84, 85, 87, 88, 90, 221
Encapsulation, 71, 89, 93, 229, 230
Endothermic reaction, 114
Energy-intensive method, 11
Enterococcus
  faecalis, 120
  faecium, 130
Environmental footprints, 70
Enzymatic activities, 19
Enzyme stabilization, 46
Escherichia coli (E. coli), 22, 24, 25, 52, 57, 81, 82, 120, 223, 225, 227, 230
Extraction, 3, 23, 24, 27–29, 46, 47, 56, 57, 69–75, 78, 79, 91, 92, 122, 227, 260

F

Faraday reactions, 46
Fat globules, 60, 221
Fatty acids, 130, 215, 217, 222
Federal communications and commissions, 11
Fermentation, 25, 37, 43, 56, 71, 203, 205–210, 212–233
    types, 206
        solid-state fermentation, 212
        submerged state fermentation, 208, 210
Fermented dairy products, 191
Fermenters, 207, 209, 210, 233
Filamentous fungi, 130
Flavonoids, 75, 83
Fluid jet
    system, 50
    unit, 50
Fluidization effect, 154
Foam
    characteristics, 163, 170
        density, 171
        expansion, 171
        factors affecting foaming parameters, 172
        stability, 170
    density (FD), 155, 170, 171
    expansion (FE), 170–172
    formation principle, 168
    history, 164
    mat drying (FMD), 163–167, 175–178
        classification, 175
        commercial application, 176
        economic importance, 176
        technology, 164
    stabilizers, 174
Foaming, 22, 85, 163, 164, 166, 168, 170, 172–174, 176–178
    agent, 163, 164, 166, 168, 170, 172–174, 176–178
        additional foaming agents, 174
        egg albumin, 173
        milk protein, 173
        protein containing foaming agent, 172
        soya protein, 173
    device, 166, 178
    operation, 166
    stabilizers, 175
    theory, 168

Food
    and drug administration (FDA), 81, 116
    emulsion, 88
    fabrication, 101, 109
    grade protein, 173
    industry, 38, 40, 41, 55, 70–73, 88, 90, 91, 101, 102, 119, 122, 125, 127, 128, 131, 142, 144, 145, 148, 158, 163, 173, 177, 203
    manufacturing sector, 5
    modification, 69–71, 83
    parameters, 113
    preservation, 79, 131
        drying, 79
        freezing and thawing, 82
        high-temperature preservation, 80
    printing, 101–104, 106, 109
    processing, 3, 4, 11, 12, 16, 20, 27, 28, 37, 38, 41, 47, 59, 69–72, 79, 82, 86, 91, 101, 104, 108, 113, 116, 118, 119, 127, 128, 131, 163, 164, 213, 214, 218, 221, 224, 231, 232, 245, 256
        applications, 11, 214
        baking, 16
        cocoa-based fermentation, 231
        dairy-based fermentation, 218
        drying and dehydration, 11
        fruits- and vegetables-based fermentation, 224
        grain-based fermentation, 214
        industry, 41, 113, 119, 214, 218, 221, 224, 231
        meat-based fermentation, 221
        sector, 37, 69, 163
        technique, 38, 47, 59
    quality, 11, 29, 47, 60, 70, 82, 84, 91, 113, 164
    safety, 28, 40, 51, 70, 72, 84, 90, 91, 108, 109, 113, 131
    sector, 38–40
    security, 104, 107
    synthesizer, 102
    warehouses, 113
Foodborne diseases, 39, 40
Fortification, 40, 245–247, 257, 259–261
Four-dimensional (4D), 108, 109
Fourier's equation, 250
Fraction collector, 50

Free
   fatty acids, 74
   radicals, 80, 81, 83, 205
Freeze-drying, 3, 147, 150, 152, 154, 155, 175
Frequency, 3–6, 9–11, 27, 45, 46, 48, 52, 54, 60, 70, 73, 79, 86, 89, 91, 145, 151, 152, 191, 204, 217, 226, 227
   spectrum, 70
Fruits- and vegetables-based fermentation
   cider, 227
   kimchi, 230
   probiotic products, 228
   sauerkraut, 229
   vinegar, 228
   wine, 224
Fumigant, 113, 121, 131
Functional heating, 42
Fusarium, 130, 215
Fused deposition modeling (FDM), 102–105

G

Gamma irradiations, 204
Gaseous treatment, 130
Gas-expanded solvents, 72
Gelatinization, 17, 18, 28, 47, 218, 246
Gelation, 173, 220–222
Gellan gel, 22
Generally recognized as safe (GRAS), 73, 74, 116
Geriatric nutrition, 106
Germination, 84, 85, 130
Globulins, 173
Glycinin, 174
Glycosides, 229
Glycosylation, 172
Grain-based fermentation
Green
   chemistry, 70
   consumerism, 69
   engineering, 70
   extraction, 74, 75, 78
   food processing, 70, 71
   solvent, 71, 72, 74, 75, 84, 90, 93
   space interstellar project, 106
   technology, 69, 70, 72, 74, 75, 91, 93
Greener technology, 131
Guar foaming albumin (GFA), 174
Guarded hot plate, 250
Gulabjamun, 183, 187, 188

Gum Arabic, 87, 90, 175

H

Hanseniaspora uvarum, 226
Hansenula anomala, 225
Hazard analysis, 69, 90, 91
Heat
   and mass transfer, 17, 79, 80, 142, 147, 152, 155, 158, 163, 164, 246, 251
   flux, 250, 251, 256
   resistance enzyme, 81
   resistant spores, 28
   sensitive enzymes, 54
   transfer
      coefficient, 250, 256
      modeling, 256
Heavy metals, 259
Heterogeneous heating, 39
Hickling and molecular segregation theories, 82
High-pressure processing (HPP), 38, 72, 204, 221, 223, 226
Homogeneous heating, 27
Homogenization, 71, 87, 88, 191, 204, 221
Horizontal spray drying (HSD), 148, 149
Hot
   air drying, 16, 27, 144
   water immersion treatment, 22
Hybrid drying, 152, 158, 175
Hydration, 84, 85, 220
Hydrocolloids, 87, 105
Hydrodiffusion, 23, 78
Hydro-distillation, 23
Hydrodynamics, 87
Hydrogen
   bond, 24
      donors, 75
   bonding, 81
   free radicals, 81
Hydrophilic colloids, 175
Hydrophobicity, 84, 85, 174, 221
Hydroxide radicals, 83
Hydroxyl ions, 5

I

Inactivation mechanism, 42
Inclined scraped surface heat exchanger, 184
Incubation, 108, 214, 225
Inexpensive technique, 78

Infrared, 21, 27, 28, 106, 144, 147,
    156–158, 217
    drying (IR), 156
Inherent qualitative properties, 39
Inlet-outlet conditions, 155
Innovative transduction, 38
Insulators, 7
Interatomic friction, 4
Internet of things (IoT), 107
Intracellular substances, 204
Inulin, 90, 219, 220, 229
Inverse technique, 256
Ionic
    fluids, 8
    liquids, 72
    losses, 9
Iron, 80, 145, 182, 183, 257, 258
Irradiation, 38, 113, 114, 204, 223, 227
Irritable bowel syndrome (IBS), 216, 217
Isomaltooligosaccharides, 230
Isotropic foods, 44

J

Joule
    effect, 41, 44
    heating, 39

K

Khoa, 181, 183–188, 192–197
Kinetic
    energy, 5, 6, 42
    parameters, 52

L

Lactic
    acid
        bacteria (LAB), 25, 49, 205, 214–217,
            219, 222, 227, 229, 231, 232
        fermentation, 230
    fermentation, 233
Lactobacillus, 204, 207, 214–218, 220, 222,
    223, 225, 226, 228, 229
    acidophilus, 220, 228
    pentosaceus, 222
Lactococcus lactis, 205
Large-scale production, 107, 259
Levy model, 254

Lignin, 56
Lignocellulosic biomass conversion, 70
Limonene, 54
Line heat source probe, 251
Lipase activity, 122
Lipid, 56, 57, 72, 86, 90, 172, 220
    molecules, 79
    peroxidation, 20
Lipolytic activity, 222
Liquid
    foods, 22, 23, 166
    liquid microextraction, 78
Listeria innocua, 24
Litopenaeus vannamei, 125
Lossy dielectrics, 7
Lycopene, 78, 147
Lysozyme, 173

M

Macro-components, 254
Magnesium, 258
Magnetic
    compounds, 7
    field, 5
    stirrer, 48, 50
Malathion, 129
Mallard reactions, 246
Maltodextrin, 90
Manganese, 78, 79
Manothermosonication, 81
Mass flow, 42
Material extrusion, 103
Maxwell-Eucken model, 252, 254
Meat-based fermentation
    novel technologies, 223
    starter culture, 222
Mechanization, 181, 182, 184, 186, 190, 191,
    197
Melting point, 117
Membrane
    filtration, 71
    fouling, 88
    separation, 69, 70, 88, 89
MeshLab, 248
Metabolic
    activities, 203, 206
    compounds, 56
    performance, 81

Metallic cations, 20
Microalgae, 86, 128
    cell disintegration, 86
Microbes, 38, 39, 52, 71, 113, 118, 125,
    130, 131, 146, 209, 222, 223
Microbial
    cells, 42, 43, 56
    death rate, 51, 52
    inactivation, 24, 25, 38, 42, 51, 55, 57,
        72, 80, 118, 124, 130
    multiplication, 19
    population, 55, 120, 230
    safety, 38, 51, 126, 222
Microcapsules, 80
Microcirculation, 83
Micrococcus, 207
Microencapsulation, 90, 219
Microextraction, 78, 79
Microfiltration, 227, 228
Microflora, 127, 191, 218, 219, 221, 230
Microfluidization, 88
Microjets, 73
Micronutrient malnutrition, 259
Microorganism
    cell membrane, 42
    degradation, 126
Microspheres, 89
Microwave (MW), 3–5, 8, 11, 12, 16–29,
    39, 40, 72, 74, 75, 78, 80, 84, 156, 158,
    175, 223, 227
    assisted extraction, 72, 78
    freeze drying, 3
    heating, 29
    hot air assisted drying, 3
    radio frequency plasma processing, 3
    vacuum drying, 3
Mild steel, 183, 184
Million tons (MT), 163, 254
Minimal processing time, 56, 58
Modulated differential scanning calorimeter
    (MDSC), 249
Moisture
    content (MC), 4, 8, 18, 20, 22, 26, 45, 84,
        121, 122, 141, 143, 150, 163, 183, 212,
        213, 248, 255
    loss, 47, 56, 251
    migration, 27, 251
Molecular
    interaction, 11
    segregation, 82

solvents, 72
structure, 85
weight, 172
Momentum, 150, 246
Monoglyceride oleogels, 106
Mycelium, 207
Mycotoxins, 121, 129, 130, 215
Myrcene, 54

N

Nannochloropsis oculata, 87
Nanofiltration, 89, 120
Nanometer scale, 90
National
    aeronautics and space administration
        (NASA), 106
    dairy development board, 184, 186
Native biomolecules, 206
Neolithic age, 205
Newton's law, 250
Nonconductive materials, 7
Nonenzymatic browning, 20, 53
Nonionic surfactants, 78
Nonozone degradation, 129
Nonpolar molecules, 5
Nonthermal, 1, 24, 28, 38, 42, 51, 56, 69,
    70, 72, 80, 131, 145, 149, 204, 205, 223,
    233
    technology, 38, 204, 205, 223, 233
Nontoxic quaternary ammonium, 75
Nonuniformity, 3, 26, 29, 47
Novel
    cultures, 205, 233
    drying techniques, 146, 158
        horizontal spray dryer, 148
        refractive window drying, 146
    ingredient, 216
    pretreatments, 143
        controlled pressure drop, 144
        infrared, 144
        poultice-up process, 143
        ultrasonication, 145
    processing techniques, 3, 60, 204
    starter culture, 204, 214, 218
    substrate, 205
    techniques, 38, 40, 81, 151, 217, 218,
        221, 225, 230
Nucleation, 82

Nucleic acids, 24
Nutrient diffusion, 207
Nutritional compounds, 52, 55

O

Ohm's law, 39, 41, 253
Ohmic
  assisted thawing, 56
  cooking, 51, 59
  heaters, 38, 40, 41
  heating (OH), 23, 37–60, 80, 83, 223
    assisted extraction, 78
    design, 43
  processing, 43, 50, 54
  sterilization, 55
  system, 43
  thawing, 56
Okra, 80
Omega-3 fatty acids, 125, 259
Open-source software, 103
Order kinetic energy, 5
Organic
  acids, 227
  compounds, 54, 56, 59
  matter, 118, 128
  media, 83
  solvents, 56, 57, 72, 73, 75, 78
  sonochemistry, 70
Organoleptic properties, 12, 19, 20, 203, 216, 221, 222, 227, 233
Osmotic dehydration, 47
Ovalbumin, 85, 173
Oxidation, 24, 46, 83, 124, 128, 154
  reaction, 46
Oxygen (O_2), 23, 113, 114, 116, 119, 120, 128, 206–209, 211, 219
  bonding, 114
Ozonation, 113, 114, 121–126, 128–131, 223
Ozone, 85, 113–131
  chemical and physical properties, 116
  dairy products processing, 126
  dissolution, 117, 120
  effect, 127
  food
    industry water treatment, 119
    packaging, 127
  generation, 114, 116
    electrical (corona) discharge method, 114

    electrochemical (cold plasma) method, 114
    ultraviolet method, 116
  grain processing, 121
  hydrocolloid processing, 123
  meat processing, 123
  odor and food waste treatment, 128
  pesticide residues, 128
  poultry processing, 124
  processing efficiency affecting factors, 117
    concentration, 118
    flow rate, 117
    organic matter, 118
    pH, 118
    temperature, 118
  radiochemical ozone generation, 114
  seafood processing, 125
  spice processing, 122
  technology, 113, 118, 123
  water treatment, 129

P

Parenchyma cells, 44
Pasteurization, 3, 4, 17, 19, 20, 22, 23, 26, 28, 29, 37–39, 55, 57, 59, 91, 144, 227
Pathogen, 25, 26, 51, 57, 120, 122, 127, 205, 227, 230
Pathogenic
  microflora, 206
  organisms, 221
Pectin, 54, 58, 74, 78, 107, 175, 231
  methyl esterase activity, 58
Peda, 186
Penetration depth, 9, 10, 27
Permeate flux, 88, 89
Peroxide value (PV), 86, 124
pH, 20, 50, 52, 58, 60, 82, 85, 86, 109, 118, 120, 121, 124, 125, 204, 206, 210, 214, 216, 217, 224, 225, 228, 229, 231, 258
Phenolic, 90
  compounds, 56, 89, 90, 130, 227–229, 232
Phenols, 78, 83, 215, 226, 227
Phosphorylation, 172
Photopolymerization, 103, 104
Physicochemical properties, 28, 38, 53, 88, 126, 150, 259
Physicothermal properties, 245, 246, 254, 255
Physiochemical parameters, 124

Phytochemicals, 23
Pigments, 56, 216
Pinene, 54
Plant matrix, 73
Plateau border (PBs), 168, 169
Polar molecules, 19
Polarity, 5, 6
Polarization, 50, 88
Polycarbonate, 48
Polydispersity, 89, 90
Polyethylene, 125, 127, 248
    pouches, 248
Polymer, 83, 90, 102, 220
    chemistry, 70
Polymerization, 89, 226
Polynomial equations, 254
Polypeptide interaction, 173
Polyphenols, 56, 74, 89, 215–217, 228
Polysaccharides, 73, 86, 89, 174, 175
Polyunsaturated fatty acid (PUFA), 217, 219
Porosity, 8, 216, 253–255
Postbaking, 12, 17
Postharvest
    aging, 45
    storage, 45
Potassium, 257, 258
Potential
    energy, 6
    oxidative damage, 83
Power ultrasound, 70, 71, 81, 82, 84, 87, 88,
    93, 145, 146, 226
Preservation, 28, 39, 40, 55, 69, 70, 79, 83,
    90, 119, 124, 125, 131, 139, 141, 142,
    145, 164, 203, 205, 223
Probiotic, 105, 204, 218, 220, 223, 228, 233
    foods, 228
    products, 204
Proliferation, 205, 206, 217
Proofing, 17, 20
Prophylactic properties, 204
Propylene
    glycerol monostearate, 174
    glycol alginate (PGA), 175
Protein, 24, 42, 59, 86, 89, 101, 130,
    172–174, 178, 220–222, 254, 258
    degradation, 222
Prototyping, 102, 103
Protozoa, 126
Pseudomonas, 25, 126
    contamination, 126

Pulsed electric field (PEF), 38, 72, 203, 204,
    221, 223, 225, 226
Pulses, 86, 113, 204, 246, 260, 261
Pycnometry, 247

Q

Qualitative integrity, 39, 40, 47

R

Radiative heat transfer, 18
Radio frequency (RF), 4, 8, 11, 12, 16–19,
    22, 24–29, 223
    heating, 29
Radiofrequency, 39, 40
Rarefaction, 73, 80
Raw material, 73, 90, 91, 101, 107, 121,
    183, 186, 191, 205, 228
Reactive oxygen species, 120
Reference index, 51
Refractance window (RW), 146–148
Rehydration, 27, 28, 80, 144, 156, 158, 164,
    177
Relative
    dielectric constant, 6, 7
    error, 256
    humidity, 83
    permittivity, 6
Relaxation frequency, 6
RepRap movement, 103
Reverse osmosis, 120
Rheological properties, 188, 219
Rhizopus oryzae, 24
Ripening, 124, 219, 223
Robocasting, 104

S

Saccharomyces, 56, 127, 204, 207, 210,
    224, 225, 228, 231
    cerevisiae, 56, 127, 207
    yeasts, 224, 225
Salmonella, 25, 57, 120, 123, 130, 222, 223,
    230
    typhimurium, 120
Scott crump, 102
Scrapped surface heat exchanger (SSHE),
    184–197
Seed replacement method, 247
Selective laser sintering (SLS), 102, 104

Semolina, 16, 130
Sensory parameters, 130
Shearing forces, 204
Sheet lamination, 103
Siemens per meter (S/m), 43
Sitotroga cerealella, 26
Small-scale units, 103
Sodium, 5, 41, 85, 175, 223, 258
Solid
    and submerged state, 233
    displacement method, 247, 248
Soluble solids content (SSC), 217
Sonication technique, 81, 88
Sonocatalysis, 70
Sonolysis, 81, 83
Soxhlet extraction techniques, 23
Soya protein isolate (SPI), 173, 174
Spray drying, 89, 90, 148, 149, 152, 158,
    164, 165, 175–177
Stabilizer, 172, 174–176, 178
Staphylococcus aureus, 81, 120, 222
Starter culture fermentations, 230
Steady-state methods, 249–251
Steam distillation, 23
Stephan processing unit, 185
Stereolithography (SL), 102, 104
Sterilization, 3, 4, 22, 23, 26–29, 37–39, 49,
    50, 55, 57, 59, 80, 81, 127
Sublethal temperatures aids fermentative
    processes, 42
Submerged-state fermentation, 206
Subtractive manufacturing, 102
Sulfur dioxide, 224, 226
Supercooling, 82
Supercritical fluid extraction, 72, 75
Superior conductive properties, 45
Supplementation, 246, 258, 260, 261
Supramolecular solvents, 72
Surface
    active property, 117
    pasteurization, 19
    tension, 74, 196
Surfactant, 78, 90, 168, 170, 172
Switchable solvents, 72

T

Tenderization, 84, 146
Thawing, 3, 12, 17, 26–28, 55–58, 82, 83,
    146, 257

Thermal
    assistance, 38
    conductivity, 8, 9, 249–257
    diffusivity, 47, 252, 257
    effect, 42, 51, 52
    energy, 6, 11, 12, 28, 40, 148, 194, 196,
        226, 248
    equilibrium, 250
    inactivation, 24, 52
    kinetics, 47
    processing, 12, 19, 39, 40, 54, 55, 70, 72,
        78, 81, 227
    resistance, 43
    resistant microorganism, 51
Thermocouples, 48–50, 251
Thermophysical property, 47
Thermosonication, 78
Thiamin, 260
Three-dimensional (3D), 101–106, 108,
    109, 123, 168, 169, 248
    printing (3DP), 101–106, 108, 109, 123
Tissue engineering, 106, 108
Toxic solvents, 91
Trace elements, 78, 79
Traditional
    heating techniques, 56
    Indian dairy products (TIDPs), 181, 182,
        191, 192, 194, 196, 197
        electric energy, 195
        thermal energy, 194
    smokehouse cooking, 59
Transient methods, 250, 251
Triatomic molecule, 113, 116
Trichloro phosphate, 174
Triglycerides, 173, 220
T-tube heater, 48, 49
Two-dimensional (2D), 104, 108, 250

U

Ultrafiltration, 89, 189
Ultrasonic, 73, 78, 79, 82, 85–89, 91, 128,
    129, 142, 146, 158, 204
    drying, 79
    emulsification, 88
    power treatments, 129
Ultrasonication, 87, 88, 93, 218, 223
Ultrasound, 23, 38, 69–75, 78–92, 145, 146,
    204, 203, 218, 221, 226

application, 71, 82, 226
assisted
    air drying, 79
    extraction (UAE), 72–75, 78, 90, 91
    green extraction, 76
    high-pressure extraction, 75
    homogenization, 87
    maceration extraction, 75
    microwave extraction, 75
    ohmic heating extraction, 78
food processing, 69, 90, 91
inactivation mechanisms, 81
technology, 72, 91
treatment, 73, 81, 84, 89, 226
waves, 70, 226
Ultraviolet (UV), 113, 114, 116, 117, 123, 156, 203, 223, 227, 228
treatment, 123
Uniform
    cell structure, 257
    heating techniques, 39

V

Vacuum
    conditions, 175
    drying techniques, 176
    freeze-drying, 154, 155
    frying, 38, 58, 84
    heating, 60
    impregnation, 53, 58, 188
Valorization, 72, 91
Van der Waals interaction, 81
Variable frequency drive (VFD), 191, 192

Velocity, 70, 79, 150, 152, 154, 155, 168, 211, 227
Viscosity, 42, 44, 45, 54, 58, 74, 84, 87, 123, 174, 175, 209–212, 215, 219–221
Volatile compounds, 215, 216, 222, 225, 231
Volumetric
    heat transfer technique, 16
    heating, 24, 41, 175

W

Wastewater treatment, 118, 128
Water
    activity, 11, 38, 59, 129, 163
    displacement method (WPM), 247, 248
    glycerol combinations, 74
    holding capacity, 84, 221
    molecules, 5, 8, 17, 81, 83
Wavelength, 9, 10, 117, 156
World War II, 4, 148

X

Xanthan gum, 175, 229

Y

Yeast cell concentration, 227
Yield stress, 221

Z

Zero effect, 121, 124
Zinc (Zn), 78, 257, 258
Zygosaccharomyces bailii, 52

For Product Safety Concerns and Information please contact our EU
representative GPSR@taylorandfrancis.com
Taylor & Francis Verlag GmbH, Kaufingerstraße 24, 80331 München, Germany